"Those with visions of vast satellite communications networks dancing in their heads would do well to read John Bloom's new book . . . Bloom . . . tells this story well . . . He does a good job of explaining the technology and the importance of the inventors who made the technology possible." —*Washington Post*

"Engaging and ambitious . . . A panoramic narrative, laced with fine filigree details, that makes for a story that soars and jumps and dives and digresses." —Jon Gertner, *Wall Street Journal*

"An enlightening narrative of how new communications infrastructures often come about: with a lot of luck, government help and investors who do not ask too many questions." —*Economist*

"Extensive . . . Sprawling . . . A detailed and entertaining history of the rise, fall, and rebirth of Iridium." —*Space Review*

"Highly engaging . . . Check it out." —*News Tribune*

"A good read." —*Marketplace*

"A prize-worthy example of the investigative genre . . . [*Eccentric Orbits*] has conflict and triumph on a Wagnerian scale . . . John Bloom has achieved in *Eccentric Orbits* an admirable balance of the human and the technological in what is at heart an age-old tale of one man's triumph against apparently insuperable odds." —*Literary Review*

"Riveting . . . I've never used the term 'tour de force' in a book review before, but if it ever belonged in one, it is this review of *Eccentric Orbits*." —800-CEO-READ

multibillion-dollar debacle, Iridium, is for you. *Eccentric Orbits* is both a novelistic thriller and a cautionary tale, a page-turner about a reach for the heavens and a business primer on a near-fatal fall back to the earth." —Julian Guthrie, author of *The Billionaire and the Mechanic*

"This book takes readers on an unusual, head-shaking investigative journey about the provocative but little-known history of Iridium, the only phone network that covers anyone, anywhere in the world—and which almost disappeared in a hell-bent suicidal incineration. Impeccably researched, and in smooth, easy prose, John Bloom interweaves fascinating historical trivia about the space race, satellites, and global communications with detail-filled personality snapshots and cringingly revealing, often disturbingly humorous, insights about the many ways big business can shoot itself in the foot." —John Brewer, former president and editor-in-chief of the *New York Times Syndicate and News Service*

"John Bloom's *Eccentric Orbits*, which tells the story of one of the most ambitious projects in the history of technology, is the most compelling book I have read in a long while. Bloom somehow coaxed the deepest thoughts and darkest secrets out of many satellite engineers, skeptical VCs, business royalty, inner-city tycoons, Italian marketers, Russian rocket launchers, Arabian princes, corporate CEOs, African leaders, Washington insiders, insurance giants, Pentagon brass, government lifers, politicians, and frustrated bankruptcy judges. This is a masterpiece of research and storytelling. If not for Bloom, one of the greatest stories of American ingenuity and bullheadedness would still lie scattered in thousands of documents and the memories of those who lived it." —Gary Kinder, author of *Ship of Gold in the Deep Blue Sea*

"This is a monumental piece of non-fiction, not just for the breadth and depth of the research, but for its audacity: Bloom seeks to make technology and marketing and high finance dramatic and funny and instructive of the human condition—and succeeds. Until I read this, I had always assumed that my cell phone was created by something like spontaneous combustion; like one day, it just appeared between my right hand and my ear, as if it had always belonged there. Bloom has given all of us—all billions of us—the back story on it, and what a strange, tangled, convoluted, fairly hilarious one it is."

—Jim Atkinson, *Texas Monthly* contributing editor

"Build a better mousetrap, and the world will erect every possible obstacle to its success. That's the sobering lesson of John Bloom's book on the progress of a reliable, cheap, encrypted, worldwide mobile phone system to supermarket shelves. The exhilarating lesson is that it can be done if you have visionary geeks, hard-boiled veterans, retired capitalists, and the occasional eccentric rebellious bureaucrat determined to do it. This is high scientific journalism, exciting business journalism, and a rattling good tale. It even includes Nazis."

—John O'Sullivan, author of *The President, the Pope, and the Prime Minister: Three Who Changed the World*

ECCENTRIC ORBITS

Also by John Bloom

Evidence of Love: A True Story of Passion and Death in the Suburbs
(with Jim Atkinson)

JOHN BLOOM

ECCENTRIC ORBITS

The Iridium Story

How the Largest Man-Made Constellation in the Heavens Was Built by Dreamers in the Arizona Desert, Targeted for Destruction by Panicked Executives, and Saved by a Single Palm Beach Retiree Who Battled Motorola, Cajoled the Pentagon, Wrestled with Thirty Banks, Survived an Attack by Congress, Infiltrated the White House, Found Allies Through the Black Entertainment Network, and Wooed a Mysterious Arab Prince to Rescue the Only Phone That Links Every Inch of the Planet

Grove Press
New York

Published simultaneously in Canada
Printed in the United States of America

First Grove Atlantic hardcover edition: June 2016
First Grove Atlantic paperback edition: June 2017

ISBN 978-0-8021-2678-8
eISBN 978-0-8021-9282-0

Grove Press
an imprint of Grove Atlantic
154 West 14th Street
New York, NY 10011

Distributed by Publishers Group West

groveatlantic.com

17 18 19 20   10 9 8 7 6 5 4 3 2 1

What we are building now is the nervous system of mankind, which will link together the whole human race, for better or worse, in a unity which no earlier age could have imagined.

> —Arthur C. Clarke, "The Social Consequences of Communications Satellites," Twelfth International Astronautical Congress, Washington, D.C., 1961

eccentric /ik-'sen-trik/] 1 *adj.* in aerospace engineering, an orbital path that's off-center. 2 *n.* in psychology, a person who is off-center. 3 *n.* in business, Iridium.

CONTENTS

Prologue

THE DEATH SENTENCE

LABOR DAY WEEKEND, A.D. 2000
MOTOROLA WORLD HEADQUARTERS,
SCHAUMBURG, ILLINOIS

And so it came to this: Dannie Stamp, bitter, brokenhearted, packed into a sterile cubicle on an upper floor of The Tower, staring off into the hazy shimmer of the oppressive heat, watching the planes taking off and landing at O'Hare Airport. The end-of-summer *sirocco* reminded him of better days back home in Persia, Iowa, where his father would worry over the corn burning up before the harvest. Even after half a lifetime doing the top secret work of the Air Force, Dannie still had his farm-boy drawl, still came across as the boy from the sticks. The corporate headquarters of Motorola, largest electronics company in America, with its crisp pressed suits and its mahogany boardroom tables—the arena where the big shots plotted against the Japanese—this was not his baili-wick. Dannie was a "Chandler guy," one of the engineering geeks from The Lab in Chandler, Arizona, and he would have preferred to get back there as quickly as possible. But he had been ordered here by the boss. Motorola chairman Chris Galvin told him he couldn't leave. He had to stay through the holiday weekend, so that he could be summoned at any moment. Dannie Stamp was the designated executioner.

Things were eerily quiet in The Tower that day. Everyone else was on the way to barbecues or baseball games, but apparently destroying seventy-four satellites was so goddamned important that it couldn't wait until Tuesday morning. Dannie was used to taking disagreeable orders, and he was used to dirty work, but this was eating him up inside. He thought back to the time he spent in the ultimate war room, with the giant map of the United States at the bottom and the upside-down map of the Soviet Union at the top, covered with blinking red dots identifying the location of every long-range ICBM in the world. It was quiet there, too, and nerve-racking, but it was never depressing the way this was. This was just . . . death. In the old war room they talked about the "hot line" that would initiate Mutual Assured Destruction, but Dannie had never actually seen one. Now he had one on his desk.

It was a sick joke. Carl Koenemann, the Chief Financial Officer, had given him a giant red button, the kind used in science fiction movies to activate the Doomsday Machine. Everyone except Dannie thought that was funny. He had stopped eating lunch in the company cafeteria because he was sick of Koenemann stopping at his table and razzing him about the "Iridiots," a play on "Iridium," a word now identified with bankruptcy and failure. Dannie had nothing to do with that failure—he was an operations guy, and his system was performing beautifully—but for better or worse he was now the official representative of the Arizona working stiffs who got the company involved in a $6 billion white elephant space project. Stamp was not a thin-skinned guy—he could keep his poker face while being chewed out by the most profane three-star general in the Pentagon—but this was humiliation of a different order. Thirteen years spent building the largest and most complex satellite constellation ever conceived was about to end with his entire staff being turned into janitors and garbagemen, supervising the disposal of some burned-up space junk.

Motorola always used euphemisms for what was about to happen: this was a "de-orbit" or a "decommissioning" or a "liquidation of nonperforming assets." Sometimes even "de-orbit" was considered too direct so executives used "*re*-orbit" since all they were doing was tilting a few satellites into a different orbital trajectory—one that would, coincidentally, strike the surface of Earth. What was surprising about it was not even the wanton destruction of Dannie's satellites, one of which now hung from the rafters of the National Air and Space Museum, but that the suits were so self-righteous about it: *Business is tough. Sometimes you have to do unpleasant things. Yes, it's a shame, but life moves on.* Chris Galvin, the third-generation Motorola boss, was normally a hands-off kind of guy, delegating authority to others, confiding in only a chosen few and managing everyone else by proxy. So why would he be so omnipresent now? He had made it clear that these were Dannie's birds and Dannie was expected to bring them down like a good company soldier. It was the "Old Yeller" principle—it's your dog, so you have to shoot it. It might happen today, it might take until Monday, but as soon as all the obstacles were removed—something to do with waiting on Lloyd's of London and a fax from a Belgian insurance company—Motorola would destroy the constellation that was whipping around Earth in low orbits, with satellites converging at the North and South Poles and then fanning out into a canopy that covered every inch of the planet. There had been only two engineering projects in history this complex—the other one was building the atomic bomb at Los Alamos—and Iridium had the kind of elegant structure that stirred the imagination of even the most jaded scientist. None of that would matter a few hours from now. It was only for the moment that the satellites soared over the Poles, synchronized and cross-linked, communicating in space like electronic cyborgs, doing whatever dance they were told to do when their daily software uplink was transmitted. Soon that transmission would say, "Die."

The organized carnage had been outlined in an e-mail. When Galvin said, "De-orbit," Stamp would place a call to the Satellite Network Operations Center in Leesburg, Virginia, better known as "The SNOC," where a handful of technicians would gather around the NASA-style monitors and watch as the satellites passed over the North Pole and received instructions from suicide software written by Dannie's staff—stark lines of computer code set down in a nine-page step-by-step memo that was reminiscent of the protocol used in the gas chamber ("Step 6.3.2. Once the apogee lowering has been completed, a final perigree burn will be executed to lower perigree as much as possible given the remaining fuel level"). The death throes would commence less than a minute after the satellites passed over the Pole, as they flew within reach of the earth station at East Iqaluit, in the remote Nunavut Territory of northern Canada, or the one at West Yellowknife in the Northwest Territories, or past a third facility in the frozen tundra north of Egilsstaðir, Iceland. Three earth stations, three chances to get the "kill yourself" message. Normally that would be called a "failsafe" system, but since its purpose was to create failure, it was more properly called "failcertain."

It takes a million fine-tuned calculations to place a satellite into its flight path, but it takes only one to euthanize it. As soon as the operational cyanide hit, the first satellite would be jolted off course by the same hydrazine thruster normally used to hold it securely in place. It would then tumble out of position and lurch into a wild elliptical orbit that would get longer and longer, like pulling a piece of taffy, until it hit the atmosphere and, if all went well, burst into flames and flutter to Earth as microscopic ash. The plan was to destroy two satellites a day until they were all on their way down, but crash orbits are so unpredictable that it can sometimes take decades to put a bird out of its misery. Then, after lethal injection had commenced, several dozen engineers and technicians at the SNOC were planning to share a few stiff drinks at Stonewalls Tavern in Leesburg, or maybe the microbrew place over in Sterling. The ten remaining scientists at The Lab in Chandler might or might not drift down the road to their local watering hole, Churchill's. Most of them would probably just head home early that day. Satellite engineers are not a sentimental bunch, but something about this seemed like sacrilege.

Dannie, on the other hand, would take his rental car back to O'Hare, alone, and call his wife to tell her it was all finally over.

Galvin had already asked several times, "Why is this taking so long?"

Half a continent away, on a high bluff overlooking the river that runs past the U.S. Naval Academy, Dan Colussy awakened at 6:00 A.M. and, careful not to wake his wife, eased down a breezeway past his lacquered Tang dynasty folding screen, centerpiece of his Chinese art collection. Settling into his wood-paneled study, he realized that, for the first time in six months, he had nothing to do. Apparently the damn fools still intended to destroy the satellites, and he had run out of ways to stop them. Outside his window faint rays of sunlight were casting long shadows on the opposite shore, and the first sailors of the day were scudding across the Severn through a light mist on their way down to the Chesapeake. If this were a normal holiday weekend, he and Helene would have plans to entertain. They would be converting the pool area into a ballroom and maybe bringing in an orchestra for dancing. But for once in their lives they had no plans at all. Colussy was subdued. He was not the kind of guy to admit it, even to himself, but he was depressed.

At sixty-nine years old, Colussy had planned to spend the whole year learning to play golf. He had never understood the appeal of the game until he retired, always thinking it was just an excuse for businessmen to goof off. But once he and Helene gave up sailing for something less strenuous, they had enrolled in a total-immersion program at the Jim McLean Golf School in Miami, and that experience had turned them into fierce competitors. Now he hadn't touched the clubs in weeks. Iridium occupied his attention 24/7. *I don't need to be running a company at my age*, he told himself. *The whole thing is too complex. There are too many egos in the way, too many bureaucracies. It's probably impossible. Motorola doesn't care what happens to the satellites, and apparently no one else does either.* Any of a hundred American companies could save Iridium with what amounted to pocket change for them, and yet everyone seemed willing to stand by while the constellation was scrapped. It looked like Iridium was going away and he could go back to the driving range.

Then he felt a pang from twenty years ago: it was the pang of losing, and he hated it. *Review the evidence one more time,* he thought.

Colussy took out the leather folder he used to hold a frayed spiral notebook, and he turned to a new page. When in doubt, make a list.

Twenty minutes later he had nine items that constituted his reality check of the hard facts on the ground. Each one was more depressing than the one before.

One: Motorola is shutting down its earth station in Tempe, Arizona, the processing center for all Iridium phone calls—a clear indication they're dead serious.

Two: The Italians have already shut down the European earth station in Fucino.

Three: Motorola just wired $5 million to its French insurance company to activate a de-orbit policy.

Four: To add insult to injury, Motorola wants Colussy to reimburse that $5 million, as a penalty for slowing them down. Colussy was used to being told that he had to pay millions for this or that delay, but he never took those threats seriously until now, when they seemed to be signing off.

Five: Motorola has shut down the production line in Libertyville, Illinois, that manufactures the actual Iridium phones. Without handsets there's no Iridium.

Six was the thorny issue of Colussy's own insurance. No one wanted to take the risk that the satellites would injure someone when and if they were de-orbited, even if they stayed aloft for another hundred years. Colussy had arranged $2 billion worth of insurance from Lloyd's of London, but Motorola insisted that his promises were worthless because his little venture-capital company was way too small to be trusted. It seemed like one of those issues that gets worked out by attorneys, but after three months of wrangling, Colussy suspected that Motorola was no longer acting in good faith, and that they were using the insurance issue as an excuse to destroy the constellation.

Item *Seven* was the only positive one: people in the government were researching an obscure statute, Public Law 85-804, to determine whether the United States government could simply insure the satellites directly. There's an enormous aversion within the halls of Washington toward anything that smacks of propping up private industry with government guarantees, but it had happened at least once before in history. Colussy worked for Pan American World Airways during the Vietnam War, and he recalled the day Pan Am canceled passenger service out of Saigon because the Vietcong were shooting at planes during takeoff and landing. The State Department wanted Pan Am to continue passenger service to and from Saigon—for symbolic reasons, if nothing else—and the Pentagon needed Pan Am's charters for rest-and-recreation flights to Bangkok and Hong Kong. The problem was solved when Secretary of Defense Robert McNamara invoked Public Law 85-804, which fully protected the airline against any damage or liability. This was obviously a different situation, but at least Colussy knew it was possible.

The last two items were the most dispiriting of all.

Eight: His own fund-raising efforts were in disarray. His investor group, after all this time, had not come together. His original backer, a well-heeled New York investment bank, had dropped out suddenly—and with embarrassing publicity—just five weeks earlier. Colussy had been so content to let more famous financiers and businessmen take the lead that many people in government still didn't know his name and, if they did, couldn't pronounce it—"Kuh-LOOS-ee," not "Kah-LUSS-ee" or "Kah-LOOCH-ee." The only real commitment he had now was from Syncom, a small African American firm based in Washington, D.C., that had some success with cable TV franchises and urban radio, but had no experience in telephony, aviation, or space, and besides, was only good for a fourth of what was needed. A Saudi prince seemed perpetually on the verge of investing, but Colussy thought his legion of advisors and handlers were difficult to deal with, tentative about their support, and slow to act. Just the previous week Kazuo Inamori, the Japanese business titan, had seemed on the verge of making a commitment, but now he apparently felt insulted by a canceled meeting. The Texas Pacific Group, which had hundreds of millions available for venture capital, had expressed slight interest. Others had come and gone and come again, talking about investing but never writing checks. The crazy thing about the fund-raising is that all Colussy needed was $100 million—1.6 percent of what Motorola claimed the system was valued at—an amount so small that any Fortune 500 firm could put it up. Anyone with the slightest background in telecommunications *should* be putting it up. But the very word "Iridium" had become poison. The name itself sent investors fleeing.

Finally, Item *Nine* made it painfully clear how badly the week had turned out. After being optimistic for most of the summer, Colussy now had to grudgingly admit that his business case—the financial projections for future success—might be all wrong. One thing he knew from a career spent acquiring and evaluating companies was that you had to guarantee the revenue side in advance. But the Iridium customers had been cut off, so the revenue side was zero. Now only one customer could possibly guarantee income—the Pentagon. Colussy had believed the Iridium system was so essential that a government contract would be a cinch, but now he had to face the fact that there were only a few supporters here and there, mostly the lower-level "war-fighters" in the Army and Marines and the special operatives in the DEA and CIA. The higher-ups at the Pentagon were lukewarm—it was not their system, after all—and their usual congressional supporters didn't want to be associated with what they were calling a bailout. The chances of getting the crucial government contract were slim.

Colussy closed the notebook and wondered if he should even bother trying to find anyone on the Sunday morning of a long holiday weekend.

It might be too late anyway. Keith Bane and Ted Schaffner, the two-man Motorola assassination team, had told him that the deadline for giving them a completed deal was 8:30 A.M. today or else they would start destroying the satellites. For all he knew, they were jerking them down already.

Suddenly the phone rang. Colussy glanced at the caller ID. It was a classified number, the kind government agencies use, so he picked it up.

"Dan, how are you? It's Dorothy."

He recognized the bright, upbeat voice of Dorothy Robyn, White House assistant to Gene Sperling, chief economic advisor to President Clinton. Thank God for Dorothy. Dorothy believed so strongly in Iridium that one of the top officials in the Pentagon had pleaded with Colussy to make her stop placing phone calls to the generals, admirals, and colonels. "That woman is driving us crazy," he said. And here she was, popping up with cheery optimism on one of the worst days of his life.

She had been working through the weekend and was calling from her White House office. *All is not lost,* she said. *This thing is not over. Keep the faith.*

What she didn't say was that she'd been on two phone calls between Motorola chairman Chris Galvin and Gene Sperling, and that Galvin was virtually begging for permission to destroy the satellites. Dorothy told her colleagues that Galvin sounded like a man pleading to be allowed to take his wife off life support. She thought he sounded weak and desperate. She thought they could stop him.

"We'll get this thing done, Dan."

He thanked her. At least one person thought so. He just hoped she had the horsepower.

Later that day Ted Schaffner settled into his seat at Chicago's Comiskey Park to watch the feisty playoff-bound White Sox face the Anaheim Angels. The weather was beautiful, he was enjoying a rare day out with his daughter, and for a few hours he could forget all the harsh words and tense meetings of the past week as he argued Motorola's position in Washington. The Iridium fiasco was almost over and he could finally start to relax. If everything went according to plan, the satellites would be de-orbited sometime over the weekend and that would be one less dirty job on his desk.

The matchup with the Angels turned out to be one of the most exciting of this or any season, starting with thirteen runs in the first inning and staying close the whole game as both teams knocked out hit after hit, run after run. About the time Herb Perry was tripling down the right field line in the eighth inning to even the score at 12–12, Schaffner's cell phone rang. Very few people had that number, just family and the top executives at Motorola. The caller ID displayed one of those codes that government agencies use.

When he answered it was a bright, upbeat female voice.

"Mr. Schaffner! I'm calling from the White House."

His heart sank. Suddenly he had a headache. He listened to the woman's spiel with growing irritation. It seemed there was a meeting at the Pentagon on Tuesday and someone at the White House wanted him to be there.

The final score was 13–12, and the Comiskey crowd was berserk with joy, but it no longer mattered. Ted Schaffner's day was ruined.

Chapter 1

THE CONSPIRATORS

APRIL 15, 2000
KENNEDY WINTER WHITE HOUSE,
PALM BEACH, FLORIDA

It was the season of the dot-com bubble. Computer jargon had come into vogue—"Y2K" instead of "2000"—which was more than ironic for the year when all things high-tech became suspect. It was the spring when the NASDAQ stock exchange crashed, bringing down half of Silicon Valley with it, but that was long after Iridium had been taken off the trading board anyway, a symbol of failure and corporate hubris of the highest order. By the time Dan Colussy started studying Iridium LLC, more out of curiosity than anything else, it was being talked about in terms reserved for iconic business failures like the Edsel, Betamax tape, and, more recently, Pets-dot-com. As a brand, Iridium seemed as shopworn as Bill Clinton, scarred by the Monica Lewinsky scandal, entangled in a Balkan war no one understood, trying to preserve his legacy in the few months he had remaining. There couldn't have been a more turbulent moment to tell a cranky President that seventy-four American satellites were about to be intentionally put into uncontrolled death spirals that would end up destroying the world's largest man-made constellation.

When he did hear about it, Clinton was reportedly incredulous. Was there a chance, he asked, that one of the errant satellites would be dangerous to the public? Demolish a bus full of schoolchildren in Kansas—something like that?

"Well, Mr. President," came the answer, "not likely, but we can't rule it out."

Clinton's response was unequivocal: *Not on my watch.*

What his lieutenants had failed to tell Clinton was that the fate of Iridium might not be within his control. The satellites had been launched by private industry, without a single government dollar being invested. In fact, there were corporations owned by sovereign nations, including Russia and China, that were part owners, whereas the United States was a mere customer. The satellites might soon be crashing to Earth whether Clinton liked it or not.

Fortunately, unbeknownst to the Commander in Chief, a meeting was taking place in a second White House, a makeshift shirtsleeve White House from another era, an oceanside villa where another Democrat had solved other crises four decades earlier. It was not entirely coincidence that when two old Harvard classmates convened to talk about the ailing Iridium system, they did it in Joseph Kennedy Sr.'s house at 1095 North Ocean Boulevard in tony Palm Beach, Florida. John Castle, the current owner, was an American history buff and, more to the point, a lover of the grand gesture. It was in this white Mediterranean getaway home where Castle and his old friend Dan Colussy—two men who had never before had anything to do with operating satellites or telephone networks or working in outer space—would have a what-if luncheon that might, just possibly, solve the President's problem.

The meeting in the Kennedy Winter White House that spring happened because the phone in John Castle's yacht stopped working.

On March 17, 2000, Castle's forty-two-foot Hinckley Sou'wester, named the *Marianne* in honor of his wife, was locking through the Panama Canal, north to south toward the Pacific Ocean, with Colussy aboard. The *Marianne* was making a round-the-world voyage, but since Castle was busy running his eponymous New York investment bank,

Castle Harlan, he could join the yacht only sporadically. The rest of the time he lived vicariously through satellite phone conversations with his captain. The previous week, as the yacht approached Central America from the Atlantic side, Castle had called Colussy to ask his friend if he'd ever traversed the Canal and, if not, would he like to come aboard for a few days. Actually Colussy *had* traveled through the Canal, but not since 1954, when he was a boyishly skinny ensign and navigator on the Coast Guard cutter *Blackhaw*, serving under a dapper captain who insisted that the ship stop at every resort and tourist spot between Charleston and Honolulu so that he could disembark in his dress whites and . . . well, Colussy was never sure what the captain did exactly, but he was a mustang officer who loved the nightlife. Colussy accepted the invitation to fly down and make the passage, as he wanted to see how the Canal had changed since it was returned to Panama in 1999. He boarded the yacht at the port of Colón, in Limon Bay, and enjoyed naval shoptalk with the captain all through Gatun Lake and the Gaillard Cut.

And then Castle's Iridium phone went dead. Fortunately the *Marianne* was close enough to Gamboa for Colussy to get a cell signal and get back in contact with Castle. Colussy had known Castle for almost four decades and sat through many tense meetings with him, so he knew Castle was crotchety and easily annoyed. On this day he was yelling about Motorola's decision to shut down Iridium. Castle was especially upset because it was the only phone that worked all the time, at all longitudes and latitudes, without any kind of delay.

Listening to Castle rant, Colussy had an epiphany.

"John, I already know all about this."

It was more or less coincidence—one of a hundred coincidences that would favor Colussy's quest that year—that a neighbor of his in Anne Arundel County, Maryland, was a midlevel Motorola executive. If he'd been a high-level executive, he never would have talked to Colussy. If he'd been at a lower level, he wouldn't have known enough. But Dennis Dibos was perfect. When you live in the moneyed suburbs of

Washington, D.C., especially the area around Annapolis, the canonical six degrees of separation becomes two or three, so maybe it wasn't such a coincidence after all. Colussy lived half the year in Florida and the other half in Glen Oban, an unincorporated area of mansions arrayed across a picturesque part of the Broadneck Peninsula, where average incomes were 170 percent of the national average and where the men who ran the military-industrial complex had established a Republican enclave in an overwhelmingly Democratic state. It was there, during the previous summer, that Colussy had first noticed the struggles of Iridium in the financial press and decided to purchase a little of the vastly deflated stock, thinking, *Someone will save this thing. It's too valuable to waste.* The following day he asked his wife, "Don't you golf with someone who's married to a guy who works for Motorola?"

In fact, she did, and Helene knew everything he needed. She knew the name of the woman in her golfing group (Jan), the name of the woman's husband (Dennis), his rank at Motorola (Vice President of Public Relations), and the reason he was able to live in Annapolis while commuting to Motorola headquarters in Schaumburg, Illinois.

"Invite them to dinner" was all Colussy said.

Dennis Dibos, it turns out, was not exactly close to the satellite division of Motorola. His area of responsibility was, in fact, golf. He flew around the country setting up golf tournaments for the company, and he was the current champion at Chartwell, the club the Colussys had just joined. But as luck would have it, another couple had already been invited to dinner that same night: recently retired four-star Admiral Chuck Larson and his wife, Sally. Since Dibos was himself a reserve naval officer, this was like a club boxer being invited to a private dinner with Muhammad Ali. Admiral Larson had been Commander in Chief of CINCPAC, the U.S. Pacific Command in Honolulu, pretty much as high as you can go in the Navy, but he was better known locally as a popular superintendent of the Naval Academy—in fact as "the man who saved the Academy," having been brought in for a second stint in the job to clean up an institution mired in cheating scandals, drug scandals, sexual-assault scandals, and even a stolen-car scandal. Larson was the kind of impressive military figure that normally exists only in fiction, having been a Top Gun fighter pilot, a nuclear submarine commander running sensitive Cold War spy missions, and, incidentally, the guy who carried the black suitcase containing the nuclear launch codes for President Nixon. But if Colussy thought he needed Larson to loosen up Dibos, he needn't have worried. The weather was warm, so dinner was served on the terrace overlooking the Severn, and the conversation was so effortless that Colussy decided he didn't need to waste any time. Before dinner was over he turned to Dibos and said, "I'm curious about Iridium. What can you tell me?"

Dibos was more than happy to expound on the subject and turned out to be a fount of information, mostly negative company scuttlebutt about what was regarded as Motorola's biggest mistake of recent years. Iridium was bad news, he said. It was a narrowband system in the emerging broadband world. It didn't have enough capacity to carry the kind of data people would need in the future. The Iridium phone was cumbersome and much too large. Motorola was anxious to get rid of the whole thing. He knew most of the history of the project. He knew the three executives directly in charge of Iridium and gave Colussy their names. He also knew why the satellites were being abandoned, or at least the official version of that story. He knew approximately how many billions had been spent on the system, and he knew where the bodies were buried.

Colussy made a few follow-up phone calls after the dinner party, but nothing too serious. He was still skeptical about Motorola's threats to force Iridium into bankruptcy, especially since he'd learned from Dibos that Motorola had only an 18 percent ownership share while the rest of the company was held by venture partners around the world, many of them tied directly to powerful countries like China, Russia, India, Japan, and Saudi Arabia. Colussy had always been a "small-cap guy"—Wall Street jargon for companies valued under a billion dollars—so Iridium sounded a little too rarefied, the kind of company that would need the global support of a true large-cap multinational like Sony or AT&T. It was certain to be attractive to anyone trying to establish a stronger foothold in the booming global telecom markets. Colussy purchased some Iridium stock, certain that once the company was bailed out, the price would soar—only to be shocked when, less than a month after meeting with Dibos, Iridium did go ahead and file for Chapter 11 bankruptcy protection. He still didn't believe it could be anything more than a reorganization. When he saw that Craig McCaw had expressed interest in buying the company, he assumed the whole saga had fallen into the category of one big organization selling its mistake to another big organization. McCaw was the original cellular visionary, having bought up the very first cell-phone licenses in the 1980s and built a company that he eventually sold to AT&T for $20 billion. McCaw was in many ways the perfect choice to figure out the next stage of telecommunications.

But now that had all turned sour. Motorola went through a six-month due diligence process with McCaw, who said he would probably buy Iridium out of bankruptcy for $600 million, but McCaw backed out at the eleventh hour, sending the Motorola board of directors into paroxysms of rage. Now apparently they'd simply thrown up their hands and shut everything down.

As Castle continued to rant about his dead satellite phone, Colussy felt an inner pang that was close to outrage. He'd been in love with aeronautics since he was a boy, watching wide-eyed as barnstormers barreled into town to star in "air circuses" at his great-uncle's dairy pasture in Coudersport, Pennsylvania, later to become Colussy Brothers Airport. He'd then gone on to become a licensed commercial pilot and airline executive, so everything about aerospace fascinated him. He couldn't believe that such a revolutionary scientific achievement was about to be destroyed.

The conversation between the two men was brief, but it brought home the immediacy of the situation: if Motorola was actually turning off the phones, then maybe this was something more than a typical Chapter 11 reorganization.

Colussy disembarked in Panama City, flew back to his winter home in Palm Beach Gardens, and resumed working on his short irons. But he continued to brood over the satellites. Why would Motorola shut off the phones of sixty-three thousand paying subscribers? Even if the birds were destroyed next month, why wouldn't they keep the revenue flowing until the last moment? More to the point, why couldn't a giant corporation like Motorola restructure the company with a new business plan? A few days later Castle called from New York to say that Iridium service had started up again as the *Marianne* neared the Cook Islands, but his sources at Iridium were warning him that as soon as his yacht crossed the 120° west longitude line, the service would be lost again as it was passed off to the Thailand earth station, which was no longer operational. *What was going on?*

Colussy decided to investigate, and a few days later he called Castle again.

"What do you think of Iridium as an investment?" Colussy asked his friend.

"At the right price, I like it," said Castle.

And so the conspiracy began. Hence on this Saturday in mid-April Colussy brought his investigative findings to the very room in Palm Beach where, four decades earlier, John F. Kennedy had written the words "Ask not what your country can do for you—ask what you can do for your country." The room was small, but Castle took pains to make it look as it had when Senator Kennedy wrote *Profiles in Courage* while recovering from back surgery. From the window you could see the six palms planted in the exact positions they'd occupied when the Kennedy boys were playing touch football on the soft zoysia grass. It was a place, Castle thought, imbued with good mojo, an incubator of ideas.[1]

John Castle lived in a realm of opulent display and First World creature comforts. An investment banker who preferred the nineteenth-century term "merchant banker," he gave the impression of a throwback tycoon straight out of a Dickens novel: Falstaffian in his girth and appetites, puffy faced and crinkly eyed, fastidious in his refined tastes for Thoroughbred horses, fine wines, society balls, and Park Avenue penthouses. Castle was the very picture of East Coast Old Money, although that picture was misleading. His money was brand spanking new, as he had grown up in the small town of Marion, Iowa, a precocious Eagle Scout who left to attend MIT and scrap his way up through the rarefied investment banking world of New York City. He was regarded, in fact, as the "savior" of legendary investment bank Donaldson, Lufkin & Jenrette (DLJ) during the free-spending eighties, an era when Wall Street's dealmakers were regarded as wizards and alchemists, existing as they did in an arcane universe of leveraged buyouts, corporate raiding, junk bonds, and exotic securities. Castle thrived in that world long after its more famous proponents had either crashed and burned or become persona non grata among their peers. His reputation—first at DLJ, then at his own boutique bank—never lost its luster, even when one of his high-flying takeovers would go unexpectedly bad.

Colussy, who drove over from his winter home facing the eighth hole at BallenIsles Country Club, was also self-made, but in every other respect Castle's opposite. He had grown up in a small Pennsylvania town, then excelled at the Coast Guard Academy, before going on to run Pan American World Airways and several other companies. Soft-spoken, slender, athletic, low-key, a fan of classic jazz and real yachts (the kind that have sails), Colussy had an affable manner reminiscent of Jimmy Stewart. Colussy's story was, in fact, almost stereotypically American, complete with a 1953 wedding to a New England beauty who became a homemaker and mother, two daughters he sent to private schools in both New and Old England, a period of military service, a rise through the corporate world as the family joined yacht clubs and gated communities, a social life full of dinner parties with other buttoned-down couples, and a modest fortune that he built through hard work and sensible investments in tax-free municipal bonds. His children sometimes made fun of him for playing Andre Kostelanetz, Perry Como, and the sound track from *The Phantom of the Opera* during dinner and for drinking the same *chianti classico* for years at a time. Coming of age during the Eisenhower era, he believed that America was the greatest country in the world and that the free enterprise system provided the best opportunities for the greatest number of people. Dan Colussy was decidedly *not* a fan of leveraged buyouts, junk bonds, mortgage-backed securities, or anything that would later be emblematic of the financial world that he called "guys in their pajamas with laptops." Racehorses bored him. But right now he had more immediate concerns: he was looking for someone he could trust from Wall Street, and someone who understood Iridium.

With the clock ticking—two de-orbit dates had already passed, mainly because the government was virtually begging Motorola to wait— the two men needed each other. Castle's merchant bank was a means of quick capital: the company was set up to invest in private equity deals, with Castle himself as the majority stockholder who could green-light purchases without any bureaucratic meetings. Colussy was a hands-on operator who specialized in turnarounds. He had taken charge of the floundering United Nuclear Corporation in the early eighties, when nuclear projects were being canceled all over the country ("thanks to Jane Fonda"), and he had transformed it into a robust, diversified aeronautics firm called UNC that ran more than thirty businesses, eventually selling out to General Electric in 1997 for $725 million, a hefty premium. He then promised Helene that he would never work again and retired to that part of the Florida coast that has more professional golfers and more business tycoons than any place in the world. Even though Castle was ten years younger, he had bonded with Colussy as an outside director at UNC. The two men were famous for their heated arguments during board meetings, sometimes so intense that the other members would take them for mortal enemies, only to be surprised when Castle would drop his withering sarcasm, vote with Colussy, and back his decisions. Their friendship would resume five minutes after the argument was over, as though they had simply been engaging in one of the famous case studies taught at Harvard Business.

Still, this was no routine turnaround. Iridium was the biggest bank-ruptcy in American history.[2] Iridium was a $6.5 billion sinkhole, larger than anything either man had ever faced. That's why, when Colussy arrived on Ocean Boulevard that day, he didn't come empty-handed: he had a thick Motorola intelligence file.

Colussy had visited the only remaining Iridium office in Reston, Virginia, and spoken to the man left presiding over a skeleton crew of just seventeen employees: Randy Brouckman. Brouckman was a Stanford-educated electrical engineer who had run the company's soft-ware systems during the years leading up to the launch of Iridium, but now he was the last man standing after every other executive had left. When asked if he was the Chief Executive Officer of Iridium, he paused for a moment and said, "Yes. Yes, I suppose I am." Brouckman made it clear that the one-month delay promised by Motorola was over and the de-orbit was next Wednesday—five days away. So it was not just the eleventh hour; it was 11:58 verging on 11:59.

In most bankruptcy situations, it would be possible for a company to lie dormant for a few months until it was revived. In this case it would take one order from the Motorola CEO and the satellites would vanish. As Castle and Colussy started evaluating their options, the enormity of what was about to happen was just starting to sink in. What was at stake here was nothing less than one hundred years of scientific and technological development that, if destroyed, might never again be duplicated. The Iridium constellation was something the public probably didn't even understand. Jaded by the word "satellite," thinking of them as passive reflectors for their TV signals and technical doodads that floated around in space spying on other countries, the average investor might never get up to speed on what Iridium really was. Even the science-minded probably put these satellites in the same class as the Defense Navigation Satellite System, better known as GPS, which had been in orbit since 1973 but only available to the public since 1988.

The Iridium engineers who worked at the SNOC actually made jokes about what a snoozefest it must be to work at the Colorado Springs control room for GPS, which consisted of twenty-four satellites positioned 12,600 miles above Earth, because the Iridium birds were the polar opposite of that—literally. They moved in low fast orbits, north and south, over the Poles. They were just 485 miles above Earth, or 2 percent of the height of a typical geostationary orbit, and traveled at 16,776 miles an hour, fast enough to pass over the North Pole fourteen times a day, or every 100 minutes and 28 seconds. The sixty-six Iridium satellites could also communicate in space, unlike the GPS system, so that when a phone call went up, it could be passed among the sister satellites and downlinked anywhere in the world, if the receiver had an Iridium phone, or through an earth station, if the sender was calling a non-Iridium phone. (If two users were both talking on Iridium phones, the call would never come to Earth. It would simply bounce from phone to satellite to satellite to satellite to phone—up to twelve satellites could cooperate on a single call, depending on where the users were located—and it was virtually untraceable.) A call passed through a traditional geostationary satellite could have up to a half-second delay, causing a jerky, disconnected effect in the signal, or creating the lengthy pauses often seen in network TV conversations between an anchor and a field correspondent. The Iridium signal, on the other hand, was virtually instantaneous, even though the actual cells were speeding across the surface of Earth at 4.6 miles per second.

Pick any location on the planet, and there was at least one Iridium satellite between the horizons, probably two or three. That satellite would be there for about nine minutes as it passed overhead, then another one would replace it. If you've ever seen a fast-moving "star" in the night skies, it was probably an Iridium satellite, not a meteor, and its silver antennas were reflecting the sun's rays, creating a phenomenon called "Iridium flare," which is brighter than Venus and can last up to ten seconds. But it was the interlinked antennas that made Iridium satellites unique, as they made autonomous decisions and carried on a constant conversation with one another, with each satellite able to pass signals to four other satellites, resulting in the almost instantaneous transmission of any call to any point on Earth. The satellites were arrayed around the globe in evenly spaced rings, eleven satellites per ring, with 2,485 miles between one satellite and the next in the chain. You need a three-dimensional model to appreciate the beauty of the system, but if you can imagine Earth as a solid sphere inside a slightly larger virtual sphere made up of six strings of satellites racing up one side of Earth and then down the other, while Earth rotates beneath them at a right angle to their planes, then you'll have some idea of its complexity. The six columns were arrayed along half of the Equator, so that they flew up to the North Pole in a staggered "herringbone" formation, with Planes 1, 3, and 5 even with one another, and Planes 2, 4, and 6 also arrayed in military rank. After converging at the Pole, they flew down the other side of Earth, converged again at the South Pole, then fanned out to the places at the Equator where they began. But because they were moving north-south while Earth rotates east-west, each satellite would eventually pass over every inch of the planet. When the satellites crossed the Poles, there were so many overlapping beams that most of them were turned off, and they also had to reorient themselves at the precise moment they stopped moving north and started moving south. When approaching the North Pole, a Plane 2 satellite had to look to the right to talk to Plane 3 and to the left to Plane 1, but after passing the North Pole the same satellite had to look to the right to talk to Plane 1 and the left to Plane 3. And since sending all the satellites in the same direction on one side of Earth resulted in two of the planes going in opposite directions on the opposite side of Earth, the spacing was not precisely even—Planes 1 through 6 were 31.6 degrees of longitude apart, leaving only 22 degrees for the "seam" between Plane 1 and Plane 6, where the satellites were traveling in opposite directions. At the North Pole, an Iridium satellite passed overhead about once every ninety seconds. In the unlikely event of a malfunction, there were several spares circling Earth at 414 miles up—71 miles below the working satellites—in what was called a "storage orbit." If one went out of service, it was a fairly simple procedure to boost the spare higher and turn it on.

Because there were so many satellites traveling so close together and communicating in such a complex way, the orbits had to be eccentric. If the orbits were perfect circles, you would have a three-satellite collision over the North Pole every 50 minutes and 15 seconds, and within nineteen hours the whole constellation would be destroyed. To avoid that, the orbital paths were slightly disinclined, from 90 degrees to 86.4, and they were offset by an eccentricity parameter of 0.003 so that they never intersected.

Of course, there was no need for the Iridium user to know or care about any of this. He didn't even have to be concerned about the direction of his antenna. Whereas all other satellite devices had to be pointed at the geostationary satellite, Iridium did the reverse: the satellite found you, locking on to the handset no matter how fast it was moving.

Colussy and Castle couldn't have known all this at their first meeting, but they had a vague sense that this was one of those deals that comes along once in a lifetime. They also had the sense that they were looking at a unique system that was light-years ahead of anything previously put into space. So what were they not understanding? Sure, the company had huge debt—six hundred claims totaling $11 billion at last count—but that was what bankruptcy was for, to get rid of debt. Sure Craig McCaw had rejected it—"We're not interested in low-bandwidth mobile services," said McCaw's huffy CEO as he backed out—but who else had sixty-six perfectly functioning satellites in the sky? Colussy thought the remaining Iridium executives did seem like profoundly discouraged men. Each passing day brought more bad news for the company. The Japan gateway was shut down and locked. The India gateway was being powered down. The Thailand gateway was closed, and that's the one that controlled the whole western Pacific. There were several businessmen who had checked in with the bankruptcy court, claiming they had the means to buy the company and run it, but most of them seemed like crazies.

The one solid name among the potential buyers was a guy named Stan Kabala. He was a seasoned telecom executive who had worked in Canada for AT&T and Rogers Cantel, and since Colussy figured he needed all the help he could get, he gave him a call. Kabala now lived in Naples, Florida, where he was running a company called Executone that was, among other projects, developing a proposed National Indian Lottery. He had already worked up an Iridium restructuring plan, so at the last minute Colussy and Castle decided to invite Kabala to the get-together at the Kennedy estate.

It took Kabala four hours to drive over from the opposite Florida coast, and he arrived energized and enthusiastic about rescuing Iridium from bankruptcy. He systematically laid out the conventional wisdom. Iridium had done a disastrous marketing job, trying to sell a narrow-band system in a rapidly changing environment where cell towers were available even in the wilds of western Pennsylvania, if not always the wilds of western Montana. As press reports had relentlessly reminded everyone, the Iridium phones were heavy, boxy handsets that looked more like World War II Walkie-Talkies than telephones. You usually had to be outdoors to use them—they wouldn't work in your hotel room or office—and a long antenna had to be extended to get good reception. For the privilege of using the ponderous contraption, you paid $3,500 for the phone itself and anywhere from $4 to $7 a minute every time you used it. No wonder Iridium never made enough money to pay the interest on its operating loans.

Kabala had come to a lot of the same conclusions Colussy and Castle had, but he had also ferreted out usage rates for the Department of Defense and the State Department. It became clear at a very early stage that there was only one big customer who was going to find it hard to live without the Iridium phone: the U.S. government. The owners of Iridium never wanted to be tied to the government and, in fact, had alienated agencies that might otherwise have become customers. Instead, Motorola planned to market the phones to an elusive creature called the "international business traveler"—only to find out that a) there weren't enough of those, and b) all of them already owned sleek, portable cell phones that worked anywhere in the civilized world at a fraction of the cost. Iridium had burned through almost $200 million just on advertising, trying to get the superrich and the supercool to adopt its handset. It would go down as one of the least effective campaigns in advertising history. The Iridium satellite telephone, as marketed, deserved every bit of its reputation as the most useless techno-toy ever created.

And yet . . . there were those sixty-six satellites, sailing around the planet in a geometrical canopy, blanketing Earth with sensors penetrating into the Amazon rain forest and the remotest parts of the oceans and every square foot of every major city, waiting for phone calls that never came. Colussy didn't know specifically what they could be used for, but he knew they could be used for *something*. Could it be that something so ambitious and powerful was just a science project? Wall Street was calling it exactly that, a "science project," and when investors used that term, it was the same as saying, "Please get out of my office." The investment community had watched as Iridium introduced a satellite phone to the entire civilized world and watched as the world rejected it, and so they had moved on.

But the key word here was "civilized." Cell phones work *only* in the civilized world. They don't work in remote areas, they don't work on any large body of water, and they don't work in the 86 percent of Earth that has no cell-tower infrastructure. When a hurricane or other natural disaster strikes a region, the first thing that disappears is cell service. Even in areas already served by satellite, there's no way to communicate if you're very far north or south of the 65th parallels. This is because almost all communications satellites are set in fixed stationary orbits above the Equator. If you live in northern Alaska and you want to get satellite TV, you have to point your dish almost dead even with the horizon so that the signal can skim just over the curvature of the earth as it emerges from the Western Hemisphere satellite. Even if you're closer to the Equator, a mountain or a tall building between you and the satellite can make access impossible. The Iridium satellites, on the other hand, covered every part of the planet with equal clarity and consistency, making Iridium the best friend of the doctor in the Ugandan field hospital, the Arctic oil-rig operator, the Marine cut off from his unit, or, as it turns out, men with secret identities who sneak across borders in the dead of night.

When Motorola cut off service on March 17, 2000, lives were instantly endangered, notably those of Jo Le Guen, a Frenchman attempting to row solo across the Pacific from New Zealand to Cape Horn, and two Norwegians on a four-month trek from northern Siberia to Ward Hunt Island in Canada by way of the North Pole, dependent on Iridium phones for reports of deadly ice crevasses on the shifting glaciers. (All three adventurers survived, but Le Guen suffered the amputation of his toes after he capsized and had to be rescued by the Chilean navy.) Iridium was revolutionary for previously unreachable regions of the planet—but alas, the reason it's called civilization is because that's where the consumers live. And that was why Wall Street thought Iridium was a colossal failure that could never be saved. As one analyst put it, speculating on what kind of "international business traveler" would need an Iridium phone, "Not many people travel by camel."

Given the spectacular flameout of the business plan, the first thing the three men in Palm Beach had to do was make a leap of faith. Without knowing how best to use the satellites, they had to assume that somehow, some way, they *could* be put to use, and at a profit. Kabala's initial inquiries indicated that Iridium had been spending $100 million a month just to keep its doors open, but the conspirators were all experienced enough to know that that number was likely to be inflated, especially since Motorola was trying to exaggerate the damage it was incurring for the benefit of the bankruptcy court. They also discussed the name "Iridium," which would probably have to be abandoned. It was a meaningless word to most people. It didn't even begin to suggest the promise and power of the product, and Wall Street regarded it as a synonym for a business debacle. Kabala suggested changing the name of the company to "Polaris," which he had already registered, and Castle and Colussy agreed that was a much better choice.

After Kabala excused himself and went back to Naples, Castle and Colussy roughed out an estimate of costs as well as sources of potential revenue. Iridium had been paying Motorola $45 million a month just to operate the system. Obviously that had to change. Kabala's suggested way of doing it was a little questionable, though: he wanted to slash salaries and give the engineers and technicians stock options instead. Colussy knew that Motorola's salaries, as carried on the official company books, were not the real salaries—that number was always padded with overhead—so pay cuts wouldn't be necessary. Besides, the last thing anyone needed was two hundred disgruntled engineers and scientists fleeing the company with irretrievable background knowledge. Some of the other ideas in Kabala's plan seemed a little flaky as well, such as his proposal to sell Iridium phones in Walmart and Costco stores. They certainly intended to charge less than $3,500 a phone and $7 a minute, but not *that* much less. This was still a specialized product for specialized uses, and it was unlikely that every deer hunter would have one in his pickup truck. Kabala also placed a huge emphasis on "worldwide alarm monitoring"—using Iridium to notify call centers when a security alarm was triggered—and he thought Iridium would function well as a luxury car phone. But Colussy was increasingly inclined to bypass the consumer market altogether. This was obviously a tool for businesses and governments.

And one of those governments was China. While making prelimi-nary calls, Colussy had spoken to Ed Staiano, the Iridium CEO ousted a year earlier by disgruntled partners, and Staiano went into a long spiel about conversations he'd had with the Chinese Minister of Communications, Huang Zhendong. China had been one of the original Iridium investors, mainly by contributing the use of its rockets and launch facilities, and in its quest for rapid industrialization, it was now focused on bringing phone service to the most remote regions of the country, regions too isolated for cell towers or trunk lines. Staiano had made several trips to China expressly to seek business from Huang, and finally the minister indicated he would be interested in buying up to fifteen thousand Iridium phones—one for each inaccessible Chinese village—hoping that the phone would serve as a sort of community link to the outside world. But Huang had one concern: he didn't want the phones if they could be used to call outside China. Oddly enough, Iridium did have a way to do that—the satellites could recognize international borders precisely and refuse to pass a call outside a designated geographical area. Staiano had gotten so excited that he started developing a prototype for a solar-powered phone booth for each village. What Colussy liked most about the Chinese idea were the big numbers: fifteen thousand village phone booths in China could be worth about $75 million in annual revenue. Kabala had also spoken optimistically of getting government clients for the phones, including all the American military allies, but his projections for that business seemed wildly optimistic.

All of these were back-of-the-envelope computations, none of them supported by research, but Colussy and Castle were not that worried, as both were old hands at estimating things that can't be estimated. At the end of the day, they decided the range of what they would need was anywhere from $100 million to $200 million, probably closer to the latter. That would be enough to buy the system out of bankruptcy and run it until positive cash flow kicked in.

Castle said, "Don't worry, Castle Harlan is good for a hundred million, and if we need more I'll help find it."

In fact, they thought maybe they'd already found the rest. Kabala said he was in touch with two "Beltway bandit" firms that fed off government contracts. He also had a relationship with a San Diego real estate company that he thought would be interested, mainly so it could go into competition against ADT and other home security companies. Less appealing were Kabala's plans for managing the new company. He said he would be CEO, but he would commute from his home in Naples to the headquarters in Virginia. The last thing Colussy wanted to do was come out of retirement to run a company, so he was grateful for Kabala's offer to be the hands-on executive. But Colussy also had extensive experience with commuting executives, and he didn't like them: they left on Thursday night and they returned on Monday afternoon. That didn't really show the level of dedication that would be required.

Still, this was an argument to have later. Colussy and Castle agreed to get some offer papers in front of the bankruptcy court as soon as possible to try to stave off the de-orbit. But the bigger question for the long run was: Where would the revenue come from? You weren't going to be able to monetize this company with Himalayan mountain climbers and Congolese diamond prospectors and Chinese phone booths. Colussy had identified several industries that he thought would want the system—oil exploration, mining, forestry, anywhere employees worked in remote areas—but all those taken together still weren't enough to put Iridium into the black. The two best bets, in his opinion, were the shipping industry, which was a proven user of satellite phones, and the aviation industry. When Colussy ran Pan Am, the airline served 121 cities in 84 countries, and its communications system was a patchwork quilt of phones, radios, cables, and telegraphs that worked most of the time but would have been a thousand times smoother if he could have parked an Iridium phone in every cockpit. There was no question in his mind that the airline industry, and anyone who owned an airplane, would adopt Iridium as the standard. And from his experience as a Coast Guard officer and lifelong sailor, he knew that the ship-to-shore system of marine communications, virtually unchanged for twenty years, was expensive and unreliable. Cell phones were worthless in the oceans. That had to be an easy-to-take market.

Then there were those sixty-three thousand people who already owned Iridium phones. They would be essential customers as a baseline to build from. The problem was that they were all angry. After paying exorbitant monthly rates, their service had been shut off without notice. But Colussy was an old marketing hand. Back in the sixties he'd suffered through a year as an account executive at Wells Rich Green, the power-house Madison Avenue agency where he successfully sold the edgy "Up, Up and Away" campaign to a skeptical board at Trans World Airlines (TWA). He eventually became frustrated working at a place dominated by long-haired creative types who smoked marijuana in their offices, telling CEO Mary Wells Lawrence, "I'm too square for the advertising business," as he left for a more straight-edge marketing job at Pan Am. Still, he knew how to build a customer list, and he was sure that by giving the disaffected Iridium users some initial free service, most of them could be retained as subscribers.

By the time Colussy and Castle concluded their meeting, both men were fairly confident that they had the rough outline of a business plan. What they didn't know was that almost all the assumptions they'd made—sources of financing, the willingness of the government to extend contracts, the mood of the Iridium users, the attitudes of foreign governments, the partners most likely to invest, the division of labor among the three men present in Palm Beach, the interest of the aviation and maritime industries—were wrong.

Chapter 2

NERDS, NAZIS, AND NUKES

MAY 2, 1945
VILLAGE OF OBERJOCH,
BAVARIAN ALPS, GERMANY

Only a scientist could use the word "love" when speaking of a satellite. The Iridium space vehicles were created and nurtured entirely within the domain of stolid, sensible engineers—not a species known for singing romantic ballads—and yet, when searching for metaphors, they would lapse into *amour*: "I love the system." "We all loved the constellation." "There was a lot of love for Iridium." In some cases it was an embarrassed love, like the young boy infatuated with the girl no one else likes. In others it was a rebellious love, inflamed by parental disapproval. But in most cases it was the inner thrill of seeing a beauty no one else was aware of, because the satellites weren't attractive in any outward way. They were gold, silver, and black—bold and tacky, like cheap gowns sold in casino dress shops—and their "wings" (actually solar panels used to charge the battery) were ungainly. They were skinny as satellites go, at barely fourteen hundred pounds, and they went around the planet at the wrong angle—over the Poles instead of around the Equator. It was not uncommon for people close to the Iridium program to refer to them, years later, as "my satellites." Yes, there were people who hated them, but their hate had a peculiar tone—it was the plaint of the jilted lover.

How did these satellites hold the fascination of so many people? How did they develop such a legion of lovers?

Iridium was, in fact, the culmination of one hundred years of sci-fi geek history.

The idea of sending an object into orbit around the planet is so odd to the nonscientist that no one even thought it could be done until 1897, and even then it remained in the realm of theory because no rocket existed that was powerful enough to thrust through the ionosphere to start the orbit. The first proof that satellites were possible came not from a scientific laboratory or a university, but from a partially deaf high school math teacher speculating from the little town of Borovsk on the bleak Russian steppe. Konstantin Tsiolkovsky was thirty-nine years old when he wrote an article called "Investigating Space with Rocket Devices" and submitted it to a Russian science journal, but it wasn't published until six years later, in 1903, the same year the Wright brothers achieved flight at Kitty Hawk. Tsiolkovsky's article was all but incomprehensible to the layman, consisting as it did of elaborate mathematical calculations, and in fact went mostly unnoticed in the West except by other scientists obsessed with space, almost all of whom were dismissed by their respective governments as kooks. Tsiolkovsky, then, was the prototype for every satellite inventor of the next century, a combination of dreamer and scientist who, from boyhood, was fascinated by stories about space travel, especially the novels of Jules Verne and H. G. Wells. He even wrote articles about how man would someday communicate with creatures from outer space, and how the eventual colonization of space was the key to human immortality. No wonder his fellow citizens in Borovsk thought of him as a bizarre eccentric.

For the next half century, most of the writing about orbital vehicles used the term "spaceships," not "satellites." It was assumed that anything put into orbit would be operated by human beings who would function as latter-day explorers. And almost all the early speculation about satellites was also concerned with rockets, since nothing could happen, space-travel-wise, until someone developed a rocket powerful enough to overcome the gravitational pull of nature. The necessary speed of the rocket, as calculated by Tsiolkovsky, was 11.2 kilometers per second, and for the next forty years this would be the holy grail of outer space visionaries. Who would be the first to develop a rocket with enough thrust to accelerate a heavy object to 25,000 miles per hour?

But human nature being what it is, the only way to get money to build rockets was to design rockets that killed people. So there was always a disconnect between the men who wanted to put things into space and the men who controlled their means to do so. The advanced rockets developed in the United States and Europe leading up to World War II were designed to travel vast distances very close to the ground and then explode on impact, destroying as much as possible. The science of telemetry—communicating with the rocket and its payload—was mostly devoted to guiding it toward an enemy target. It was an American, the physicist Robert H. Goddard, who solved the principal technical problem of an outer space rocket in 1926 when he mastered the use of liquid fuel, but he would die in 1945 without seeing his invention used for any purpose except war. When scientists would ask the military for rockets to be used for peaceful purposes—firing them into the stratosphere, for example, to get information about cosmic rays or geomagnetic storms—they would inevitably be turned down. And it didn't help that many of the scientists were also writers of science fiction.

Herman Potočnik, a Slovene who worked as a rocket engineer in the 1920s in Vienna, wrote a book under a pseudonym called *The Problem of Space Travel: The Rocket Motor*, which was mostly ignored by his colleagues in the Austro-Hungarian scientific community but was reprinted in serial form in the American pulp magazine *Science Wonder Stories*. Hermann Oberth, a Transylvanian high school math teacher who first developed the concept of the multistage rocket, wrote a doctoral dissertation in 1922 called *By Rocket into Planetary Space* that was rejected by the University of Göttingen as "utopian," its author nothing but a "romantic futurist." Oberth would become the scientific consultant on the first film to feature scenes from outer space, Fritz Lang's *Frau im Mond* (*Woman in the Moon*), which premiered in 1929 and inspired a whole generation of amateur space-flight enthusiasts, most of whom joined the Verein für Raumschiffahrt (Society for Space Travel), a Breslau-based club that stirred the imagination of its youngest member, seventeen-year-old Wernher von Braun. Oberth would turn up again in the 1950s, serving as advisor to the Hungarian producer George Pal's space-travel film *Destination Moon* and contributing articles to *Flying Saucer Review*, where he traced alien visitations back to references in Pliny the Elder—but then again, there was not much separation at the time between science fiction writers and science *fact* writers. When John Robinson Pierce, Director of Research at the prestigious Bell Laboratories in New Jersey, first wrote about the potential benefits of orbiting communications satellites, it was in *Astounding Science Fiction* magazine in 1952. His article appeared right alongside the work of George O. Smith, whose system of communications-relay satellites between Earth and Venus was first proposed in 1942 and then expounded in pulp paperbacks such as *Operation Interstellar* and *Lost in Space*, all of which were taken seriously by the satellite community even though Smith had no standing as an academic. And the man who first calculated the orbit at which a satellite would stay in the same position above Earth—26,199 miles from the center of the earth, or 22,237 miles above mean sea level at the Equator—was not a scientist at all but Arthur C. Clarke, science fiction writer, retired Royal Air Force electronics officer, and proud member of the British Interplanetary Society, in a letter to *Wireless World* magazine in 1945. (Today a geostationary orbit is called the "Clarke orbit," and there are more than 450 satellites that use it.)

The geeky scientific dreamers went on publishing their stories about space travel, waiting on the technological means to lift their visionary vehicles into space, but it was the Nazi war machine that finally made it possible. The Peenemünde Army Research Center, based on the tiny island of Usedom in the Baltic, was where Wernher von Braun and his team of four thousand engineers used slave labor to build the most powerful rocket the world had ever seen, the V-2 "Vengeance" missile used to terrorize London from six hundred miles away. (Painted on the base of the V-2 was the logo of *Frau im Mond*, the Fritz Lang movie.) It was typical of rocket scientists that on October 3, 1942, the day the V-2 achieved its first successful launch, Nazi General Walter Dornberger's celebratory toast at Peenemünde was devoted not to winning the war but to "making space travel practicable." In fact, von Braun spent two weeks in a Gestapo prison after being overheard telling someone he would prefer to be building spaceships instead of military rockets—a comment ruled disloyal to the war effort. Von Braun's design center, the Heeresversuchsanstalt, also developed plans for an intercontinental missile that could reach the United States from Germany, but that technology would have to wait until von Braun switched teams.

As Nazi Germany started to collapse in early 1945, von Braun feared the savagery of the Red Army, which was approaching Peenemünde from the south, but after what he'd done to London, he feared the British army even more. He had a half-dozen conflicting orders on his desk, including orders to fight to the last man, orders to preserve his research, orders to destroy his research, and orders to retreat to various places in the mountains. He finally decided to evacuate all five thousand scientists and engineers and their families, taking the top hundred officers with him as he crossed Germany north to south and made his way toward the American lines. Von Braun's insurance policy was the blueprint for the V-2 missile—fourteen tons of paper that he loaded into trucks and stowed in an abandoned iron mine in the Harz Mountains while his team made its way toward Bavaria, using his status as an SS officer to avoid being detained. The German rocket scientists ended up waiting for General Patton's army at the resort village of Oberjoch, on the Austrian border, and it was von Braun's brother Magnus who rode a bicycle out to the edge of town and accosted the first American soldier he could find.

That turned out to be Private First Class Frederick Schneikert, who was perplexed by Magnus von Braun's message and told him frankly, "I think you're nuts." But von Braun's broken English and Schneikert's broken German were eventually sorted out, and passes were issued so that the von Braun team could leave their hotel and surrender to the First Army, headquartered at the nearby Austrian village of Reutte, on May 2, 1945. It was at that moment that the entire history of space exploration, and of satellites, was altered.

Von Braun was moved to "P.O. Box 1142," the notorious secret prison in Fort Hunt, Virginia, where his membership in the Nazi Party and his career as an SS officer were erased from his biography so he would qualify for amnesty. All the Nazis were then scrubbed and debriefed in Operation Paperclip (code-named for the way their false biographies were attached to their personnel files) so they could be sent to Aberdeen Proving Ground in Maryland, then White Sands Missile Range in New Mexico, where sixty-four V-2s were reassembled for testing. For years afterward, V-2s were fired vertically into the skies, eventually achieving altitudes of 244 miles, and the design of the German rocket was used for many of the earliest American missile launchers. Von Braun was later assigned to the Redstone Arsenal in Huntsville, Alabama, and put in charge of rocket development, but once again he bristled at being forced to work on military matters instead of his first love: spaceships. In his spare time he became an advisor to Walt Disney, and in 1952 he wrote an article for *Collier's* calling on America to immediately start building "cargo rockets" and a wheel-shaped Earth-circling space station to be used as a terminal for all the ships that would soon be shuttling to and from distant planets. Von Braun had wanted to bring 500 scientists to the United States, but the Americans accepted only 177. The others were captured by the Russians, so both countries used the technology of Peenemünde throughout the Cold War to build rockets, ICBMs, and, almost as an afterthought, satellites.

Still, no one had thought of bouncing a radio wave off a satellite until, in the middle of World War II, the radio intelligence agents of the Federal Communications Commission encountered a strange and scary phenomenon: radio messages from outer space. Every day these agents "cruised the ether" using aperiodic receivers scattered across North America to search for Nazi fifth column radios as well as signals from enemy submarines on both coasts and Nazi agents in South America and Africa. ("Aperiodic" simply means capable of responding to every radio signal, no matter where it appears in the spectrum, without having to turn a dial.) Each officer would identify suspicious transmissions, then send his data to the Radio Intelligence Division in Laurel, Maryland, where technicians would triangulate the signals with a device called an Adcock Direction Finder. Once they had a general idea of where the signal originated, an unmarked Hudson sedan would be sent to the suspected area. This prowl car had a short-range direction finder that could be extended through a hole in the sedan's roof, allowing officers to further narrow down the location. Finally, an undercover agent would be outfitted with a "snifter" (a direction finder concealed under his clothing) so that he could get close enough to identify the building and then the room, and presumably destroy the Nazi transmitter or turn the agent. At any rate, this was how the system was supposed to work, but the FCC was perplexed by certain signals that seemed to come from nowhere and could never be triangulated. These transmissions were truncated, appearing sporadically, only at certain times of the day or night. So James H. Trexler, an engineer at the Naval Research Laboratory's Offboard Countermeasures Branch, started studying these "anomalous propagation" signals and eventually discovered that they were coming from . . . the moon. If the moon happened to be in the pathway of a radio signal, the signal would reflect back to Earth at the precise angle it glanced off the moon's ionosphere—which was news to everyone, since until then no one knew the moon *had* an ionosphere, and so it was assumed that any signals striking the jagged lunar surface would glance off at random angles.

After the war, both the U.S. Navy and the U.S. Army Air Force came up with plans for an orbiting spaceship—the word "satellite" was still not in common use—that would be used to reflect radio-wave messages as well as listen in on Soviet transmissions, but the Pentagon killed every proposal as so many speculative pies in the uncharted skies. The Army even went so far as to prove it could be done by sending and receiving the first moon bounce on January 10, 1946, when a 3,000-watt pulse was sent from a laboratory at Fort Monmouth to the lunar surface and back down to New Jersey in 2.5 seconds. Later that same year, the Douglas Aircraft Company published a feasibility study called *Preliminary Design of an Experimental World-Circling Spaceship*, but it failed to impress the Pentagon, mainly because the satellite wouldn't be able to carry any weapons. Various pro-satellite forces inside the Pentagon managed to get development funds placed in the 1948 defense budget, only to have the money stricken when there was a public outcry about using scarce public resources on what were regarded as fanciful playthings for scientists.

The only way communications satellites were ever going to be built was to find some warlike purpose for them, so most of the continuing research went on at the dreary top secret Naval Research Lab in Maryland. It was there, on July 24, 1954, that Trexler sent his own voice from the Stump Neck Radio Antenna Facility to the moon and back, amounting to the first satellite phone call. (For security reasons, all he transmitted were vowel sounds.) This breakthrough led to Operation Moon Bounce, a program to use the moon for real-time communications, and a year later the first moon-bounce telegraph message was sent from Stump Neck to the Navy Electronics Lab in San Diego. By 1956 a recorded message by President Eisenhower had been sent from Stump Neck all the way to Wahiawa, Hawaii, to the same military communications center that would be used by Iridium forty years later. Throughout the early fifties, the Navy was establishing leadership in the science of manipulating radio signals in outer space, having set up a complicated system that allowed its ships to send and receive messages via the moon.

But obviously a moon-based system had major drawbacks—namely that the moon was not always on the side of the planet where you needed it and, depending on its trajectory across the skies, would be useless if you couldn't get the right angle on it. The self-evident solution would be to replace Earth's natural satellite with an artificial one, but the only scientists who seemed to be focused on that project were at Bell Labs, the research arm of AT&T, and they couldn't do anything without the military loaning them a rocket. Fortunately the upcoming International Geophysical Year (IGY) gave everyone an opportunity to claim rockets. The scientific community had known for a long time that the eighteen months between July 1, 1957, and December 31, 1958, would be a period of maximum solar activity, creating perfect opportunities to use new technology—especially rockets, radar, and computers—to study outer space and the oceans. So in 1954 the International Council of Scientific Unions met in Rome to propose the equivalent of the world's biggest science fair, challenging countries to put aside their differences and contribute projects that would map and describe the expanding universe. Since President Eisenhower had made several speeches about making sure outer space was used for peaceful purposes, the handful of satellite men in the armed services saw this as their big chance.

First in line for IGY funding were the moon-bounce guys at the Naval Research Lab, who wrote a proposal in 1954 for Project Vanguard, an experimental satellite that would orbit Earth sometime during 1957 or 1958. The Glenn L. Martin Company, creator of the Viking rocket developed at White Sands using Nazi technology, was given the contract to build the launcher. The Naval Research Lab would create the satellite itself, an orbiting laboratory that would be carried aloft in order to chart what might lie beyond Earth's atmosphere. The project would be run by the National Academy of Sciences—to make it clear that it was a civilian, not a military, project—and the whole thing was announced on July 28,

1955, at a White House press conference full of speeches about how important it was to move away from the emphasis on building ICBMs and start building space vehicles that would benefit mankind. Whether "freedom of space" was something Eisenhower genuinely believed in is a matter of some historical debate. The Air Force had taken that 1946 study by Douglas Aircraft and used it to develop plans for a spy satellite—and the very first satellite authorized for production, in 1954, was called Weapons System 117L, which doesn't exactly sound like a peaceful use of space. So what's more likely is that the Pentagon wanted to test the right of countries to protect their airspace from spy satellites. Having satellites circling the planet for purely experimental purposes would establish the principle that there was no such thing as outer space airspace.

Not so surprisingly, idealistic notions of a scientific United Nations in the sky died quickly. Less than two weeks after getting the contract for the Vanguard rocket, the Martin Company got hired to build the new Titan missile for America's fledgling ICBM system. As any satellite man could have predicted, all the top Martin engineers were pulled off the Vanguard project in Baltimore and sent to Denver to design the airframe of the Titan. The Vanguard program then puttered along with very little notice, greatly annoying the Air Force generals who had to make room for it at the Atlantic Missile Range, part of Patrick Air Force Base near Cocoa Beach, Florida, later to be called Cape Canaveral.

To say that Vanguard was the most neglected space project in history is no exaggeration. First the Vanguard team was forced to share Launch Complex 18 with Douglas Aircraft, which was developing the intermediate-range Thor missile. Then, because they were so chronically underfunded, the Vanguard builders had to scrounge for materials, at one point being reduced to ordering Victor mousetraps from the manufacturer in Pennsylvania to harvest the cadmium-plated springs, then sending engineers to a Sears Roebuck department store to buy screw jacks. They asked the U.S. Army Corps of Engineers to send a salvage team out to White Sands to bring back the ninety-five-foot gantry they used for their Viking missiles, believing they were doing the country a favor by saving the cost of a new one, but Major General Donald N. Yates, commander of the Air Force Missile Test Center, barged into a Vanguard planning session to fulminate that "the structural monstrosity proposed for your launch complex will be erected over my corpus delicti!"

Fortunately the Vanguard scientists had been assigned an equally profane troubleshooter, Captain Winfred E. Berg, who told Yates that if he interfered with their plans, they would just get permission to use the Redstone launch complex instead and mess that up worse. Eventually Yates backed down and became fond of the Vanguard people, mainly because they were such underdogs—not just low-priority people but no-priority people. When they would request special equipment, he would complain they were "clobbering up my range," but privately seek to find what they needed. Still, even with all the cost-saving measures and the sharing of facilities, the Vanguard budget ballooned from $9.7 million to $110 million, causing the Pentagon and the National Science Foundation to start fighting over who should pay for it.

By the summer of 1957 work on the Vanguard satellite had slowed to a crawl, causing Rear Admiral Rawson Bennett, Chief of Naval Research, to pay a personal visit to Launch Complex 18A. Pulling a couple of engineers into a room, he said, "What's going wrong down here anyway?"

"Just one thing," answered Daniel Mazur, head of telemetry. "Instead of rockets, Martin is sending us *shit*."

And indeed by that time Vanguard engineers had come to think of the Glenn L. Martin Company as careless and sloppy, an impression reinforced by a visit to its Baltimore offices, where designers worked out of an old plant with broken windows and would sometimes arrive in the mornings to find that sparrows had left droppings on their architectural drawings. Vanguard was just not a priority—not for the government, not for the military, and not even for the lead contractor. Because of the escalating cost and the blasé attitude of the government, there was even talk of scrapping the program entirely—until October 4, 1957, when a reception for scientists at the Soviet embassy was interrupted with the announcement that "Sputnik," the Russian word for "satellite," was already in orbit.

Sputnik was pretty much the lamest excuse for a satellite ever launched—an aluminum-alloy sphere, twenty-three inches in diameter, weighing about 184 pounds—but it was passing over the United States every ninety-six minutes, making the *beep beep* sound that would forever be appropriated by filmmakers depicting objects in outer space. Sputnik was actually a last-ditch effort by Sergei Korolev, the chief designer of Russia's R-7 missile, to get a satellite into space before America did. Like von Braun and his American counterparts, Korolev was a science fiction buff who had read Jules Verne and been an amateur rocketeer as a boy, but his main job in the fifties was working anonymously inside the sexily named Design Bureau Number 1 at Scientific Research Institute Number 88, a secret installation near the elite Moscow suburb of Podlipki. Korolev was known as the most brilliant Russian rocket designer, but times being what they were, individual talent was subsumed under a morass of acronyms and vague official-sounding organizations with purposes known only to a few. Fortunately Korolev was a tough man—he had been imprisoned for six years in the worst part of Stalin's gulag, the frigid Kolyma gold mines—and he knew how to work the bureaucracy.

Most of the Russian satellite men had worked with gliders—or what America called cruise missiles—during the early thirties, and one of Korolev's close friends from that era was Mikhail Tikhonravov, a rocket scientist who figured out the liquid-fuel problem for Russia's Katyusha rockets and was one of the Soviet officers who searched through the forests of Poland, looking for remnants of the V-2s launched from Peenemünde. After the war Tikhonravov was assigned to a "closed city" called Bolshevo to develop applications for ICBMs. The two friends stayed in close touch and thought it would be possible to use Korolev's version of the V-2—called the "Semyorka," but always referred to by its official name, the R-7—to put a vehicle into space. All through the fifties Korolev tried to get the Soviet Union to back a satellite. The Academy of Sciences of the USSR gave lip service to his plans, but when he would submit designs, he would be told that rockets were too expensive and too necessary to the national defense to be wasted on an orbiting science experiment. Korolev asked his friend to do the technical studies anyway, and in 1954 Tikhonravov completed a detailed paper called *Report on an Artificial Satellite of the Earth*. Thanks to President Eisenhower's announcement of the American Vanguard program, it was just enough ammunition to allow Korolev to go to the forbiddingly titled Minister of Medium Machine Building and get minimal funds to do the initial design work. Without much enthusiasm, the USSR Council of Ministers finally approved a small "Object D" to be launched by an R-7 rocket, sending some scientific instruments into orbit during the International Geophysical Year. (*D*, or *Д*, is the fifth letter in the Cyrillic alphabet. The four previous "objects" authorized for R-7 usage were nuclear warheads.)

Then, on September 20, 1956, Korolev got word through KGB spies that the Americans had attempted to launch a satellite from Cape Canaveral—and were almost successful. This report was false. What the spies had detected was Wernher von Braun's test of his Jupiter-C rocket, which did set an altitude record of 682 miles and a distance record of 3,350 miles, but was strictly intended for use as a weapon and had nothing to do with Project Vanguard. This was an easy mistake to make, since von Braun had been begging Eisenhower to let him launch a satellite, but the President was determined to leave the satellite program in the hands of civilians. Von Braun was so upset by being blocked, in fact, that he told an Army general he was going to launch a satellite anyway and pretend it was a "mistake"—a notion that he was dissuaded from only after threats to his funding and position.

Back in Russia, alarmed by what he thought was an almost-successful satellite launch, Korolev scaled back his design to virtually nothing—the engineers called it Prosteishiy Sputnik, or "bare-bones satellite"—and went to the Ministry of Defense with a bold presentation that made it politically dangerous to oppose him. "We made it in one month," said future cosmonaut Georgy Grechko, "and with only one reason, to be first in space." They actually wanted to launch it *before* the beginning of the IGY, to emphasize how far ahead of everyone else they were, but the R-7 had three failed launches during the summer of 1957, bringing into question whether the Soviets' main rocket would ever work at all, much less be available for Prosteishiy Sputnik. Finally, in late August, the R-7 launched an ICBM 6,500 kilometers and hit a target on the Kamchatka Peninsula. Oddly, the announcement of this achievement—which really did threaten world peace—was hardly even noticed by the Western press. Not so thirty-eight days later, when the same rocket was used for Korolev's "toy satellite." The act of launching Prosteishiy Sputnik from Tyuratam—later to become known as the Baikonur Cosmodrome—at 22:28:34 Moscow time on October 4, while the Vanguard was still doddering along, shook the American public to its foundations as a seeming symbol of Soviet superiority in outer space.

Yet everything about Sputnik was primitive to the extreme. The Russians didn't have access to a computer, so they used trigonometric tables and arithmometers (mechanical calculators) for the whole project. They had no tracking stations, so they couldn't even be sure at first that the satellite had achieved orbit. The nervous engineers at Tyuratam had to wait an excruciating eighty-six minutes from the time Sputnik left Russian airspace over the little village of Klyuchi, on the Kamchatka Peninsula, then passed over North America, the Atlantic Ocean, and Europe, before appearing again on the western horizon. The last two minutes were truly harrowing, but when the satellite did make its appearance, the bunkers at Tyuratam broke into chaotic cheering. Soviet Premier Nikita Khrushchev was the first person to be notified. He said, "Fine," and went back to bed, assuming it was yet another routine project to measure cosmic rays or whatever. The next day's *Pravda* gave a brief, dry description of the launch, attaching little importance to it. It was not until the Russians saw the alarmist headlines in the Western press that they realized they had scored a propaganda coup.

The Vanguard team was devastated. Admiral Bennett proved incapable of diplomatic word choices when he told a reporter, "Sputnik is a hunk of iron almost anybody could launch." Secretary of State John Foster Dulles was so apoplectic he couldn't even bring himself to congratulate the Russians. "What has happened," said Dulles, "involves no basic discovery and the value of a satellite to mankind will for a long time be highly problematical." The first half of his sentence was correct, but the second half was decidedly unprescient. Long after ICBM silos had been turned into museums, long after manned space vehicles had become passé, hundreds of satellites would still be circling the planet, revolutionizing our knowledge of Earth, the stratosphere, and the universe on the one hand, and becoming essential to our daily lives on the other. The communications satellite, of which Sputnik was the earliest primitive prototype, was in fact the most useful and valuable result of all space programs, past or present.

Almost alone among elected officials, President Eisenhower was not that worried about it. After being informed of the satellite's size, he asked CIA Director Allen Dulles if someone had misplaced a decimal point. By the time the National Security Council convened a week later, Eisenhower told the assembled officials that "I'm beginning to feel somewhat numb on the subject of the earth satellite." To make his point, he posed a rhetorical question to Detlev Bronk, president of the National Academy of Sciences.

"Were we Americans the first to discover penicillin?" Eisenhower asked.

"You know the answer to that, Mr. President," said Bronk.

"And did we kill ourselves because we didn't?"[3]

Ultimately the chief benefit of Sputnik was to launch the American space program for real. John P. Hagen, the bespectacled, mild-mannered head of Vanguard, had been at the Soviet embassy when Sputnik was announced and was plunged into an immediate maelstrom of "How could you let this happen?" second-guessing. Hauled before Congress and derided by the same politicians who had held up his funding, the hapless Hagen was now showered with money and forced into a launch before he was quite ready. So on December 6, 1957, two months after Sputnik, the Vanguard satellite was fitted into the nose cone of a Juno rocket at Cape Canaveral and prepared for launch. The rocket rose about four feet off the ground, lost power, sank back to the launchpad, exploded, and toppled toward the ocean shore, engulfed in flames. The satellite detached from the nose cone anyway, rolled over on its side, and started beeping like a defective smoke alarm. It beeped for much of the day, in fact, until workers could get close enough to shut it off. Hence the first attempt to launch an American satellite of any kind became known in the press as "kaputnik," "flopnik," and "stayputnik."

Stock in the Glenn L. Martin Company fell so sharply the day after the failed launch that the New York Stock Exchange had to stop trading. At the United Nations, a Soviet delegate asked American delegates if they would be interested in receiving aid from the USSR's program of technical assistance for backward nations. Debate would go on in the engineering world for a long time about what caused the failure—the most likely culprit was the engine itself, the responsibility of General Electric—but the nation didn't care what the reasons were. Secretary Dulles told the President that the United States had become "the laughing-stock of the whole Free World," and could they please stop announcing the precise times and dates of launches until the launches were successful? Not likely, Dulles was told, when the United States was the country saying all this stuff should be publicly available to the world scientific community and *not* shrouded in military secrecy.

By then Project Vanguard had become the whipping boy for everything thought to be wrong with America, and the powers that be felt so humiliated by the experience that they finally asked Secretary of Defense Neil McElroy to go ahead and send up a military satellite. McElroy had been at a Redstone Arsenal cocktail party the night of the Sputnik announcement, so Wernher von Braun, gloating and licking his chops, had gotten in his face and made an impassioned plea for reviving his Project Orbiter, which had been rejected for the IGY in favor of Vanguard. "Vanguard will never make it!" pronounced the Prussian. "We have the hardware on the shelf. For God's sake, turn us loose and let us do something. Just give us a green light and sixty days." As much as Eisenhower despised the idea of using an ICBM rocket for the launch— the same rocket that Hagen had asked permission to use, only to be told by the Army that it was needed for more important things—von Braun was given the go-ahead. So it was the Army, not the Navy, that ended up having the first successful American launch on January 31, 1958, when the satellite Explorer, designed by the Jet Propulsion Laboratory at Caltech and built in eighty-four days, was placed on a Jupiter-C rocket and launched into an elliptical orbit ranging from 224 to 1,575 miles above Earth.

A week later, Hagen was given a second chance with Vanguard, but the results on February 5 weren't much better. His rocket reached an altitude of four miles but then exploded—and it didn't help public perception when Hagen seemed on the verge of tears at the post-launch press conference. The Vanguard's third try, on St. Patrick's Day, March 17, 1958, had to be briefly postponed because—in a one-in-a-million coincidence suggesting the Fates were angry—von Braun's Explorer was passing overhead at the moment planned for liftoff. A few minutes later, the long-awaited journey of Vanguard finally succeeded, reaching an apogee of 2,460 miles in an orbit so perfect, with all scientific instruments working so well, that it should have exonerated Hagen and his team. Vanguard is still in orbit today and expected to stay there for two more centuries—unlike Sputnik, which lasted only ninety-two days before crashing. Project Vanguard had accomplished every single goal of its original mission—building a civilian rocket from scratch, putting up a satellite during the IGY, pioneering the use of miniature circuits and solar batteries, developing the world's first global tracking system—and this had all been accomplished while working in public view, unlike the projects of von Braun and Korolev. The Delta rocket, which would become America's most dependable launcher, evolving into the rocket used for most of the Iridium constellation, took its second and third stages directly from Vanguard. Most important of all, Vanguard showed the world that civilian space projects had value.

But virtually no one made these points at the time, because it was thought of as, if not too little, certainly too late. In fact, the German press called the Vanguard mission "spaetnik," *spaet* being the German word for "late." The Russians had put two more satellites into orbit in the meantime, including one carrying Laika, a stray dog from the streets of Moscow. America was the country whose rockets always exploded. The Soviet Union's Baikonur Cosmodrome, which would send the Luna 2 spacecraft to the surface of the moon the following year, was clearly the space-race leader. The Soviets kept sending requests to the United States to join them in their peaceful scientific efforts by contributing measuring devices and monitors to the Russian satellites, and it was embarrassing for the State Department to keep coming up with excuses as to why the United States didn't want to do that. As Congress debated, the Army fully expected to be given responsibility for all future space programs, and Secretary of Defense McElroy established the Advanced Research Projects Agency (ARPA) to handle that job alongside its work with ICBMs. But history would go in a different direction. A few months later Congress passed the National Aeronautics and Space Act, creating NASA and establishing the firm principle that space belonged to civilians. Hagen had won out over both the Soviets and von Braun, even if the public wouldn't ever see it that way.

Not so surprisingly, it was the American satellites that brought back real scientific discoveries, whereas the Russians contributed virtually nothing to the IGY. Explorer discovered the Van Allen radiation belts, regions of trapped plasma surrounding Earth that are especially hazardous to space vehicles if not properly shielded. (They were named after America's leading astrophysicist, James Van Allen of the University of Iowa, who designed Explorer.) Vanguard discovered the bulge on Earth and brought back data on cosmic rays, the gravitational fields of the moon and sun, meteor dust, and the intensity of solar radiation. The Russians, by contrast, didn't even have enough earth stations to track their own satellites; it was the Vanguard technicians at six American tracking stations who told them where their birds were. In the very early days of Sputnik, before the American system was able to home in on the Soviet radio frequencies, the *sputniki* were tracked with the help of seventy thousand amateur ham radio operators who would log their shortwave signals and send them to the National Academy of Sciences. On one mission the Soviet tape recorder failed to work, rendering all its scientific equipment useless.

In fact, after that first victory, the Russians would never again lead in satellite technology. By 1960 the Pilotless Aircraft Research Division at NASA's Langley Research Center had collaborated with Bell Labs to put up Echo, the first true communications satellite, a huge Mylar "satelloon" that was launched from Wallops Island, Virginia, and passively reflected microwave signals. (The first satellite telephone call was from Goldstone Dry Lake in the Mojave Desert of California to Crawford Hill in Holmdel, New Jersey.) And it quickly became apparent that satellite technology thrived when in private hands and didn't do so well when the Pentagon and Kremlin attempted it. The first advanced communications birds were put up on an experimental basis by AT&T (Telstar), RCA (Relay), and Hughes Aircraft (Syncom), and at one point these companies were so far ahead of Washington that President Kennedy became alarmed that space might become privatized. Kennedy was a fan of the technology, using Syncom to make the first satellite phone call between heads of state when he dialed up Nigerian Prime Minister Abubakar Tafawa Balewa so the two men could chat about the career of Nigerian middleweight boxing champion Dick Tiger. But Kennedy pointed out that the budget for NASA was just $300 million per year, while AT&T had $7.4 billion in sales for the year 1959 alone. Concerned that AT&T could use its position as a protected monopoly to outspend everyone in space, the President insisted on the formation of a public satellite corporation that would be jointly owned by the public and the major government contractors. The result was COMSAT, formed in 1962 and widely feared by the rest of the world as an incipient American monopoly.

To assuage the complaints of Europe, it was agreed that a "single-world" communications satellite system would be built jointly, with 61 percent owned by COMSAT and 39 percent by the western European nations, plus Canada, Japan, and Australia. The resulting company, INTELSAT, was responsible for the most famous launch of the sixties: Early Bird, the first commercial geosynchronous satellite (GEO) and the celebrity satellite of the West. Early Bird was positioned in a Clarke orbit over the Atlantic Ocean, and together with its sister GEOs, Lani Bird and Canary Bird, it accomplished the first live worldwide television broadcast, a 1967 show called *Our World* that was intended to unite the nations—or at least the nineteen nations that chose to participate. (Moscow toyed with the idea, then pulled out.) The 150-minute show featured superstar celebrities from the countries where people were still awake on a Sunday evening—the artist Pablo Picasso, the soprano Maria Callas, the Vienna Boys Choir, the Canadian media theorist Marshall McLuhan—but in Japan, where it was 4:20 A.M., a segment on workers building the Tokyo subway system had to suffice. The show was a propaganda coup—probably the closest the American satellite makers came to the excitement over Sputnik—mostly because of the closing segment in which the Rolling Stones, guitarist Eric Clapton, drummer Keith Moon, singer-songwriter Graham Nash, and singer Marianne Faithfull all joined the Beatles on the final chorus of "All You Need Is Love," composed by John Lennon especially for the occasion as a tacit Vietnam War protest.

Early Bird was the first satellite that actually produced revenue—but there should have been more of them. Kennedy administration policies barring the privatization of space had served as a giant stop sign to AT&T, ITT, General Electric, GTE, Lockheed, RCA, and Hughes, all of which had been poised to spend hundreds of millions on satellite systems. It was more than ironic that Congress formed COMSAT to avoid "a monopoly in space" when there were at least seven competitors ready to go, and that didn't count the British, French, Canadian, and West German companies. It was even more ironic that the government, to use the most polite term, *appropriated* patents owned by Bell Labs and others, because without them the first government satellites couldn't have been built.

In the 1960s, COMSAT and INTELSAT promoted the "open skies" policy of the Eisenhower, Kennedy, and Johnson administrations, but after that the two big satellite companies started to act like monopolies themselves, and America was increasingly perceived as the bully trying to claim all of outer space for itself. The Nixon administration did the most damage by insisting that foreign countries adhere to various American standards if they wanted to use rockets. The Europeans were rightly suspicious that they would never be allowed to launch their own communications satellites on American rockets, thereby endangering the preeminence of COMSAT, and so the European space industry was born, especially in France, which had always been annoyed by the "technology gap" between the United States and Europe. France had had nuclear capability since 1960, when it tested its own atomic bomb, and it had dropped out of NATO in 1966 so that, among other defense goals, it could develop its own ICBMs. So it's no accident that the first European rocket able to launch communications satellites was French—the Ariane, first used in 1979 from the French space complex in Kourou, French Guiana. Thus the fascination with satellites came full circle, since the original space-geek bible—*De la terre à la lune,* read by Tsiolkovsky, Potočnik, Oberth, Goddard, Pierce, Korolev, and von Braun—was written by a Frenchman: Jules Verne.

The Soviets, on the other hand, never got the hang of communications satellites. Three weeks after Early Bird went up, they launched a vehicle called Molniya that had one of the weirdest elliptical orbits ever devised, an orbit so strange it's been called the "Molniya orbit" ever since. Inclined at 63.4 degrees, Molniya took twelve hours to circumnavigate Earth and had such a high apogee (twenty-five thousand miles) that the United States naturally assumed its real purpose was to monitor ICBMs at high latitudes. The Soviets made one brief effort to copy INTELSAT with a 1968 cooperative satellite called Intersputnik, but the eight-nation consortium was not exactly stable after the Soviets invaded one of its members—Czechoslovakia—just a month after the venture was announced.

By the dawn of the eighties, the United States was still the acknowledged leader in space, with eighteen civilian satellites in orbit, but no longer the only country that could design systems. Three French and three West German companies, working together, had put up two experimental satellites called Symphonie in the early 1970s. Telesat Canada had put up seven Anik satellites, becoming the first country to set up its own commercial GEO system, and as time went on the rest of the developed world would become competitive. The Palapa satellite in Indonesia (1976), the Sakura satellite in Japan (1983), and the Olympus satellite in England (1989) were state-of-the-art communications projects. The Indian National Satellite (INSAT) was the most complex satellite ever launched prior to Iridium. Called "a crowded Indian bus shot into space" when it was introduced in 1983, it combined television transmission, weather forecasting, remote sensing, voice, data, telegraph, and video.

And for thirty years that's how satellite projects were done—they were too big, and too important, for mere corporations to attempt. But as the Cold War started to wane, all that was destined to change. In 1988 a maverick American television producer named Rene Anselmo inaugurated the first privately owned satellite, PanAmSat, which was launched from French Guiana on an Ariane rocket decorated with a cartoon dog that had its leg raised—Anselmo's message to what he called the INTELSAT monopoly. By then the action in the United States had abruptly turned back to the military side anyway, as billions were once again being poured into outer space projects—specifically one project called "Star Wars." All the science developed over the previous four decades was about to be concentrated in a bizarre stranger-than-fiction system that President Reagan would use to move the Soviets toward peace.

"Star Wars" was originally a term of derision. When Reagan proposed the Strategic Defense Initiative (SDI) in March 1983, his loudest critic was Carol Rosin, the schoolteacher turned spokesperson for Werner von Braun in the years leading up to von Braun's death in 1977. Rosin appropriated the title of the George Lucas space-war movie to make fun of SDI as science fantasy—based on mere promises from the Lawrence Livermore National Laboratory—but the media picked it up, and eventually even the engineers involved in its design started using the term. SDI was so complicated that there were only two or three people who truly understood all its aspects. As originally drawn up, Star Wars was to be an arsenal of satellites equipped with X-ray lasers and particle-beam weapons that would blow up Russian missiles by turning their own nuclear warheads against them. There were three programs within the program, and they all involved aggressive, fast-moving satellites that could make split-second decisions without human control. The first was a system of what were called "kinetic kill vehicles," to be housed in giant satellite garages called "gunracks." These KKVs were designed to deliver explosive warheads to Soviet missiles almost immediately after the enemy rockets were launched. The gunrack was eventually ruled out as being too large a target—too easy to shoot down—and it evolved into its virtual opposite, a system called "Brilliant Pebbles." Brilliant Pebbles was a constellation of four thousand mini-satellites that would fire watermelon-sized canisters ("pebbles") at high velocities in configurations that allowed each canister to destroy up to ten Soviet warheads at a time. If for some reason one of the pebbles failed to eliminate the missile during its initial launch phase, the canister would burst open like a replicating techno-creature and a dozen smaller kill vehicles would fly into action. These "Genius Sand" interceptors were designed to blow up the Russian rockets when they were flying through space on their way to North America.

The third Star Wars system—and the one that would make Iridium possible—was Brilliant Eyes, a constellation of several hundred light-weight low-earth-orbit satellites (LEOs) carrying infrared sensors that would track the Soviet missiles at a later stage so they could be destroyed, either by Genius Sand or by a ground-based missile arsenal similar to the ones housed in silos that had peppered the northern regions of America and Canada ever since the 1950s. No wonder Carol Rosin and others made jokes about it. Star Wars had a massively complicated architecture that sounded so far-fetched that even its designers at the Livermore Lab were sometimes skeptical about whether the systems could interoperate successfully. But its key elements were that a) it carried no nukes, and b) the whole system was entirely defensive—none of the vehicles could hurt anything except a nuclear device that had already been launched. For some reason it frightened the Soviet Union as no previous space program ever had.

It was the peculiar logic of the Cold War that the agreed-upon concept of Mutual Assured Destruction (MAD) was considered safer than one country having an advantage over the other, even if that advantage was merely the ability to knock incoming missiles out of the air. The Soviets condemned Star Wars as an act of aggression, claiming that it violated the 1972 Anti-Ballistic Missile Treaty that limited the United States to one hundred defensive missiles stationed at one facility: Grand Forks Air Force Base, just north of Emerado, North Dakota. The ABM Treaty also forbade the two countries from deploying "mobile" systems, and Brilliant Pebbles was about as mobile as any system ever devised, but the Americans chose to interpret the word "mobile" to mean transportable via railroad cars or trucks. So the Russians were right—each tiny Brilliant Pebble and Genius Sand was, in actuality, a miniaturized antiballistic missile—but Reagan continued to build Star Wars anyway, defending it as an instrument of peace, which meant hundreds of contractors, including Motorola, continued to design it. The common denominator of all the satellites used in Star Wars was that they could talk and think (hence "brilliant" and "genius"). They were, in effect, autonomous, computerized outer space cyborgs—the first time anything like that had ever been attempted.

The nation spent $44 billion on this system before eventually abandoning it in 1994, mostly due to withering attacks by Senator Sam Nunn, chairman of the Armed Services Committee. Democratic opposition to Star Wars was based partly on the desire to honor the ABM Treaty, but primarily on the concept first put forth by General Eisenhower that there should be no weapons in space. It was always a strange conceit, since there were at least twenty thousand active nuclear warheads and all of them were positioned on vehicles that would fly through space to reach their targets. The very first ballistic missile, the Nazi V-2 used to bomb London, flew up to fifty miles high, which qualifies as outer space by virtually any definition. The only nuclear warheads not delivered from space were carried on bombers. The Strategic Air Command had plans to scramble B-52s that would rendezvous at the North Pole and wait for windows of opportunity to dip down into Soviet airspace and cause damage, and presumably the Soviets had a similar plan, but those maneuvers would be used only after the space missiles had already been launched. In other words, the military aspects of the Cold War were *all* conducted in space. The only differences between Star Wars and the other systems were that it was satellite based instead of rocket based and it was a destroyer of nuclear weapons but not a nuclear weapon itself.

Technology developed for the Star Wars satellites, most of it classified, was what would be built into the Iridium system. Once again, it took a war-fighting purpose to create the technology that could be used in peacetime. In the estimation of many analysts, Star Wars was what led Mikhail Gorbachev to seek peace with the West, but it had another, purely technological legacy, and that was Iridium.

Chapter 3

THE SPOOKS

APRIL 28, 2000
SENSITIVE COMPARTMENTED INFORMATION FACILITY,
PENTAGON, ARLINGTON, VIRGINIA

After the meeting at the Winter White House, Dan Colussy canceled his golf lessons and flew back to his summer home near Annapolis to be closer to Iridium headquarters. Then, in the early-morning hours of April 28, he grabbed his leather notebook and eased his black Porsche 911 Cabriolet down the steep, narrow lanes of Glen Oban to the John Hanson Highway for the hour drive across Maryland to the Pentagon. It was an overcast Friday, with a slight chill in the air, one of the last breezy days before the muggy D.C. summer. Crossing into Arlington, he pulled off the Jefferson Davis Highway into the world's biggest parking lot athwart the Potomac River. Colussy felt energized. Today's mission: to find out just exactly how badly the U.S. government needed Iridium phones.

Colussy was nothing if not methodical. He was the kind of manager who incessantly made lists and took notes. He kept a copy of every document that ever crossed his desk. He liked to know things in advance, so he was fond of background checks and "What do you know about this?" phone calls. He was a hoarder of facts, always willing to pay for expert opinions and number crunching and every kind of data that can affect a business decision. When members of his large Italian family were casting about for an appropriate birthday gift, they decided to present him with the century-old business ledger of his grandfather's, a first-generation immigrant from the Alpine foothills north of Venice. In meticulous longhand it recorded every penny the old man had spent on turn-of-the-century construction projects and payments for mechanical work to a brother and son who in 1916 founded Colussy Motors, the oldest Chevrolet dealership in the world. The Colussis, who changed the spelling to "Colussy" so as not to suffer anti-Italian discrimination, had been builders and businessmen, bringing to America the masonry techniques of the Friuli region on the Adriatic coast. Dan was a third-generation Colussi and the first family member to marry a non-Italian, but he had the same methodical scrimp-and-save style that his grand-father and uncles had used when building the little town of Bridgeville, Pennsylvania. Every time Colussy started a new project, he christened a notebook and a phone list, and he kept them for years in metal file cabinets that eventually required a room of their own adjacent to his home library. While running UNC he purchased thirty-two companies in an effort to move away from the nuclear business, but his takeover style was less like J. P. Morgan summoning the railroad barons and more like an artisan sewing the last square on an elaborate quilt. Colussy's takeovers were low-key affairs, a matter of drawing up papers and hav-ing a celebratory cocktail at the local hotel. Colussy was the guy in the back of the classroom who always did his homework a day in advance. He believed in information and common sense and, above all, guaran-teed sources of revenue. Most people making a cold call at the military headquarters of the free world would arrive with lawyers and PowerPoint presentations and aides-de-camp. Colussy arrived with his notebook. He got to the Pentagon early.

Colussy was well versed in Pentagon ritual, since many of the com-panies held by UNC had sold products or services to the military. As he walked across Honor Guard Plaza, he was expecting to pass through the metal detectors at the River Entrance, then sit in a waiting area until a polite enlisted man in a crisp uniform came to escort him the half mile to whatever office he was scheduled to meet in. The routine was time-tested and inevitable; the only thing you never knew was when the polite enlisted man would show up. The wait could sometimes last all day, as the foyer would fill up with nervous salesmen in expensive suits, world-weary lawyers with battered briefcases, well-scrubbed officers seeking reassignment, and the occasional group of schoolchildren, all waiting on audiences with decision-makers. The Pentagon moved very slowly at every level.

But on this particular day, nothing would go as expected.

"Are you Dan Colussy?"

He heard the voice as soon as he walked through the doors. It came from a smiling civilian, slightly built, wearing thick glasses and a slightly rumpled shirt. He seemed to appear out of nowhere, boyish and bouncy, like the water boy for the football team, offering to bring fresh towels.

He extended his arm. "I'm Mark Adams."

Colussy had heard the name somewhere but couldn't immediately place it.

"I'll be helping you. I'm an advisor to the Department of Defense."

Colussy thought, *When did the Pentagon start using advisors who look like waiters at TGI Friday's?*

The two men shook hands, and Mark Adams eventually revealed a bit more of his résumé. He was about forty years old, not fourteen, and he was an employee of MITRE Corporation. Now Colussy started to understand. MITRE was a Cold War think tank on Dolly Madison Boulevard in McLean, Virginia, three miles down the road from CIA headquarters. It had grown out of the same Lincoln Lab at MIT that helped design rockets and telemetry for the early ICBM systems in the fifties, and its main mission had been to "automate war" as the United States entered the computer age. One of MITRE's first achievements had been the huge SAGE units—the largest computers ever built—that were used by NORAD and featured in *Dr. Strangelove*. But now that the

Cold War was over, MITRE was called a "private nonprofit consulting firm" for the government, although it tended to have ex-CIA officers in senior management positions. MITRE, in fact, had long been a favorite bogeyman of conspiracy theorists, as it was linked to evil plots by the RAND Corporation, the Council on Foreign Relations, ill-intentioned Zionists, the Trilateral Commission, and, of course, the New World Order. In the year 2000 it was being talked about as one of the sinister forces behind DARPA, best known for creating the Internet.

Colussy didn't subscribe to any of those theories—he knew MITRE simply as a place where engineers and scientists did classified research—but Mark Adams was nonetheless triggering his "spook detector." Colussy had dealt with quite a few "black ops" guys, especially while President of Pan Am, when his planes were being regularly hijacked to Cuba and, on one memorable occasion, blown up by terrorists on the Cairo runway. In Mark Adams Colussy recognized a type of all-American boy who was favored by the intelligence services. Even the name "Mark Adams" sounded like the hero in a juvenile boy's book, or perhaps a comic-book character who has a secret superhero identity. Colussy chose not to be curious about exactly where Mark Adams came from and what he was doing here. If Mark Adams wanted to "guide you through the process," then Colussy wasn't going to argue. This might be the beginning of a beautiful friendship.

As Colussy was led through the 6.6-million-square-foot maze toward the inner sancta of the Air Force command and the Joint Chiefs of Staff, he reflected that he now had two road maps he could follow. One was his designated spook, Mark Adams, looking like an undergraduate planning a party for homecoming weekend, so effusive and effervescent about Iridium that Colussy kept speculating as to whom he might be taking orders from. The other was a private checklist Colussy had been given by his old friend Beverly the day before.

Beverly Byron was a former congresswoman from Maryland who had served on Colussy's UNC board and was known for having major influence at the Pentagon while on the House Armed Services Committee. Colussy had asked for her advice, knowing that, even though she'd been out of office for seven years, she knew how to find the soft underbelly of the Pentagon. She was not a traditional politician, having entered Congress by accident in 1978 when her husband, Goodloe Byron, died suddenly, seven days before his certain reelection to a fifth term, and Democratic Party leaders begged the grieving widow to fill his shoes. She won easily and inherited all her husband's committee assignments, including Armed Services, where she became a thorn in the side of Chairman Les Aspin, who didn't like Byron's tendency to vote with the Republicans. Byron had a huge advantage, though. Her father had been aide-de-camp to General Eisenhower during World War II, so the generals and admirals loved her. She was, in fact, the first woman to fly in the SR-71 "Blackbird" spy plane, which could reach Mach 3 at ninety thousand feet and was the only plane that, if attacked by a surface-to-air missile, could simply outrun the missile. Byron endeared herself to military men forever when she convinced Barbara Mikulski, the liberal Maryland senator and military critic, to take a ride in an F-15 fighter plane. The resulting pictures of the stocky senator in a flight suit were collector's items.

And sure enough, Beverly Byron had lots of information. *Don't waste your time talking to NASA,* she told Colussy. *NASA has no money for this. Beware of a retired Vice Admiral named Bill Owens.* Owens had been the commander of the Sixth Fleet during Desert Storm and the Vice Chairman of the Joint Chiefs of Staff, but now he was a hired gun for Craig McCaw, skulking around the Pentagon, trying to extort millions in operating fees from the government so McCaw could restart his bid for Iridium. Whoever liked Owens, or included him in meetings, would be no friend of Colussy's. *Be nice to Admiral Nutwell,* she said. Bob Nutwell was part of the Office of the Secretary of Defense, so anything he said could conceivably be heard by the "SecDef" himself, Bill Cohen. The man to start with, though, would be General Jack Woodward. *Jack will*

*be on our side, and he'll get you down the road, but he won't be able to close
the deal. The man you want to ultimately get to is Rudy deLeon. He's the
Deputy Secretary of Defense and controls most of what goes on in the DOD.
He also ran the staff of Les Aspin back in the day, so don't mention the name
Beverly Byron, but the only person more powerful at the Pentagon is the SecDef
himself. Good luck*, she told her old friend.

All of this advice was racing through Colussy's head as he was led
down the corridors toward their assigned SCIF (pronounced "skiff").
A SCIF is simply a secure room where nobody can hear you, but the
Pentagon, in its acronymic wisdom, calls it a "Sensitive Compartmented
Information Facility."

Waiting for them in the SCIF, right on cue, was three-star Air Force
General Jack Woodward, who turned out to be animated, bald, beefy, and
friendly as all get-out. Woodward had been in the Air Force since 1968,
when he went straight from the ROTC program at little Hobart College
in Upstate New York to a combat commission in Vietnam, and since
then he'd held so many titles and ranks and assignments that it required
four single-spaced pages to list them all. At the moment he was Director
of the Command, Control, Communications and Computer Systems
Directorate of the Joint Chiefs of Staff, as well as the Deputy Chief of
Staff for Communications and Information for Headquarters U.S. Air
Force, as well as the Deputy Chief Information Officer, Headquarters
U.S. Air Force, Washington, D.C. These were titles too daunting even for
his fellow military bureaucrats, so he was known as a "senior communicator," an information specialist, and the man in the J-6 with influence
over the J-3. The J-6 is the communications division of the Joint Chiefs
of Staff, including the command that runs the "secret Internet," and
the J-3 is the part of the Joint Chiefs that can move troops around the
world at a moment's notice, but is also charged with briefing the highest commanders on everything going on in the field. It's a quick-action
crisis-response position that involves fighting wars, but also providing
humanitarian assistance and disaster relief, as well as the tracking down

of international drug cartels any time the military is involved. Knowing enough about Woodward to see that he was a major player at J-3, Colussy was starting to realize Iridium was a higher priority with these men than he'd first thought. Who could be a better prospect for an Iridium phone pitch than somebody who tracks down drug traffickers and parachutes into earthquake-stricken refugee camps?

Joining them in the meeting that day was Rear Admiral Bob Nutwell, the key connection to the SecDef singled out by Beverly Byron, and he turned out to be a former aviator with 700 aircraft carrier landings on his résumé. (The record is 2,407, but still.) Admiral Nutwell had his own impenetrable title: Deputy Assistant Secretary of Defense for Command, Control, Communications, Intelligence, Surveillance and Reconnaissance and Space Systems. Nutwell had just come from a conference in San Diego, where he'd told his colleagues that the GPS system was so vulnerable to enemy jamming that the technology used in microwave ovens could take it out, so the fragility of satellite systems was at the forefront of his thinking that day, and Iridium was something that might help him patch things up. Nutwell was accompanied by Colonel Rick Skinner, from the Pentagon Space Office, while Captain Steve Slaton attended for the Navy Space Projects Department. There were also assorted assistants and deputies and administrative staff on the Pentagon side of the SCIF table, but Colussy was all alone on his. Or almost all alone. It seemed that Mark Adams was on *his* side of the table, and that he was going to do whatever possible to help Colussy pass the first exam.

Woodward had a set speech for anyone who came to see him about Iridium. He called it "prekindergarten for due diligence." The speech had one overriding purpose: to make it clear that the government wanted and needed the Iridium satellite telephone system. He was sticking his neck out to tell people this, since, if the truth were known, he hadn't found much support for his position within the Building. A decision had been made long ago that the government didn't want to own Iridium, so all he could vouch for was that the Pentagon wanted to continue as a customer. That meant somebody had to find a buyer they could live with.

General Woodward made small talk about how he'd just gotten off a plane from Thailand and how helpful Iridium had been there and in Mozambique. Colussy had no idea what had gone on in Thailand or Mozambique, but knew that Mozambique was one of the poorest countries in the world, unstable, with no reliable communications, full of child soldiers and corrupt leaders. Whatever happened there must have been life-threatening to one or more American operatives. Woodward kept using the term "disadvantaged user," meaning soldiers and other personnel who worked solo in remote parts of the world, implying that Iridium was the perfect solution for covert action. He had first monitored the phone, he said, when it was in a testing stage during the U.S. engagements in Bosnia and Herzegovina, and he found it "awfully sound." Since much of the action in the Balkans was secret and in small units, the military needed a "quiet" means of communication, so Iridium calls—untraceable and requiring nothing more than a handset—turned out to be just what they were looking for. Woodward himself had taken an Iridium handset up in a C-12 Huron turboprop jet, the kind used for medical evacuations, and executed a series of figure eights to test its reliability while flying all over Bosnia, Herzegovina, and Kosovo. The plane wasn't even fitted for the phone—he just jammed it up against the window—and he couldn't find any maneuver that would throw it off its signal. So Woodward was a fan of the phone, but he wasn't sure how to pay for it.

Woodward went on to say that Secretary of State Madeleine Albright always traveled with an Iridium phone, that there were at least two in every embassy, that every flag officer had one, that commandants carried them, and that all of America's English-speaking allies used them as well. (Federal Reserve chairman Alan Greenspan was another big fan of Iridium service. During the weeks leading up to the Y2K scare, Greenspan became nervous that all computer systems would fail on

January 1, 2000, so he had Iridium transceivers installed on the roof of every Federal Reserve bank and the central banks of twenty-seven foreign countries.) Woodward used the term "war-fighters" more than once, indicating he thought of himself as the spokesman for the front-line troops, and at one point he looked directly at Colussy and said, "We have a real interest in preserving Iridium." He also said he was pissed off at Motorola for dragging its feet on supplying the "secure module," a tamper-proof sleeve that qualified the phone for top secret usage. Motorola had promised the module and then cut off all work on it. If the government were going to use these phones, the secure module was not only necessary but indispensable. He could foresee a long-term use of Iridium and the purchase of anywhere from twenty thousand to fifty thousand handsets, but only if he got his encrypted module and only if the system could eventually be "netted," so that everyone with a handset could listen in on the same transmission.

Finally, Woodward said that it was important to the military to keep operating the top secret Iridium gateway in Hawaii. This was where all satellite calls on military phones came to Earth, regardless of whether the phone being used was in Sarajevo, Singapore, or the next Hawaiian island over. The United States had already spent $150 million building the earth station there, and they were not anxious to write that off. The place he referred to was actually well known to any senior communicator, a high-tech complex located on the island of Oahu in a little redneck town called Wahiawa, full of tattoo parlors, used-car dealerships, cheap bars, and Filipino fast-food joints.[4]

Colussy couldn't have known it at the time, but Woodward had extended his "prekindergarten" talk into slightly more arcane areas once he became comfortable with Colussy's background. Most of the early visitors interested in buying Iridium had been outright nutjobs, in Woodward's opinion, and he'd finally told his aides to start vetting people first to see if they had not only money but a functioning cerebellum. As Woodward and Nutwell both knew, getting support for Iridium within the Pentagon was already going to be a tough assignment, despite the fact that the Navy's own satphone system, to use one example, was over-subscribed by 250 percent and was likely to remain that way well into the future.[5] This meant that, whenever a crisis occurred, 60 percent of the users were knocked off the system and forced to fend for themselves. Logistics units were the first to go, causing logjams in places like Kosovo where too many government agencies were trying to use the same satellites. The problem was the "damn stovepipes," Woodward said—the idea that each branch of the service needed its own satellite system—and those systems couldn't even communicate with one another.

It was a strange time in the history of American military planning. No one was happy with the lessons of the Balkan wars, which had shown how ill prepared the American military was to deal with enemy objectives like "ethnic cleansing." And the Pentagon was still in psychic shock from the disastrous engagement of American troops by Somali militia fighters in what was called the "Black Hawk Down" conflict in 1993. It was widely regarded as a humiliating defeat after the bodies of U.S. soldiers were mutilated and dragged through the streets of Mogadishu, followed by a hasty withdrawal ordered by President Clinton. Ever since then the buzzwords at the war colleges had been "asymmetrical warfare" and how to engage the "bottom-feeders." In the world of combat planning this was called "chaos in the littorals," the idea that future conflicts would not be on the plains of Normandy or on the open seas, but in the inlets, estuaries, creeks, and ravines of the constantly changing shoreline—in other words, in unstable places such as Rwanda, Chechnya, and Iraq—against "savage, determined, and intelligent" militias. These were scenarios in which anything built for massive nation-against-nation fighting was worthless. In the communications world, what would be needed was "more rapid sensor-to-shooter information flow" as the various military services became "network-centric." The Marines would need "high-capacity, over-the-horizon, continuous on-the-move communications to provide a full tactical picture." And the ultimate goal for all the armed forces was "interoperability," meaning that everyone needed to be able to communicate with everyone else. The outdated idea of an army that operates on land, a navy on water, and an air force in the sky had been blown away by war scenarios in which all branches of the service were being funneled toward the same tiny speck on the planet—so it had become crucial that everyone use the same communications devices. Unfortunately, it was something that only the theorists believed in. Jack Woodward believed that the damn stovepipers would try to keep their systems separate.

Woodward was able to discuss these things more openly with Colussy than he would have with some other visitor. Colussy's background check had turned up a pro-military businessman who had served in the Coast Guard, the only branch of the armed forces that no other branch felt a rivalry with. And since the Coast Guard Academy was the only military academy that had both a rigorous exam to get in and even more rigorous exams to get out—with no congressional appointments—Academy graduates like Colussy got credit for being eggheads. Colussy also knew all about stovepipers. He had been involved with the Navy's nuclear submarine program, famously founded by Admiral Hyman Rickover, but he wasn't going to mention that to Woodward, since that particular program had always been a sore point with the Pentagon. Rickover had seen to it that his budget was controlled by the Department of Energy, giving him carte blanche to do whatever he wanted without consulting the Navy command, and it was still set up that way. Colussy knew this because for years, as chairman and CEO of UNC, he had overseen construction of the nation's nuclear submarine reactors from an isolated site on the Thames River near New London, Connecticut, that was semisecret: the plant had an entrance on a dirt road hidden by a sign for DEAN'S CHICKEN FARM.[6] Colussy's company had also trained all the Army helicopter pilots at Fort Rucker, Alabama; run all the flight simulator training at the Pensacola Naval Air Station; and managed the nation's four-hundred-square-mile nuclear weapons facility in Hanford, Washington. All of these programs were affected by inter-branch jealousy and intrabranch careerism—in other words, stovepiping.

Still, Colussy knew that he was not the first choice of the Pentagon. Beverly Byron had told him the brass really wanted Eagle River, the company owned by Craig McCaw, but McCaw and his rent-a-general were poisoning the well by asking for $15 million a month—too rich for everyone. McCaw's plan was to buy Iridium, combine it with another satellite system he'd recently bought (a system called ICO Global Communications that had yet to be launched), and then fold both of those into a 288-satellite "Internet-in-the-sky" system called Teledesic that he'd been developing since 1994 with Microsoft chairman Bill Gates. All of McCaw's plans were expensive and grandiloquent. Colussy couldn't match McCaw's bankroll, so when he got the chance to speak, he played up the financial stability of his friend John Castle's bank, Castle Harlan. He briefly outlined his plans for running Iridium—not that they were all that well formed yet—and he made it clear that this was probably a quid pro quo: he could possibly put together a deal to buy the system, but he would need a contract from the government in advance.

What ensued was a conversational dance during which neither man said what he was really thinking. Woodward knew Colussy was asking for money. Colussy knew Woodward wasn't authorized to promise money. Woodward let it be known that it was theoretically possible, at some theoretical date in the future, under some set of theoretical circumstances, for there to be a government contract. And that was all Colussy needed. He understood the halting, tentative noncommitment and treated it like money in the bank. It was only a matter now of finding out how much money, and something told him that his new friend Mark Adams would know that number.

When Colussy left the Building that day, he was certain of two things:

One, there were at least a few powerful men at the Pentagon who were desperate to see these satellite phones survive.

Two, there would be no more hard questions. They trusted him.

The question now was: Would the Iridium creditors sell him the system and at what price? Motorola didn't seem desperate to find a buyer, and once the easiest solution had evaporated—the checkbook of their friend Craig McCaw—they seemed content to destroy everything. But surely Motorola didn't have sole control of the company. There were banks that had loaned money to Iridium—in fact, thirty banks from around the world, including Chase Manhattan, Barclays, and Citibank—so maybe it was time to head for the pawn shop known as the federal bankruptcy court of the Southern District of New York and find out whether those bankers would like to get a few pennies back on their dollars.

A few days later, Colussy drove down the Beltway to the Reston, Virginia, offices of Iridium—a ghostly place now that the hundreds of IT employees had been let go, leaving only a few cubicles occupied—and met a straight-arrow attorney named Joe Bondi. Bondi was what's known as a "workout guy," a specialist sent in by the banks to rescue assets. Bondi told Colussy straightaway that "these guys have deal fatigue." Several attempts had been made to save the company, and if it wasn't done quickly, the banks were going to just shut it down and take whatever cash was lying around. In fact, he said he needed a firm commitment from Colussy within nine days or else Motorola was likely to de-orbit. Since Colussy was just getting his feet wet, he had no idea how he could make that happen. He wasn't going into this deal without a contract from the government, and the transfer of a government services contract from one company to another was, for a federal bureaucrat, a *very* serious matter. It was the kind of thing that could take weeks, if not months. But Bondi was adamant: *We don't have that kind of time.* The only way the deadline could be extended would be for Colussy to put up $20 million as a due diligence fee.

Meanwhile, inside the Pentagon, the debate about Iridium went on, but at what Jack Woodward considered maddeningly low levels. The Navy SEALs were telling him, "We need these phones." The people involved in combat search and rescue were saying the same thing. The military men charged with polar communications—staying in touch with vehicles and outposts and operatives working within the Arctic Circle and at the research facilities in Antarctica—were virtually begging for the phones. The first responders in war, the First Marine Division at Camp Pendleton near San Diego, carried everything on their backs, including three types of mobile communications devices. Yet none of those three devices worked "on the move," when the unit was first deployed to a location. The radio operator in each platoon had to stop moving, set up the device—which took anywhere from twenty to forty-five minutes—and *then* establish contact with his commanders.[7] In Mogadishu, small units on the ground had become separated, with no radio contact, right before they were massacred. All these units would have needed was one Iridium phone per platoon and the whole tactical problem would be solved. Woodward conferred with Mark Adams: *Isn't this thing a no-brainer?*

And Adams agreed. But what he was starting to realize was that no one in the Pentagon trusted the Iridium phone for a simple reason: it wasn't secret. Any guy on the street could buy one. How could it possibly be worth anything when it was privately owned, easily available, and relatively cheap? The Air Force Space Command at Peterson Air Force Base in Colorado Springs, better known as NORAD, was where Jack Woodward had been the "chief communicator" for much of his career, and it was there that he spent thirty years putting communications satellites into space that were secure, protected from jamming, and secret. These were what soldiers and fliers and seamen used, not some handset that a bunch of geeks created for international playboys in Monaco.

Fortunately for Colussy and Iridium, Mark Adams could go places no one else could enter, and he knew people in the government who were using Iridium phones and loved them. Adams left his office at MITRE and drove the eleven miles out the Dulles International Airport corridor to the Iridium offices, showing up at the Colussy-Bondi meeting unannounced. And then he told Colussy exactly what he could do. He could go to the DISA, which operated the GIG, and get an IDIQ.

What? Say that again.

Colussy had no idea that he had just stepped into an international drama that had been going on for thirteen years.

Chapter 4

THE DREAMERS

MARCH 27, 1987
SCRUB DESERT, NEAR DUGAN'S DAIRY FARM,
CHANDLER, ARIZONA

When Ray Leopold first laid eyes on the rock-strewn patch of desert waste along Price Road, he was more than a little skeptical. The bleached suburbs of Phoenix looked like an ancient race of giants left behind a burned-out campfire, an open rock quarry, and some flimsy kindling that people had somehow turned into windowless bomb shelters. The address Leopold was given in the little town of Chandler was twenty-two miles from downtown and had no signs of life except for a ghostly dairy farm and, if you squinted into the eastern sun, past the creosote and rattlebox bushes, beyond the gnarled mastic trees and exposed gas pipelines, a sand-colored two-story unmarked building that could charitably be described as bland. Someone had put up a sign warning random visitors to stay away—the building was "Department of Defense vetted" and required top security clearances—but otherwise you would never notice it was there. So this was where Motorola put its engineers. Well, that was okay, Leopold was used to it. He took a left onto a winding driveway past an artificial duck pond. Oddly, there was a hot-air balloon floating over the building. This was going to be a quick interview.

Two hours later, Leopold's life had changed. That summer he would move his family to Phoenix and take an office in the isolated building that Motorola called "The Lab." On his first day his boss would say, "Ray, what title do you want? We don't care about titles in this division. Pick one."

"Chief Visionary," Leopold would reply.

The executive would think for a moment, then say, "I would, Ray, but most people would think of the occult."

Motorola was about as buttoned-down and corporate as America gets, not normally the kind of place where people invent their own titles. Motorola was, in fact, a multinational company with hundreds of divisions and subdivisions and working groups and task forces and an organizational chart so complex that it had to be updated daily. But Leopold had shown up at an exceptional time. On the day he arrived in Chandler, he hunted down his old boss from the Air Force, Dave Carroll, who was now a section chief at Motorola. Carroll had lured him to the interview by dangling the carrot that Motorola was "about to do something very special" and needed brainy engineers to accomplish it.

After twenty-four years in the Air Force, Leopold was about to retire as a Lieutenant Colonel and was ready for something different. He was only forty-one years old and felt like his best years were ahead of him, so in his usual systematic way, he'd made a list of twenty-five factors to consider before making up his mind about where to work. Then he'd spent six months going to interviews, some of them in the business world, some of them in academia, always assuming he would end up in a teaching job, since his fondest memories were of his student days at the Air Force Academy and then the years he'd taught there. Leopold's Air Force résumé also had extensive "headquarters experience"—two tours at the Pentagon, the first on the Air Staff and the second working for the Secretary of Defense—but he had grown disenchanted with the military, especially because of what he considered its shabby treatment of his mentor, legendary fighter pilot John Boyd, who never made general because he was considered a hothead. So Leopold put in for retirement at the earliest possible age, and now he was on the short list to run one of the most prestigious engineering schools in the country, the South Dakota School of Mines and Technology. Carroll had talked him into visiting Motorola before making a final decision, but Leopold considered this a "throwaway interview."

The recruiter that day was Durrell Hillis, a forty-seven-year-old Motorola veteran who headed the Strategic Electronics Division, and he got right to the point. Hillis was putting together a twelve-man brain trust of systems engineers who would spend all their time dreaming up new projects. Six of them would be chosen from within Motorola, and six would be hired from the outside. The reason: Hillis was sick of being a subcontractor. Boeing, Lockheed, and TRW got all the good government contracts—aircraft, space vehicles, satellites, missiles—and Motorola was then hired to create the electronics infrastructure that made those projects work. Motorola was the company you called when you needed a radio, or a radar system, or a transponder. The contracts always read "providing the secondary payload." Hillis had come up through the semiconductor division—the cash-cow part of the company that made solid circuits and microprocessors and wafers and other devices that ran the computers of the world—but he thought Motorola was big enough now to be building its own projects, not just supplying gear for others. What he didn't say, but the rest of the staff was beginning to sense, was that the future of Motorola in Arizona probably depended on finding something new. The Cold War was winding down, government contracts were getting smaller, and even after forty years of building top secret products for the military and NASA, the company operations in Scottsdale and Chandler were still considered "Motorola Siberia." They accounted for less than 10 percent of the company's business; the real action was back in Chicago, where the consumer electronics guys were taking the fledgling cell-phone business by storm. Once the fear of the Soviet Union went away, their jobs might be at stake.

"I want people who can think big," Hillis told Leopold. "I want system-level opportunities for Motorola. I want people who will go figure out something that no one has ever thought of before."

Leopold was not a shy man. In fact, he was so ebullient and social by nature that normally he would fill up any room with his excited, high-pitched voice. But Hillis took him by surprise. He was silent for a moment.

"I don't believe you," he said.

"What do you mean?" said Hillis.

"Jobs like this don't exist."

Later that day Leopold became the first outside hire for the new unit, and the first fellow engineers he met were two Motorolans he would soon be sharing offices with—Ken Peterson and Bary Bertiger. Peterson was a soft-spoken, wavy-haired Iowa farm boy who wore gold-rimmed glasses and was known as a pure mathematician. Bertiger was a compact, mustachioed microwave engineer from Hillside, New Jersey, who came out of the prestigious Stevens Institute of Technology and had a master's in electrophysics from New York University. Bertiger had started his career at Bell Labs but left after four years because "it was too academic and I wanted to be building stuff." He'd found his place at Motorola in 1972, developing a stellar reputation with the business managers back in the Chicago headquarters as an engineer who respected the bottom line. The three men bonded right away, and Leopold got that rush of adrenaline you feel when you've joined a winning athletic team.

And he had. If you go to the National Air and Space Museum, part of the Smithsonian Institution, housed in a cantilevered glass building facing the National Mall in Washington, D.C., and if you go up to gallery 213 to check out an exhibit called *Beyond the Limits*, you will find enclosed in a glass case a one-page handwritten document dated July 14, 1988, describing the "Global Personal Satellite Communications System" and signed by four employees of Motorola. One name belongs to the designated Motorolan who, for legal purposes, signed every patent document the Systems Engineering Group ever produced. The other three names are Leopold, Peterson, and Bertiger, the three men who invented Iridium.

How that document came to be is still a matter of some controversy among the people who were there. According to internal Motorola mythology, the idea was born when Bary Bertiger took a vacation on Green Turtle Cay, a small island in the Bahamas, and while there his wife, Karen, had trouble getting a phone connection so that she could complete a real estate deal back at her office in Scottsdale. Exasperated, she asked Bertiger how, "if you're such a smart guy," it could be possible to design devices that could communicate with the Voyager probe in deep space, but there was still no cell phone that worked everywhere in the world. Bertiger then started toying with the idea of a "world phone," and Iridium was born a few weeks later.

Leopold told a slightly different version. He said Bertiger's wife was reluctant to go to the Caribbean at all, worried that she would miss a call at her real estate business, so the vacation never happened. Peterson, on the other hand, thought the whole vacation story was a myth concocted by the Motorola public relations department, and that everyone at the little in-house research group was already thinking about innovative cell-phone solutions well before Bertiger took his vacation. Regardless of whether there was a eureka moment, the idea Bertiger proposed was that instead of trying to build cell towers spaced one to five miles apart throughout the whole planet, why not just put towers in the sky? It was much more than a theoretical suggestion. Spotty cell service was a huge problem at the time, and Motorola had a bigger stake in solving that problem than virtually any other company.

Motorola was, in fact, the inventor of what we now call the cell phone, having created the first car phone in 1946 (installed in the trunk, it worked on radio frequencies and required an operator to put the call through) and then the first handheld portable cell phone, the legendary DynaTAC, known then and now as "the Brick." (The name stood for Dynamic Adaptive Total Area Coverage, one of Motorola's typically complicated acronyms.) Considered a specialized luxury item at the time it was introduced, the DynaTAC was demonstrated for the FCC as early as 1973. (The first call—placed by Martin Cooper, General Manager of Motorola's Systems Division—was a wrong number.) Ten years later the commercial version, the DynaTAC 8000X, was introduced to the public, retailing for $3,995, and even then the phone was not really noticed. Since the company had thirty-five years of experience marketing car phones, the sales force assumed cell phones would be used by the same people, and that market was not large. How many people, they reasoned, needed access to a phone twenty-four hours a day? Then came the first media attention for the DynaTAC, when a Pacific Bell executive went into the stands at the 1984 Los Angeles Olympics and passed it among the crowd so that everyone could make a complimentary call. At twenty-eight ounces, the DynaTAC was one of the largest and heaviest cell phones ever produced—but demand for it was instantaneous and never abated. The first time most people saw the DynaTAC was in 1987, when Michael Douglas, as Gordon Gekko, kept one with him at all times in the movie *Wall Street*, and the only reason it wasn't better known before that was because Motorola couldn't produce them fast enough.

The early cell-phone customer was plagued with so many dropped calls and dead zones that he had come to expect them. In 1987 the problem of building enough towers to eliminate spotty service seemed insurmountable even in the United States, much less the rest of the world, so when Bertiger finally said, "Is there a way to put cell towers in the sky?" he got enthusiastic responses from Leopold, Peterson, and a young communications engineer named Greg Vatt—and no one else in the Systems Engineering Group, which is what the twelve-man brain trust had been rather unmemorably named. In fact, when the idea was first presented at the weekly meeting of the group, it was dismissed as fantasy. "'This is garbage' was their attitude," recalled Leopold. "'We have to go work on real things.'"

And yet creating a constellation of communications satellites that would cover the whole planet was not an entirely new idea. When AT&T launched Telstar in 1962, they expected it to anchor a system of forty satellites in polar orbits, fifteen satellites in geostationary orbits, and twenty-five ground stations positioned around the globe. The system would have depended on ordinary trunk lines to complete calls, but it was still acclaimed around the world as the most sophisticated satellite system ever launched and the future of communications for the planet. That promise was snuffed out by President Kennedy, who was alarmed by the very fact that AT&T was prepared to spend $500 million on the system, and the result was COMSAT, which ended up not being interested in voice communications at all. Now Bertiger was saying, "But why not?" And most of his colleagues were saying, "We don't see the relevance."

Besides, there was already a company in that business. The American Mobile Satellite Corporation (AMSC) had been formed in the mid-eighties to provide "mobile satellite services" to North America. Faced with the emerging technology, the FCC decided satellite phones would use up so much bandwidth that the agency refused to issue more than a single license. Then, after two years of fierce fighting for that license, the commissioners couldn't decide on a winner, so they ordered all seven of the applicants—including Hughes Aircraft and McCaw Cellular—to form a consortium and work together. In other words, they forced them into the COMSAT and INTELSAT model, thinking that the demand would be so small that it wouldn't matter. The problem with the AMSC business plan, though, was that the phones wouldn't really be mobile. They were what came to be known as "briefcase phones," since the user had to carry a terminal that would receive calls only after you spent a few minutes setting it up and getting a fix on a GEO. They were really more accurately called "suitcase phones"—the "mobile terminals" were bulky and heavy, like the suitcase transmitters carried by Nazi spies in World War II. So obviously the plans for AMSC were not what Bary Bertiger's wife was talking about.

As history turned out, all twelve members of the new Systems Engineering Group would have other "real things" to work on anyway, mostly things involved with the Strategic Defense Initiative. Halfway through Leopold's first day on the job, Bertiger came to him with an urgent request: "We need your help with something outside this division. Something is hanging loose." What was hanging loose was an impending decision by McDonnell Douglas as to what kind of communications system would be used on the hundreds of Star Wars satellites about to be built. Bertiger was a veteran of Cold War military systems, having designed most of the microwave communications for America's spy satellites, and Leopold had been acting director of the Milstar Terminal program at Hanscom Air Force Base in Massachusetts. Milstar was the $5 billion super-satellite warfare system being built by the Pentagon to link bombers, nuclear missiles, submarines, fighter planes, and troops on the ground. Given "highest priority" by President Reagan to support Star Wars, Milstar had cross-links among eight satellites—ten-thousand-pound "switchboards in space" that cost $800 million apiece—so Leopold did know a little about linkage issues. But these Star Wars birds were even more complex than Milstar. Motorola had security clearances for all three Star Wars systems—Brilliant Pebbles, Genius Sand, and Brilliant Eyes—and all of these satellites were going to need to communicate autonomously in space. It was something that had never been done before, so McDonnell Douglas hired Motorola to study the issue of how the satellites should talk to one another, especially since they would be moving at high rates of speed and the distances between them would be constantly changing.

It came down to two choices: laser links or radio frequency links. Laser links had enormous bandwidth, but they were not always reliable. Yet McDonnell Douglas didn't seem to like the "low-tech" solution of simply using radio linkage. Bertiger was convinced that radio frequency linkage was safer and better, but he wanted Leopold to back him up. Leopold quickly immersed himself in the project, helped by data from two top secret experimental satellites that had been launched by Lincoln Lab in the seventies, and came to the same conclusion that Bertiger had: radio frequency linkage was better than lasers. "We're getting on a plane to Washington," said Bertiger. "We've got to convince them not to go with the lasers."

And so, from the first day, Leopold was deeply involved in all of Motorola's existing communications work, much of it involving Star Wars. (Motorola lost the argument about radio communications, by the way. If Star Wars had been built, it would have used lasers. But the study became the basis of what the three inventors would use for Iridium.) Unfortunately, after the initial enthusiasm over "cell towers in the sky," Leopold, Bertiger, and Peterson were forced to work on the project in their spare time. For much of the first year after the Bahamas vacation, Bertiger was working in restricted areas of a building full of SCIFs, while Peterson was working out calculus problems on "unmentionable" portions of Star Wars. Leopold kept trying to bring everyone together for brainstorming sessions, but it was difficult to get all three men into the same room on the same day.

The main thing Leopold wanted to talk about was making the orbiting cell towers autonomous. What made the Star Wars kill vehicles so scary was that they were like cyborgs; they could make decisions and act without human control. At the time, the American public thought this was due to some massive secret operation that had developed new technology unavailable to the Soviets, but actually it was the opposite. Lowell Wood, the principal designer of Star Wars, had used off-the-shelf devices that were available to everyone, and it caused some of his colleagues to make jokes about how he was buying stuff at Radio Shack to defend the free world. After the demise of the Star Wars program, the engineers and scientists who had worked on it were curious about whether all that "autonomous action" technology would have really worked, so to find out, they packed it all into the Clementine lunar mission—and it performed beautifully. During its 303rd orbit around the moon, the Clementine space vehicle operated autonomously for the full circuit. Nevertheless, the reason it was possible to do that mission at all was that everything built into Brilliant Pebbles had been so cheap. The engineers at NASA who worked on Clementine called it "a desktop computer hooked up to some camcorders and a mobile phone." Space projects had entered a new era in which you could cobble together parts from an electronics store and still compete with the big boys. And all these eye-opening developments—the cheapness of the technology, the ability of satellites to function autonomously, the fact that a huge constellation could communicate internally and make complex decisions—would percolate through the minds of Leopold, Bertiger, and Peterson as they waited for their chance to work on the new space-based cell-phone system.

Finally, one day in early 1988, Durrell Hillis said, "Okay, guys, I like it, let's get moving on it, what can you do with $25,000?"

And so, with what amounted to the cost of a few rivets on the typical military satellite, the Iridium project was launched. The $25,000 was really just enough to allow the three men to get together and start working out the initial problem of how best to position the satellites. They assumed they would need at least four of them, since Arthur Clarke had shown years before that that was the number you needed to cover the planet. But then they backed up when Leopold said, "We shouldn't be starting with the answer. Maybe we don't *need* satellites for this. Maybe there's a way to do it from the ground, or at least within the atmosphere." Leopold had been a hot-air balloon enthusiast ever since he heard boyhood stories about his flamboyant great-great-grandfather Enrico Capece, known all over Sorrento as the "Little Duke," who was a professional balloon pilot before being killed in a duel over a woman. Leopold himself had stood in a wicker basket at 14,250 feet over Pike's Peak and knew that sturdy Mylar balloons, similar to the ones used in the Echo program, could be functional up to nineteen miles. Obviously balloons were incalculably cheaper than satellites. The problem, as they worked out the calculus, was that the balloons might bunch up over time, leaving gaps in cell coverage. But there was another substratospheric solution. What if you simply kept a fleet of planes in the air twenty-four hours a day? This would still be cheaper than satellites, although it was subject to weather events. (Ten years later, the idea was tested successfully by a St. Louis company called Angel Technologies in a project called HALOSTAR.)[8] But there were political problems. With planes and balloons you would need permission to operate within the airspace of every country in the world, whereas any satellite constellation would fly well above the Kármán line—one hundred kilometers above sea level—which was generally accepted as the legal threshold of outer space. Ultimately, they decided that satellites were the only answer.

But what kind of satellites, how many satellites, and in what orbit? These simple questions all led to complex math problems that took months to compute. So the three engineers—Bertiger the businessman, Leopold the dreamer, and Peterson the mathematician—made a sort of informal pact to keep working on it until they had the answers. And Durrell Hillis gave them another rule: "Keep your mouths shut. This is a bootleg project."

Hillis was creating what's known in engineering circles as a skunkworks—secret, autonomous, fenced off from bureaucracy. After all, this was Motorola.

In retrospect, the Motorola Corporation in 1988 was one of the few places in the world, if not the only place, where such a project could have taken shape and been pushed to fruition. Bob Galvin, who had recently retired as Motorola CEO after thirty years but stayed on as chairman of the board, liked to preach two principles to his 150,000 troops:

Do not fear failure!

Recognize the signs! (Meaning that, when failure does appear, get out quick.)

Both principles would eventually be applied to Iridium—the first led to the invention of the system, and the second led to its abandonment—but a more important principle was the one Galvin applied in private meetings. "The first step in the execution of any strategy," he would tell his lieutenants, "is to rewrite the rules."

He meant that literally. His chief lobbyist in Washington was Travis Marshall, who also happened to be America's ambassador to the International Telecommunication Union (ITU), the United Nations agency that assigns radio frequencies and works out technical standards for every communications system in the world. When Marshall was arguing before the ITU, he was representing the United States, but he was also arguing for Motorola. Galvin always believed that Motorola should be a close personal friend of the most powerful person in the room and that Motorola's information should be better than anyone else's information. Motorola's Intelligence Group was created by Galvin to mine information he obtained from the CIA, and then handed over to a full-time staff of researchers who presented "external environment" reports to senior management. As a result, Motorola often knew things before the government did.

There was no better example of the Galvin way of doing things than Motorola's complete domination of the Chinese market. All the groundwork for the move into the People's Republic of China had begun in 1979, when Patrick Choy, the company's senior finance executive in Hong Kong, introduced Premier Deng Xiaoping to the game of bridge. Deng loved the game so much that in 1981 he agreed to send a Chinese team to the Motorola-sponsored Hong Kong Inter-City Bridge Tournament. It doesn't sound like that big of a deal until you realize that it was the first time in history that Chinese and Taiwanese officials had gathered together in public at *any* event. Just as Dennis Dibos was expected to play golf with the right people, Patrick Choy was expected to practice "bridge diplomacy." Galvin then decided to beat the Japanese into the Chinese market, not by going head-to-head in China but by taking away Japan's foreign partners. Motorola aggressively developed business among the "four little dragons"—Hong Kong, Taiwan, South Korea, and Singapore—so that, when the time came to move into mainland China, it was the company full of "sincerity and love" for Asia (yes, they actually used those words), unlike the Japanese, who were perceived by the Chinese as rip-off artists. Once they gained the trust of the right people, Galvin told his Asian executives that their next goal was to rewrite the Chinese government's rules, including—and this is no exaggeration— the Chinese national constitution. China required all foreign companies to form joint ventures with Chinese-held entities and outright forbade wholly owned subsidiaries. Galvin said Motorola wouldn't be playing that game. It took years to get what he wanted, years spent cultivating friendships with Chinese officials high and low, but when Motorola finally opened a semiconductor factory and a two-way radio factory in the city of Tianjin, it was the first foreign-owned company allowed to operate in China without Chinese partners. And in a country where trade unionism was part of the official state constitution, Motorola became the first corporation to operate union-free. The result was that within five years Motorola dominated every telecommunications category so completely that the Chinese word for cell phone became "Motorola." An $85 million, nineteen-story Motorola Tower rose in the Chaoyang District of Beijing. Sales in China rose to $10 billion by 1997, and the company eventually represented 25 percent of the gross annual product of Tianjin, a city of eight million.

Bob Galvin was actually the second shrewd, feisty Irishman to run Motorola. The first was his father, Paul, who founded the company in the 1920s and was legendary in Chicago for his Horatio Alger story. Growing up in the small town of Harvard, Illinois, Paul Galvin was a hustling street kid who would sprint onto passenger trains when they stopped at the local depot and sell as many ham sandwiches as possible before the train pulled out. The thirteen-year-old was considered a pest by the railroad since what he was doing was not only against company

policy but dangerous, as he almost fell under the train once while leap-
ing off. He was charming enough, however, to grease the skids with
conductors and evade the train detectives as he enlarged his offerings
to include popcorn and ice cream. He went on to become an artillery
officer in France during World War I, then tried college but gave it up
before starting his first company in Marshfield, Wisconsin, in 1921.
The Marshfield Chamber of Commerce was so eager for development
that civic leaders gave Galvin all kinds of help for his little factory to
manufacture storage batteries—but within two years he was bankrupt.
Marshfield was on a branch line of the railroad, which meant his shipping costs were higher than battery companies more centrally based.
Chastened but not defeated, he learned the importance of being near
a transportation hub and moved his family to Chicago. While raising
capital for his next venture, he worked as the personal assistant to Emil
Brach, the elderly founder of the Brach's candy company, and finally
managed to start a second battery-manufacturing company in 1926. This
company also went broke, the victim of a product defect that caused its
creditors to shut it down.

By the time he and brother Joe founded Galvin Manufacturing
Company in September 1928, Paul Galvin's business career was hanging by a thread. He had two bankruptcies, making it difficult to raise
any more capital, and the only way he could manage to get back into
the market was by taking a chance on a niche product called the "dry
battery eliminator." In 1928 many people were getting alternating-
current electricity in their homes for the first time, but they still owned
appliances—especially radios—that worked on battery power. The dry
battery eliminator was simply a way to plug your old radio into the
wall socket instead of relying on the bulky rechargeable batteries that
sometimes dripped acid onto your carpets. It was a device that rectified
alternating current and turned it into direct current. Obviously this was
a product that would be obsolete in a short period of time, especially
since Galvin was making more money repairing the eliminators than

selling new ones. No matter. When the company opened at 847 West Harrison Street, Galvin's only concern was making payroll of $63 a week for his five employees.

The dry battery eliminator sold so well at first that Galvin decided to start making his own radios. America was in the midst of a radio craze. The first broadcast had been November 2, 1920, when KDKA announcer Leo Rosenberg read off the Harding-Cox presidential election results from a transmitting shack thrown up on the roof of a Westinghouse Electric building in east Pittsburgh. But radios didn't start to become a mass-market product until the end of 1922, when more than six hundred radio stations had been licensed. Between 1923 and 1930, 60 percent of American families purchased at least one radio, but the market was split among dozens of brands, mostly forgotten now, like Pilot, Federal, Patterson, Rogers, Stewart Warner, and Armstrong. There were also many "private" brands: companies that essentially built furniture but now decided to build furniture that housed radio receivers. Galvin started manufacturing radios for these private brands, and by the end of his first year in business he was just starting to break even.

Then came the 1929 crash, and the market became so glutted with used radios that all the private brands disappeared. To survive, the Galvins started installing radios in cars. In 1929 this was considered a luxury product. The installation could take up to two full days and required bolting three separate units with a combined weight of twenty pounds. The speaker, mounted on the dashboard, was a foot across and six inches thick. The receiver unit, installed under the hood, was the size of an automobile battery and required five large vacuum tubes that had to be periodically replaced. And the tuning mechanism, mounted on the steering column, was the size of a slice of bread. Once this bulky device was installed, it was subject to interference from the generator and distributor, and it would run only when the engine was turned on. No wonder it took a team of engineers four years to perfect it. Demonstrating Galvin's lifelong nose for technical talent, one of the men working on that team was William Lear, who would go on to invent the autopilot system in airplanes, the eight-track tape player, and, of course, the Learjet.

Lear, Galvin, and an engineer named Elmer Wavering finished the prototype for their signature car radio, the ST71, just a few days before the Radio Manufacturers Association Convention in Atlantic City in June 1930. They then installed one in Galvin's Studebaker and drove it from Chicago to promote what he billed as the "Motorola," a word chosen to combine notions of an automobile in motion and the sound created by a Victrola, which was a brand trademarked by the Victor Talking Machine Company and so never explicitly referred to. The ST71 retailed for $120, a substantial sum considering it was often installed in cars that themselves retailed for only around $600. The promotional trip paid off, though, and Galvin was soon swamped with enough orders to show a small operating loss of $3,745 in 1930 and then substantial revenue gains for every subsequent year. In 1932 his engineers figured out how to augment the power of the car battery to run the radio, making the unit much smaller, and by 1934 the interference problems had been solved. The breakout moment for the company came when Galvin signed a deal with B.F. Goodrich to sell and install car radios out of Goodrich tire stores coast to coast. Galvin launched a nationwide advertising campaign that featured Motorola billboards and led to some of the first public protests against highway sign clutter. (Motorola's signs were so big that Seminole Indians in Florida started stealing them to use as flooring in their homes.) The ad campaign also spawned laws passed by various city councils making it illegal to operate a car radio while driving, under the theory that the radio was distracting and unsafe. Even the bad publicity turned out to be good for Galvin, though. When the Philco Company underwent a devastating strike by its labor force in the late thirties, it was forced to satisfy contracts by making a deal with Galvin to produce its radios as well. Increasingly Galvin Manufacturing was referred to by the public as "Motorola," and so the name change was made official in 1946.

By then the Galvins had established themselves as the dominant players in the creation of any device used to receive an AM radio signal, but especially devices used in cars. As early as 1930, Paul Galvin had built radio receivers into police cars used by the village of River Forest, Illinois, but it was a one-way system and subject to disruption during high-speed chases and any rough roads that jarred the frame. The first Motorola radio built exclusively for police departments was the 1936 Police Cruiser—with "Magic Eliminode," whatever that meant—which Galvin promised was the most rugged car-based receiver ever built. Quickly it became the standard of most police departments, until it was replaced by an even more ambitious Motorola product, the T6920, which became the first two-way police radio in service when it was adopted by the city of Bowling Green, Kentucky, in 1940. Motorola continued to make products sold directly to the consumer, including home radios, but Galvin had a lifelong distrust of anything too dependent on the American public. Motorola ended up making car radios not for rich individuals, but for Ford, Chrysler, and American Motors, while developing radios for police and ambulance services and other first responders that resulted in the company dominating that business for the next several decades. When an NYPD dispatcher summoned Officers Muldoon and Toody by squawking "Car 54, where are you?" in the 1961 sitcom, he said it over a Motorola radio.

Motorola was a manufacturing company, but during the Great Depression manufacturing companies didn't have shiny factories with automated equipment. Like many another Chicago start-up, Motorola hired barely educated immigrants—Poles, Italians, Irishmen, Greeks—who built radio units on wooden benches in poorly heated industrial buildings on Chicago's West Side. They were the kind of guys who would spend their paychecks on liquor and sometimes have to be bailed out of jail for carousing or worse, but Galvin loved them, even when he had to command their respect with his fists. He was a hard-scrabble hustler himself, the kind of guy who, when faced with a large shipment of defective radios returned to the factory in 1934, took out his frustration with a sledgehammer. So it was typical that in 1959, the last year of his life, when he was frail and sickly, he was still telling his managers that he would take them out to the alley and teach them a lesson if they didn't toe the line. His braggadocio was regarded good-naturedly in the blue-collar trenches, and everyone knew that once Galvin liked you, you had a lifetime position with the company. When Motorola took off after World War II and became one of the leading manufacturers of television sets, radios, transistors, and the very early versions of what would become the microprocessor, it still retained its working-class ethic: overwhelmingly male, aggressive, and tribal.

Ray Leopold came from one of those same working-class Chicago neighborhoods, growing up on the streets around Belmont and Austin Avenues, where he was constantly reminded of his Italian heritage—Neapolitan on his mother's side, Calabrian on his father's—while playing back-alley sports with Greek, German, Polish, and Irish kids. His father and grandfather were both master machinists who worked for the Automatic Electric Company, making parts for the rotary dial phone. He also had a grandfather who was a master electrician and a grandmother who was a milliner. Leopold himself was an animated, precocious kid who became a licensed ham radio operator in the eighth grade, an Eagle Scout, and an athlete who delivered the *Chicago American* after school. But the turning point in his life came in 1959, at age thirteen, when his father showed him a *Parade* magazine article about the first class to graduate from the Air Force Academy. To Leopold—a born science geek who struggled in every other academic subject—it seemed like the most exciting place in the world. Through the Boy Scouts he met an older kid whose father was in the Air Force Reserves, and that resulted in a ride on a C-119 troop carrier. The C-119—nicknamed the "flying boxcar"—was about the most unglamorous plane ever built, but from that moment on, Leopold was in love with flying. He played poker to earn the $8.50 an hour it cost to take lessons in a Piper J-3 Cub, but his ultimate goal was to attend the Air Force Academy, learn to fly jets, and become an astronaut.

The problem was that, after completing the tenth grade, his grade point average was only 2.35, which meant he had to get straight A's his last two years if he wanted to have any chance of getting appointed to the Academy. For chemistry class, the highest grade you could get was a B unless you had a science fair project, so Leopold conceived the idea—mostly from an article in an electronics magazine—of building his own computer. This was at a time when even the word "computer" was new to the lexicon, and most people had never seen one. He figured he needed about $100 for his project, so he took the Austin Avenue bus and the Lake Street "L" to Newark Electronics, lied about his age (fifteen), and got a job that earned him enough money to buy the parts. He salvaged old TV sets that had been thrown out in the alley near his home in order to rehab vacuum tubes, and the resulting ungainly machine used a rotary phone dial for computations, had lights flashing on and off when it was "thinking," and could do three things: add, subtract, and multiply. He outfitted it with a cabinet and painted the wood to make it look more presentable—and got the necessary top score from the science fair judges. Then, at the age of seventeen years and five months, Leopold achieved the highest score in a competition among thirty-seven high school students sponsored by U.S. Representative Roman C. Pucinski and was rewarded with an appointment to the U.S. Air Force Academy. It was the turning point in his life and the inspiration for much of what he would do in the future. On his first day in Colorado Springs he stopped at the granite wall where the Academy had inscribed the opening lines from Sam Walter Foss's poem "The Coming American":

> *Bring me men to match my mountains,*
> *Bring me men to match my plains,*
> *Men with empires in their purpose,*
> *And new eras in their brains.*

Leopold would commit the poem to memory and quote these lines in most of the speeches he would make over the next fifty years, especially when he spoke to young people, encouraging them to find "the new eras in your brains." He was an idealist, an optimist, and a believer in the future. He thought in monumental terms.

And so did Motorola. While Leopold was starting his Air Force career and getting his Ph.D. in electrical engineering at the University of New Mexico, Motorola was becoming one of the largest electronics corporations in the world. One of the curious things about Paul Galvin's legacy was that, even though he was a college dropout himself, he always had a great respect for intellectuals. He believed in the genius who could figure out a better way of doing things, as if by magic, and his favorite egghead in the world was a University of Connecticut professor named Daniel Noble. Galvin first noticed Noble in 1940 when the professor developed an FM mobile communications system for the Connecticut State Police. Motorola's system used AM frequencies, which have a more limited geographic reach, and Galvin was so determined to retain Motorola's dominance of the police radio market that he sought Noble out. The two men were wary of each other at first, but Noble would soon become the first academic to lead Motorola into areas never dreamed of before he arrived. Noble's first act after joining the company was to change Motorola's signature Police Cruiser radio from an AM unit to an FM unit that was almost immediately successful thanks to a huge contract with the Philadelphia Police Department. Even more significantly, he invented the single most valuable piece of communications equipment used by the American military in World War II. At the beginning of the war, every Army unit carried a Motorola radio called the Handie-Talkie. The handset was a little larger than a normal telephone receiver, but it required a huge backpack carried by the communications officer. It was an AM system that Noble converted to FM and reduced greatly in size to become the Walkie-Talkie, beloved of military men in that war and also in Korea.

Recognizing a whole new market, Galvin asked Dan Noble to stay on after the war and develop new military electronics. But the Army Signal Corps made a special request: *If you guys are going to do government research, please do it somewhere besides Chicago.* Chicago was high on the nuclear target list, and all postwar planning involved keeping sensitive operations away from major cities. Noble needed a fairly isolated place that was near a major university, and he ended up choosing Phoenix, mainly because he'd vacationed there and was a devotee of horseback riding. Motorola's professor spoke grandly of the new research facility that started in a rented space on Central Avenue, but Phoenix at the time was a backwater by anyone's standards—a desert town of barely a hundred thousand people—and the consumer electronics guys back in Chicago openly made fun of him. The new research facility was called "Noble's Resort" and "Noble's Folly," even after Noble hired eminent Purdue physics professor Bill Taylor to develop a new product called the "transistor." It was Bell Labs that announced the first transistor on December 24, 1947, but it was Noble's little lab that first started building them in the kitchen of a house on 56th Street in Scottsdale. Noble believed Motorola was at the threshold of a new era in which solid-state electronics would change the world, and that world would involve more than just engineers. He would also need mathematicians, physicists, metallurgists, and chemists, all working together to do basic research into the next generation of communications devices.

And he found them. Not only did Motorola do basic research on the germanium transistor, but Noble's Folly quickly proved itself in the Korean War when Motorola became the first producer of power transistors. When the company opened what amounted to the first semiconductor factory in the world, it was in a small building at 52nd Street and McDowell Road. There they turned out glass sleeve diodes, signal diodes, rectifiers, zener diodes, and all kinds of other products that the public might not understand but would end up not being able to live without. In 1955 Motorola opened the first mass production facility for semiconductors, and that division would generate hundreds of millions in profits for the company even as Noble's Folly continued to get zero respect from Chicago.

Meanwhile, Motorola started to become too big a company for the patriarchal management of one man, so in 1956, when revenues were $216 million, Paul Galvin's son, Bob, was installed as CEO. By the time Bob Galvin passed the baton thirty years later, revenues would be $6.7 billion, as not only Motorola but the world underwent a revolution in electronics, communications systems, and, not so coincidentally, our understanding of outer space. In 1956 Motorola introduced the first pager—for use in hospitals—and in 1957 they introduced the world to pay TV through a system called Telemovies in Bartlesville, Oklahoma. When Wernher von Braun put the first American satellite into space, it carried Motorola communications equipment. Sometimes the consumer concepts and the government concepts would even overlap, as with Motorola's "Astronaut" television set, the first cordless "big screen" at a whopping nineteen inches, introduced in 1960. Television had been one of the few fields where Motorola was late to the game, as RCA captured the market shortly after television became commercially available in 1946. Looking for a way to get part of that revenue, Paul Galvin told his engineers he wanted to smash RCA's $300 price for a set. The result was the Motorola Golden View VT71; it had a smaller screen than the RCA set, but it retailed for $179.95, allowing Motorola to move into fourth place among TV manufacturers within a year of entering the market. Motorola would continue to compete through the sixties, creating the first rectangular color TV tube in 1963 and introducing the Quasar, the first all-transistor color TV, in 1967. Eventually, though, the company saw that the Japanese were starting to develop insurmountable advantages in terms of price and quality control, so Bob Galvin made the decision to sell the very successful Quasar franchise to Matsushita and get out of the TV business entirely around 1974.

By that time Dan Noble's operation in Scottsdale had become much more than a transistor factory and a think tank. When Apollo 11 journeyed to the moon in 1969, it carried thirteen Motorola communications units, including the "lifeline" transponder, a two-way radio weighing thirty-two pounds, which was the only link between the astronauts and Mission Control after they went beyond the thirty-thousand-mile point in space. Motorola also built the Up-Data Link, which had sixty-seven functions and operated as a "fourth astronaut" on all the Apollo missions. Any time a launch was aborted, it was because of an automatic detection device on the Saturn rocket that was designed and built by Motorola. At the last minute before Apollo 11 launched, NASA became worried that American citizens wouldn't be able to see the moon landing on their television sets, so Motorola rushed into production some FM demodulators that sharpened the outer space transmissions from the Motorola S-band transceiver that Motorola had already installed in the lunar landing vehicle. A Motorola microphone, a Motorola transceiver, and internal Motorola components made it possible for Neil Armstrong to say, "That's one small step for man, one giant leap for mankind." (Unfortunately, that wasn't what Armstrong actually said. The astronaut later said his precise words were "That's one small step for *a* man, one giant leap for mankind," but apparently the Motorola communications system hiccupped on the word "a," thereby rendering his expression more poetic.)

By the 1980s Motorola had become a national institution. It was not only the top brand for semiconductors and cutting-edge communications equipment, but it was the only organization that could consistently beat Japan Inc. It was known to consumers through products like the eight-track tape player—a standard feature in most cars in the late sixties—but, more important, it was known to manufacturers as the most efficient provider of solid-state electronics in the world. The ostensible reason for that success was a management system called Six Sigma that was sort of the ultimate "efficiency expert" philosophical system. Developed by Bill Smith, a Motorola engineer who noticed the relationship between manufacturing defects and customer assistance costs, Six Sigma started out as an application of the common-sense notion that if you build things that don't break, you don't have any costs of fixing them later. As codified in various mathematical formulas, Motorola determined that "if one has six standard deviations between the process mean and the nearest specification limit, practically no items will fail to meet specifications." What that means to the layman is that you will have only 3.4 defects per one million products or processes in a factory. In the context of the 1980s, achieving Six Sigma in manufacturing meant that Motorola products were more error-free than those of Sony, Ericsson, and their other competitors.

But Motorola didn't leave it there. Managers at the Schaumburg headquarters started applying "process capability studies" to every aspect of the company, essentially trying to make human beings more efficient. In popular culture, the efficiency expert had been a staple of comedies ever since Eddie Foy Jr., as the factory timekeeper, sang "Think of the Time I Save!" in the 1954 Broadway musical *The Pajama Game*. Thirty years later, Motorola took efficiency to its ultimate level, organizing mandatory Six Sigma sessions for all employees in which "empowered teams"—led by "Six Sigma Black Belts"—learned "aggressive clock management" and "critical scorecard metrics" in an "action learning framework methodology" that would live within the breast of every sentient being laboring in any Motorola building anywhere in the world. The company even created Motorola University, which sold Six Sigma training to companies all over the world, including IBM, General Motors, Ford, Johnson & Johnson, Eastman Kodak, NBC, Texas Instruments, and—two clients they were especially proud of—Sony and Toshiba. Motorola was dubbed the "American Samurai" by the business press, and it was all because of Six Sigma. It was the kind of rah-rah employee program that loners might bristle at, but it worked remarkably well at Motorola, where most employees regarded the company as family. Motorola's working-class roots in west Chicago had long ago passed into history, but its paternalism remained. "We called it being Galvinized," said Leopold. "Once you'd worked there ten years, you could never be fired. The Galvins wouldn't allow it. It wasn't always the best policy."

And that's why the little Systems Engineering Group set up by Durrell Hillis in 1987 was so out of place. Six Sigma worked fine for products that were already in production, but there was very little you could do with it when you were creating new products from scratch. Hillis needed new markets and new products. "I knew the government group was in trouble," said Ken Peterson, the mild-mannered mathematician who, unlike Leopold and Bertiger, preferred to work alone in the abstruse world of theorems, algorithms, and equations. Peterson had worked on top secret projects in Motorola's Advanced Technology Center in the early eighties—most of them early Star Wars stuff—and then he'd seen contracts canceled. In 1984 he was asked to move with another senior engineer to the marketing department. The work was so easy that he could do it while teaching graduate math at Arizona State University.

"But that's when I knew they liked me," recalled Peterson. "They were hiding me. When business was bad, but they didn't want to lose you, they would hide you in marketing. So during all of 1985 and 1986 I worked in classified areas of marketing, selling systems to the government. And that's when I saw how bad it was. The old military programs were going away. There was going to be nothing left for Motorola unless we moved on to something new."

When Bertiger, Peterson, and Leopold started plotting the outline of Iridium in 1987, they were working against the whole culture of Motorola. Motorola sold hardware, not software. Motorola supplied systems but didn't own systems. Motorola provided outer space electronics but had never so much as launched a single rocket of its own. Even when Motorola did get involved in a major project, it was always government funded, ensuring that the money could never run out. Now these three guys in the obscurest part of the company, the Chandler Lab, were talking about building satellites, launching satellites, operating a satellite constellation without government participation, and then designing a new kind of phone that could be used in every country of the world. Maybe every country in the world wouldn't *want* this new phone operating in its airspace and connecting to its ground systems. Had anyone ever thought of *that?* There were a hundred objections to the idea—the size of it, the cost of it, the sheer originality of it. "You'll never get the spectrum," they were told. "You'll never get the licenses," they were told. But, more immediately, "You'll never get this past Corporate."

"I was constantly told, 'Go to Chicago, get advice, talk to the terrestrial cellular guys,'" said Bertiger. "But the truth was Chicago was not very good at systems. They were good at radios. And when we did go to them, they simply said, 'You can't do this.' It was Chicago pride at work. We were handling switching payloads for NASA, we had twenty years' experience with complicated big-system military projects, but they didn't think that counted for anything. They were still building radios for police and taxis—it had nothing to do with systems engineering. I thought that, because of the company politics, no one would ever actually build the thing."

And so the three men worked in a sort of hopeful but cautious mood. How high or low should the satellites be? The closer they were to Earth, the more satellites were needed. The farther from the ground, the more difficult it would be to "close the link" between the satellite signal and the handset. Then there was the problem of motion. Not only would the satellites be in motion in relation to Earth, but the handset would be in motion as well. In a normal wireless system, either the transmitter or the receiving device was stationary.

Fortunately they didn't have to figure out all these issues in a vacuum. Motorola was a subcontractor to every satellite company in America, so the three inventors had the chance to pick brains without ever revealing what they were working on. Engineers from Orbital Sciences, Fairchild

Industries, and Ball Aerospace visited The Lab frequently—and it was actually a Ball Aerospace engineer who found an article in an astronautical journal that Leopold would cite as inspiring his breakthrough moment. Leopold had been asking everyone, "What's the most efficient way to cover the entire Earth with cells?" *Voilà!* Two scientists at Aerospace Corporation, William S. Adams and Leonard Rider, had recently published a theoretical article in the *Journal of the Astronautical Sciences* called "Circular Polar Constellations Providing Continuous Single or Multiple Coverage Above a Specified Latitude." Everything clicked for Leopold when he saw it. What they needed for this project were polar orbits, not equatorial. And the satellites needed to be close to the surface of Earth so that the signal delay would be as brief as possible. Unlike GEOs, which circle the globe in the same direction as its rotation, these low-earth-orbit satellites, or LEOs, would be streaking across the North and South Poles while Earth rotated underneath, crossing the Equator at a perpendicular angle. The whole purpose of a GEO orbit is for the satellite to remain still in relation to Earth so that receivers and transmitters can find it, but what if the satellites were racing around the planet and had the ability to find the receivers and transmitters on their own? But this was going to be a complex dance. A typical Iridium call might involve two handsets in motion—extremely fast motion if they happened to be in airplanes—and both of those handsets would be communicating through a herringbone-patterned chain of as many as twelve satellites that were also in motion, meaning that every millisecond would change the "talking distance" between whichever two satellites were trying to pass the call. How could the signal jump that many times that fast? The Milstar satellites used outer space cross-linking, but nothing like this. After spending several months on the problem, Leopold eventually brought in Greg Vatt to help design a "floating time slot" east-west cross-link architecture that constantly recalculated distances and times to make sure the signal was never dropped.

But there was another basic engineering problem. When the inventors first talked about using LEOs, they envisioned an orbit of six hundred to eight hundred miles up. Satellites at that altitude would have to fly through the inner Van Allen Belt, a mass of charged particles discovered by Explorer 1. In fact there were two radiation belts at those altitudes, the other one having been created artificially in 1962 when both the United States and the Soviet Union exploded nuclear bombs in space. Once again Aerospace Corporation had the data needed. George Paulikas, the Aerospace engineer who stood by at all NASA launches to describe "solar particle events" in real time, had studied every aspect of radiation belts, and his research indicated that small satellites constantly flying through that plasma would take enormous punishment every day, regardless of what material they were made of. Eventually Leopold decided to go lower—420 nautical miles—using a seven-by-eleven constellation. There would be seven orbital paths converging at the poles, and in each of those orbital paths there would be eleven satellites. This meant that they would need seventy-seven satellites to make sure the entire Earth was covered. No one had ever launched that many satellites, certainly not for a single constellation. They had already entered uncharted territory.

In retrospect, it's entirely fitting that Aerospace Corporation was the source for so many Iridium workarounds, because Aerospace had maintained for thirty years that work done in the highest reaches of space should be performed by machinery and computers, not human beings. Aerospace was a Cold War brain trust that functioned as the secret research-and-development arm of the Air Force. Working out of nondescript buildings in the bleak, concrete-encrusted suburb of El Segundo, California, thousands of scientists had been employed there for three decades to solve the thorniest problems of outer space. It was at Aerospace that Dr. Ivan A. Getting had proposed the first GPS system in 1961. It was rejected by Washington because "priority must be given to manned programs," then revived in 1973 as Project 621B, amid great skepticism by Air Force brass. More significantly for Iridium, any time Aerospace would say, "This procedure can be done with robotics, you don't need an astronaut or a pilot," their engineers would be overruled by the Air Force or by NASA, both of which seemed single-mindedly focused on putting human beings into space. Many of the engineers at Aerospace thought this was folly, since in some cases 70 percent of the operating equipment on a space vehicle would be devoted to keeping human beings alive, leaving precious little resources for actual scientific exploration.[9]

Aerospace's biggest project in the sixties was the Manned Orbiting Laboratory, a system employing four hundred scientists and engineers that was almost ready to launch when it was abruptly canceled by President Nixon in 1969. The Manned Orbiting Laboratory would have been the largest exploratory vehicle ever put into space, and much of the technology developed for it would later be used for the Space Shuttle, but most Aerospace scientists believed it should have been the *Un*manned Orbiting Laboratory. To show just how good they were in terms of running equipment from a distance, they proved to NASA and the Air Force that they could do a blind landing of a Boeing 737 within two inches on three axes—by doing it 110 times in a row. They also developed technology that created bombing accuracy within one meter. They had developed the specs for fully functioning drone aircraft decades before they were ever used. Nevertheless, most of their major projects were turned down by Washington. If a human wasn't aboard, no one was interested. That's why so much of the Aerospace research eventually filtered its way into Iridium. Aerospace engineers had long since proven that almost any computerized system could be operated in space as though a human were in space with it, and in many cases it was the human who would cause errors and ruin your attempts to achieve Six Sigma. Of course, Motorola was already the "fourth astronaut" on the Apollo missions, so Aerospace would have been preaching to the choir.

One of the more esoteric problems when you're building a wireless communications device is what kind of antennas to use. For the handset, the Iridium inventors wanted something small, of course, but everything used by the military was either too bulky or too primitive for the sort of fast system they envisioned. Fortunately Ken Peterson was a subscriber to *Boating World,* and he happened upon an article that mentioned the quadrifilar helix, also called the "volute" antenna, which looked like an upside-down electric cake mixer and could receive signals from all directions. (It was used in boats to stay in contact with weather satellites.) Like many Iridium components, low-tech won out over high-tech—they never found anything better.

It's hard to say exactly when a theory becomes a proposal and a proposal becomes a project, but Leopold would always remember a day in the winter of 1988 when he, Bertiger, and Peterson all happened to be leaving the Chandler Lab at the same time. They casually started talking about the satellite phone project, and by the time they got to the parking lot they were intensely involved. "Look, let's settle this," said Bertiger, and he went into the attendant's office and commandeered a whiteboard normally used to sign up for parking spaces. Peterson grabbed a grease pencil and quickly wrote out a few algorithms. The three of them stared at what appeared to be the sole remaining math problems nagging at them. After discussing it for a few minutes, the whole thing was done. In their minds, it was a reality. It could be built, and it would work.

The next day they told their boss, Durrell Hillis, that they were ready to proceed.

"Where's your lab notebook?" he said.

They didn't have one. They'd used a grease pencil on the parking lot whiteboard. For several weeks they'd been doing all this work like guys discussing it in a bar.

"You've got to write it all down in a lab notebook. We have to make a patent disclosure before we can get the money."

After a few more days writing down what they knew about the system they wanted to build, Bertiger was designated to go to Scottsdale and present it to the Patent Review Committee. If they got a yes from that committee, they could then start presenting it to their superiors, including Dave Wolfe, head of the Government Electronics Group in

Arizona. Wolfe was a communications engineer who had joined Motorola in 1964 to work on the Apollo program, but they were not anxious to show him the project since his job, after all, was all about government contracts, and this had nothing to do with government. But Bertiger came back with bad news.

"They turned us down?" asked Hillis.

Not exactly. The Patent Review Committee said, essentially, "We don't know what to do with this. We won't approve it, and we won't disapprove it."

But the bottom line was: no money.

Fortunately Motorola had a system in place to end-run the committee. If a manager was turned down for funding at any level, he could write a "minority report," and once a year the three men who ran Motorola would convene to listen to all those reports. That triumvirate was Bob Galvin and his two field generals: John Mitchell, a strapping, beefy, jovial son of an Irish cop who had served in the Navy, played the tin whistle, and was beloved by the rough-and-tumble workingmen, and Bill Weisz, a brilliant electrical engineer who had worked at only one company his whole life, coming directly to Motorola out of MIT in 1948. Weisz idolized Vince Lombardi and was considered the brainy guy who worked on things like spectrum issues, but he took his meals with the troops in the company cafeteria every day and was a great believer in the minority-report system. Galvin, Mitchell, and Weisz were scheduled to hear minority reports in August 1988, and the three inventors considered Mitchell their best ally, since he had run the communications division and been the chief engineer for early mobile products like the transistorized pager and the car phone. Also attending would be the newly installed CEO of the company, George Fisher.

The whole team attended the presentation—Hillis, Bertiger, Leopold, Peterson—with Bertiger and Leopold handling most of the showmanship, using charts, graphs, and three-dimensional renderings to demonstrate how the "Satellite Cellular Personal Communications

System" was not just a business opportunity for Motorola but something that would eventually be done by someone, somewhere, somehow—so why not be first with it? The price tag to design the system was $6.5 million—a lot of money even by Motorola standards for something that had no proven customers—and the atmosphere in the room was not that convivial. "All of these division heads were there," said Peterson, "and they all had money needs of their own. They were not gonna be happy to see six million go to three guys in Chandler."

At the end of the presentation, there was a moment of silence as all heads turned toward Galvin. Galvin was still feeling out his new CEO, so he turned to George Fisher and said, "Well, George, what's your decision?"

He was obviously putting Fisher on the spot, and Fisher didn't like it. He was brand-new. He'd visited the Chandler Lab one time. The project was not only expensive and ambitious, but it looked like something that would stretch over many years and had elements that were somewhat bizarre. How could he possibly listen to a ninety-minute presentation and make a decision about it? Welcome to Motorola, where people were expected to think on their feet.

"I think," said Fisher, "that I'll assign it to John Mitchell to decide."

Technically Fisher had the ability to "assign" things to Mitchell; Fisher was CEO and Mitchell's only title these days was vice chairman of the board. But Mitchell was an elder statesman who had nothing to gain or lose by backing the satellites. Maybe Fisher really wanted to know what Mitchell thought, but that's not the way attendees read it. Fisher thought the whole thing was a hot potato to be avoided. Let it be Mitchell's baby.

"I'll be happy to follow up," said Mitchell.

Secretly, the three inventors were thrilled. Mitchell liked innovators and he liked new products. Unfortunately, he was all but inaccessible during the next few weeks. He was trying to shepherd the development of a new microprocessor, the 88000 series, and that was keeping him too busy to pay any attention to the little research team in Chandler. Occasionally someone from Mitchell's office would call and say, "Do you have the patents written up so we can file them?" So patent writing became the principal endeavor of the team. The inventors decided they would write one patent per day, five per week, until they were done.

Finally, in October, Mitchell called down to Chandler to say, "Okay, I want everyone, the whole team, at a meeting. Have you turned in the patents?"

The patents had been turned in, Mitchell was told.

Still on the phone, Mitchell conferenced in Tony Sarli, the company's Senior Vice President for Patent Law. "Tony," he said, "I just wanted to make sure all these satellite phone patents are filed."

"Well, no," came the response. "We're suing the Japanese on all those other patents, and we have 150 attorneys working on those lawsuits."

"Then stop doing that," said Mitchell. "Work on nothing else until these patents are filed."

Everyone in Chandler breathed a sigh of relief, and over the next few days the story had spread through the company: Mitchell was crazy about the satellites.

Mitchell essentially told the guys in Chandler to put together a business plan and he would get them the money. The way this was done at Motorola was to create "tiger teams"—committees that met in intensive sessions until a problem was solved completely. It was a term popularized by NASA in 1970, after a ruptured oxygen tank threatened Apollo 13's ability to return from the moon. The personnel at Mission Control had split up into special-purpose units called tiger teams—some of which included Motorola communications engineers—in order to manage the crisis. Since then Motorola had used tiger teams for anything that required expert decisions and was likely to involve conflicting opinions among smart people. The team members for this project were broken into groups that would identify business opportunities, possible competitors, and potential technological problems, and a special tiger team was even devoted to Motorola internal politics. The cellular phone division in Schaumburg was riding high and dominating the wireless markets worldwide, so the inventors were steeling themselves for intramural fighting. The cell-phone executives looked at satellite phones as potential competition to their land-based systems and, more important, something that would scare their distributors. No matter. If Mitchell wanted the system, no one's opposition could possibly prevail against it.

The decision to proceed didn't sit well with everyone at Motorola, especially not Ted Schaffner. Schaffner was a Harvard Law graduate from Columbus, Ohio, whose varied career included such arcane matters as real estate tax shelters, soybean-processing issues, and oil field equipment leasing. He'd been hired at Motorola after seven years at the Cleveland headquarters of TRW, the defense contractor, and, prior to that, the Staley Grain Corporation, the biggest clearer of trades on the Chicago commodities exchanges. "I was the obstreperous geek you had to get through to get your project done," Schaffner recalled. "In any organization you have people who believe you should just salute the flag and move forward. My job as head of the Corporate Development Group was to say stop, no, you can't do that, you have to prove what you're doing is the correct decision." And Schaffner worried that all the usual safeguards were being bypassed with Iridium. "Bob Galvin was the person most enamored of the Iridium project, and he was *the* senior guy, there was no one more senior, and so it was a little bit of a special case. The problem with new technology companies was always the same: What is the market? It's very difficult to predict the demand for a technological innovation that doesn't exist yet."

Hillis and the three inventors started shuttling back and forth between Chandler and Schaumburg, walking a tightrope between keeping the project secret (fewer than a hundred people knew about it for the first three years) and continuing to promote it internally. "We were a little confused," recalled Peterson later, "because we would get wild enthusiasm at every meeting, and then when we got back to Arizona, our phones would never ring. There was all this activity, and yet the project seemed stalled for at least a year." They started to suspect that the fly in their ointment was company President George Fisher, who remained nervous about a $6.5 million commitment to what seemed like a speculative project. "We had reports that he wanted to trash-can it."

One morning, during a break in a Chandler Lab technology meeting, Leopold was accosted by Jim Williams, who headed up the cellular ground station division in Arlington Heights, Illinois. Williams motioned to him across the room.

"Ray! Come over here! I've got a name for your system."

Leopold walked over to him, and Williams grabbed a vellum Leopold was using for his presentation. The title on it was "Satellite Cellular Personal Communications System." Williams scratched out the lengthy name with a pen and wrote in the single word "Iridium."

"What is this?" said Leopold.

"Your constellation. It reminds me of electrons circling an atom. You have seventy-seven satellites. The element with seventy-seven electrons is iridium."

Iridium is a fairly obscure element—a hard, dense, brittle metal found in tiny quantities, mostly in igneous deposits in South Africa, and used for esoteric products like the spark plugs in high-performance race cars. Still, Williams' observation created an elegant image, and the word had a certain ring to it. Leopold grabbed Peterson, who was nearby, and they both collared Bertiger and showed it to him as well. By the end of the day, everyone agreed: the system would henceforth be called Iridium.

The last internal presentation the inventors had to make was for an annual Motorola event called "Technologies of the Future" in the fall of 1989. "Old Man Galvin," as they had started calling Bob, came down to see what was going on, even though he had mostly stopped coming to meetings. Most of the afternoon was devoted to a presentation by Government Electronics Group head Dave Wolfe, but Wolfe was circumspect. He had to be careful about expensive research and development.

His unit represented $685 million in annual sales, but that was only about 6 percent of Motorola's business. In the grand scheme of things, he couldn't press too hard against the much more lucrative divisions of the company. But after Wolfe's less-than-enthusiastic description of the proposed system, he allowed Bertiger and Leopold to speak.

Galvin listened to both presentations and noted that this Iridium thing had been in the works for a while now and he suspected that the bureaucrats were getting in the way of the technology guys.

"Okay, listen," said Galvin. "If one of you guys doesn't write a check for six and a half million, I'm gonna write one out of my personal account."

And with that, Galvin left the room.

That night, Leopold took his family to Lin's Grand Buffet, a popular Chinese restaurant. His fortune cookie read, "You can build it."

The next day—the same day the Berlin Wall came down—Iridium was funded.

And now, overnight, it seemed like all things were possible.

A few weeks after they got the green light, Mitchell called to say that he needed the full Iridium team in Schaumburg.

"The Canadians are coming in," he said with a chuckle, "to present *their* satellite phone plans."

Canada had always been the most aggressive nation after the United States when it came to satellite projects. The vastness of the country and its sparse population made cellular coverage impractical over 80 percent of its landmass, so Canadian phone companies were always looking to the heavens for any kind of system that would connect its far-flung territories. Eldon Thompson, CEO of Telesat Canada, had been through several joint ventures with Motorola, most of them involving cable television for the remotest parts of northern Canada, and he was anxious to talk to Mitchell about what he was touting as a revolutionary new product for automobiles. He had gathered together all the Canadian telecommunications companies—Rogers, Bell Canada, Teleglobe—in a project to create a factory-installed satellite car phone. At a morning meeting in Schaumburg, with about thirty people present, Thompson outlined his plan for a constellation of GEOs that would serve as a better solution for the often undependable car phones then in use. An antenna would be installed on the roof of the car, and the user could dial landlines through the one or two satellites orbiting overhead. It was sort of a very primitive version of Iridium, extremely limited in scope, but the main problem with it, Mitchell already knew, was that the Big Three automakers would never agree to it. Motorola had dealt with Ford, Chrysler, and General Motors for decades—the automakers hated any new process or part, especially elaborate ones that required changes in the design of the car itself.

"Eldon, that *was* a very interesting presentation," said Mitchell. "Now we've got something to show you. If you could have all your people sign a nondisclosure agreement, we would like to make a demonstration."

Leopold and Peterson then did a slick presentation of the still-secret Iridium system and, as Leopold recalled, "they about fell off their seats."

Mitchell waited for the excitement to die down, then said, "I'll sell it to you, Eldon, for $2.6 billion."

Mitchell probably knew that Telesat didn't have $2.6 billion, and neither did any of the other Canadian phone companies, but when the Iridium team got back to Chandler late that afternoon, Leopold was bouncing for joy. "I can't believe that we created something that's already worth two-billion-plus dollars."

"That actually made me mad," said Peterson. "It's worth a *lot* more than that."

And apparently Motorola thought so, too. Within a few weeks Mitchell made it clear that Iridium would be the last big project of his storied career. They were not only going to build it, they were going to make it the biggest mobile communications system ever conceived, designed, or dreamed of. To do that, they were going to assemble private partners from around the world in a sort of ultimate Ayn Rand gathering of global capitalists, and they were going to show the clumsy, slow-moving, government-operated outer space systems how things should really be done.

On June 25, 1990, Leopold, Hillis, and several other Motorolans were taken to the offices of Burson-Marsteller, the company's public relations firm in New York City, so they could take a quickie course in "how to speak to the media." Still, they could have had no idea what would occur the following day, when the Iridium global satellite system was announced at a press conference at the Hayden Planetarium on Central Park West, while other press conferences were being held simultaneously in London, Melbourne, and Beijing. In his official remarks, John Mitchell emphasized two things: that Motorola had always been the innovator in personal communications, and that this was the first step toward a true "global village." The *New York Times* and Associated Press had been given stories to break that morning, and by that night Iridium was part of Johnny Carson's *Tonight Show* monologue. Iridium also caught the attention of DC Comics writers a few blocks away, who would feature it in the next *Batman* comic book. Ray Leopold, who loved big public events even when he wasn't the center of attention, was happy to do interviews with any reporter around. Ensconced in his suite at the Plaza Hotel, holding forth for the benefit of a writer from *Scientific American*, he was suddenly interrupted by a slightly panicked John Mitchell.

"You'll have to do this later," he said. "Come with me. Right now. It's urgent."

Within minutes Leopold was being shoved into a taxi and the two men were speeding toward LaGuardia Airport.

"All hell is breaking loose in Washington," Mitchell said. "I'll explain on the way."

In its zeal for secrecy, Motorola had told only five people at the FCC what it was doing—the five actual commissioners, who had signed nondisclosure agreements stating that they couldn't reveal anything to their staffs. Now Travis Marshall, the Motorola lobbyist, was calling to say that important people at the FCC were coming unglued. They had been fielding calls all day about the new "cell system in outer space," and they didn't know what to tell people.

Mitchell and Leopold were met by a Motorola corporate plane at LaGuardia, and on the short hop down to D.C., Mitchell said, "You're gonna have to give a presentation."

Leopold panicked. "For the FCC? I have no charts! I have no pictures! I have no way to explain what it looks like or how it works!"

"Don't worry," said Mitchell, "just talk about it in general terms. When you've talked long enough, I'll give you the hook."

Less than ninety minutes later the two men ended up in a limousine winding through the crowded streets around Dupont Circle. When they arrived at the FCC hearing room, Leopold couldn't believe what he found. The place was so packed that people were sitting on the steps in every aisle. Before he could begin to reflect on what might be happening, Mitchell was being introduced by Tom Tycz, chief of the satellite division of the FCC. Mitchell made a few brief remarks about Motorola's investment in Iridium, then said, "Ray Leopold will now tell you about the system."

Later Leopold would say to Mitchell, "I thought you were gonna give me the hook!"

The presentation had gone on for forty-five minutes, followed by questions from the assembled experts. Leopold was still hyper when they left to go to the Motorola government relations office on K Street to unwind.

"Don't relax," said Mitchell. "We have to get busy. We have a strategy session tomorrow morning. Everybody has been told to get to Washington because, among other things, we don't have frequency for this."

"We don't?"

Without a radio frequency, assigned by the World Administrative Radio Conference (WARC), Iridium would be so many useless satellites.

But the news got worse. The next time the WARC was passing out new frequencies was January 1992. And the deadline for requesting those frequencies . . . had already passed.

By the next day forty more Motorola engineers and executives would be flying into Washington for what was considered a crisis meeting. It was time once again to *rewrite the rules*.

"Don't relax" indeed. Leopold suddenly realized that all his clothes were still in a room at the Plaza Hotel.

No one would relax for the next decade.

Chapter 5

TREASONS, STRATAGEMS, AND SPOILS

MARCH 4, 1992
PALACE OF CONGRESSES AND EXHIBITIONS,
TORREMOLINOS, SPAIN

The ten years that began with the announcement of Iridium in June 1990 were full of treachery, deception, and espionage worthy of the Roman Senate at its worst, penetrating across borders, arousing the ire of nations, and often resulting in outright violations of the law. On the same day that John Mitchell, Ray Leopold, and Durrell Hillis were holding their press conference at the Hayden Planetarium, Bary Bertiger was standing at a podium in London next to Olof Lundberg, Director General of Inmarsat, the quasi-governmental satellite system that supplied phone and data service to all the major ships of the world. Bertiger and Lundberg pretended to be jointly proclaiming a new era in satellite phone systems that would benefit all mankind, but secretly Lundberg was working against Motorola, trying to make sure Iridium never saw the light of day, while Bertiger had invited him to the event only because Inmarsat was based in London and it's best to keep your enemies close. On the surface the whole world loved Iridium as the first truly global communications system, a revolutionary engineering marvel that would finally connect every point on the globe to every other point. In reality, Motorola's enemies were legion.

Iridium was perceived as a game changer, and people didn't like game changers in the 1990s any more than the owners of the Erie Canal had liked the first transcontinental railroads in the 1860s. New technology always leaves a battlefield littered with bodies, so this was a time for stealth, deceit, misdirection, and what Ray Leopold, a student of fighter-plane tactics, called "maneuver warfare." Every enemy could quickly be converted into an ally, and vice versa. A hostile warplane on your tail could be transformed, with a tight loop-the-loop, into a target in your gun sights. All the alliances that would bring Iridium to life were shifting and unreliable, but it was a war of attrition—one company at a time, one nation at a time—that Motorola never shrank from. There were no government contracts involved, so the whole operation could be conducted in secrecy, and everyone was on high alert for enemy agents. But of all the mind-shattering moments leading up to the launch of Iridium, the most surreal occurred in February 1992, in an elegant French restaurant called Place Vendôme, a few steps from the historic harbor of Málaga, on the Mediterranean coast of Spain.

Motorola had rented the restaurant in order to do what the American government had been unable to do: inveigle information out of the KGB. The nations of the world were gathered in the neighboring town of Torremolinos to allocate radio-wave spectrum, an event called the World Administrative Radio Conference that is held infrequently—this was the first major one in thirteen years—but has a lot to do with what the electronic future of the world will look like. For Iridium, this was the Super Bowl, the World Series, and the World Cup all rolled into one. A decision had been made in 1990 to "proceed on the assumption that we'll get the frequency we need," but the world had changed greatly from the days when big American and British companies could tell everyone else how the airwaves would be used. Almost as soon as Iridium was announced, a chorus of "You can't do that" had gone up from several other corporations, including every national phone company in Europe. As a result, the WARC was now a live-or-die situation. No frequency meant no Iridium. That's why Motorola had spent the previous sixteen months lobbying the entire world to make sure it got what it wanted: radio frequency bands that could be used to operate the first point-to-point global telephone system, not to mention the first commercial switching system in outer space. The army of Motorola employees sent to Torremolinos far outnumbered that representing the U.S. government. The United States sent teams from the FCC, Voice of America, Department of Commerce, Pentagon, State Department, NASA, National Science Foundation, Coast Guard, U.S. Information Agency, and FAA—but all of those delegations combined were still smaller than Motorola's team.

Since Motorola had offices around the world, the company was able to identify political allies in advance, but the company's war plan went one step further and made sure that Motorola employees were named as actual voting members: the United States, Canada, France, and Australia all had Motorola employees sitting in their official delegations. Add to this the fact that Travis Marshall, Motorola's chief lobbyist, was the U.S. ambassador to the International Telecommunication Union, which administers the conference, and you start to understand why many of the WARC delegates were resentful of the pressure, regarding the Motorolans as crass salespeople determined to hold them hostage, treating them like reluctant participants in the world's biggest time-share presentation. *We'll be seeing you on the Costa del Sol in February,* came the siren song of Motorola. *Are you bringing your wife? Do you play golf? Do you enjoy casinos? I hope you'll have time for our little presentation, and, by the way, there will be a six-course dinner afterward.* Motorola had five hundred of the best "gadget guys" in the world—salespeople who hawked the latest electronics devices to service providers and systems owners on every continent—and they were all focused on Torremolinos. A few months hence, the respected Algerian historian Mohamed Harbi would complain to the ITU that he had "observed unprecedented commercial pressure" as a delegate to the WARC, and everybody knew exactly who he was talking about. One thing Motorola understood was closing a sale.

The WARC is technically a nonpolitical meeting of the ITU, a UN agency based in Geneva. The process of agreeing to stay out of one another's way on the radio-wave spectrum had been going on for almost a hundred years—since 1897, when the first commercial radio signal was transmitted from London by the Wireless Telegraph and Signal Company, founded by Guglielmo Marconi. The first meeting of nine countries took place in Berlin in 1903, and ever since then the WARC had been a boring assembly of engineers and bureaucrats, most of them working for companies like AT&T and the British Post Office and Deutsche Bundespost, who would "clean up" the spectrum, identifying parts of it that were unused and then reassigning them as technology changed and new inventions (radio broadcasting, radar, etc.) required ever more specific frequency bands. By 1992 that had all changed. The WARC had become a place where billions of dollars were at stake, not just for private corporations, but for nations that depend on telecommunications for their economic survival. There were more than 1,400 delegates from 127 nations at the 1992 WARC, but the Iridium team had such a large transnational presence that it might as well have been the 128th: the sovereign state of Motorola. All 193 countries who were signatories to the ITU had been visited by a Motorolan during the previous year, and most of them had been wined, dined, or otherwise persuaded that the upcoming conference was an opportunity to advance civilization—and the fortunes of their own country—by leaps and bounds. Iridium would be the greatest thing to happen to the Third World since . . . well, since the United Nations was formed. At last every country, and every village in every country, could be connected to the worldwide grid.

In actuality the Iridium business plan would not be focused on the Third World at all, but on the well-heeled executive travelers in North America, Europe, and Japan, but for the time being it was better to talk about straw huts in Papua New Guinea, not ski chalets in Gstaad. A promotional video for Iridium featured the President of Mali, his wife, and his staff in acting roles. After a while, Motorola's incessant statements of love and affection for the outcasts of the world started to wear thin. An observer for the U.S. Office of Technology Assessment drily remarked that the average citizen of the Central African Republic would have to work for four years to earn enough money to purchase an Iridium phone, then work seventeen hours more to pay for a one-minute call. But Motorola managed to brush off such impertinent observations as mere details. The reality on the ground was that most of the countries in the world were poor and "developing," and wouldn't it be better to get this infrastructure established? It wouldn't cost any Third World countries a single penny, and it would be permanently available whenever they established a way to use it. Besides, while Ray Leopold was working the "technology for the Third World" angle, Iridium's man in Paris, Leo Mondale, was working the darker side of human nature. Does your country have an entrenched phone company that fears Iridium? No problem—we'll tack a dollar on to the "tail charges" of every Iridium phone call made from your country and send it back your way. If idealism didn't work, maybe greed would.

"The fifty-four African countries were used to getting checks every month from AT&T," said Mondale, "so we had to agree not to undercut that direct-dial service." This stratagem would come back to haunt Iridium in later years, as national telephone companies routinely asked for kickbacks disguised as fees—tiny Madagascar wanted $500,000 a year—just to keep the Iridium license in place.

The politics of the WARC were so byzantine that new frequency allocations had to pass by unanimous consent—obviously impossible in any UN agency—so nothing was ever formally voted on. Instead, there were votes on nonbinding "recommendations" and "temporary assignments" of frequency, and then those recommendations were further refined by identifying primary and secondary users, and then further limited by adding footnotes identifying countries that would not be using that frequency for the purposes agreed to by the majority. The footnotes to the International Table of Frequency Allocations amounted to thousands of pages of fine-print text, resulting in so many exceptions to general usage that some bands were chaotic, cluttered, and highly unreliable. The unofficial motto of a WARC delegate was "What the allocations giveth, the footnotes taketh away." This was why Motorola's uphill climb would have been difficult even in the best of times—when companies were simply scrapping for underutilized frequencies—but Iridium was a whole new category. There was no frequency allocated for low-earth-orbit satellites at all—it was a totally novel use of the spectrum—and the "Big LEO allocation," as it came to be called, was not even on the agenda, since Motorola had announced Iridium too late for the 1992 conference. Fortunately this would not be a problem in the United States because the FCC had a rule allowing for expedited processing when a company qualified as a technological "pioneer"—and Iridium certainly qualified. Still, the ITU had no such category. Motorola was essentially crashing the party.

Two months earlier a huge contingent of Motorolans had flown to Thailand for a meeting of all the ASEAN nations, hoping to round up their votes in advance, and Ray Leopold's speech about what the system could do for remote tribal peoples had the Sri Lankan delegate in tears. "You have no idea what this means for my country!" she sobbed. Leopold beamed with pride. And Travis Marshall, the Motorola lobbyist, knew they were headed in the right direction.

Now they wanted all 166 voting nations to know that all Motorola needed—well, all the *world* needed, since it wasn't a Motorola application, it was for the benefit of all, and Motorola welcomed any other satellite systems that could provide this level of global coverage—all we global citizens needed was a tiny portion of the L-band, the spectrum between 1610 and 1626.5 Megahertz that nobody really needed anyway. Yes, it was true that some armies and navies and air forces were using that part of the spectrum for weapons testing and other top secret purposes, and yes, it was true that it was supposed to be used in the future for a European version of GPS, but Motorola would only need a portion of it, and look at all the benefits to mankind.

And so, on this night halfway through the conference, at the Place Vendôme restaurant in picturesque Málaga, tuxedoed waiters bustled around serving champagne and foie gras and other delicacies to two high-level KGB officers and the Russian Minister of Telecommunications. While they were dining, the Iridium Chief Operating Officer and three Motorola executives regaled the Russians with stories about what wonderful things Iridium could do for the remotest parts of Siberia, not to mention the Moscow tycoons who would soon be carrying Iridium phones as they exported Russian ingenuity around the world. The backstory to the dinner was that John Mitchell had flown in from Chicago two days earlier, only to discover that things were looking grim for the frequency allocation. Inmarsat, the monopoly satphone supplier for the past decade, had been busy stirring up opposition. Various delegations had been influenced by enemies of Motorola to find complex technical reasons why the application should be denied. Some of those objections were easy to deal with—if Botswana needed part of a frequency for a regional airport, the exception was made and a compromise worked out—but Russia was an entirely different matter. Russia was apparently going to oppose the allocation, and Russia still had clout with most of the nations that were formerly part of the Soviet Union. It took only one vote to squelch everything. But the Russians weren't saying why they needed that part of the frequency spectrum—because they didn't have to. Any nation could cite "intelligence uses" of the spectrum without specifying exactly what it was being used for. Mitchell had taken a typically Irish approach. He bought several bottles of Jameson whisky, distributed them among the troops, and said, "Keep fighting! We'll give you whatever you need!"

The objectives on this alcohol-fueled night were many, but the over-riding goal was to find out what the Russians were using the spectrum for and, more important, which Russian officials were refusing to give it up. As the evening unfurled at the Place Vendôme, as the vodka bottles were drained and the multiple courses came and went, the Russians started to loosen up. Especially attentive was Mark Gercenstein, who was not only Motorola's top foreign sales expert but spoke fluent Russian. One of the guests finally told Gercenstein, "You know, these older Russian generals, they don't like to change their ways." It seemed like an opening, so Gercenstein waited for a while and then asked whether they were speaking of any Russian generals in particular. They eventually came up with two names—somber, uniformed bureaucrats who were well known to Motorola's Russian specialists—and after a while an executive discreetly got up from the table and called Bob Galvin back in Chicago. He gave Galvin the names of the two Russian generals, who turned out to be official delegates to the WARC.

Then, while the Motorolans and the Russians continued to make small talk, Ray Leopold—holding forth on the virtues of the Iridium system—suddenly did a double take. The waiter refilling his champagne glass seemed familiar. He waited for a moment until he got another glimpse: yes, it was Dale Grimes, the low-level guy in marketing who organized conferences back at the Chandler Lab. Dale was *disguised as a Spanish waiter* in order to gather additional intelligence on the Russian delegation. He was trying to eavesdrop on the Russians every time they left the table so he could find out what they wanted to find out from Motorola. In other words, he was in place as a secret spy on the professional spies who were already being spied on by Motorola.

The following day, the two Russian generals were simply . . . gone. They disappeared, called back to Moscow. Bob Galvin had given the names to his old Texas friend Bob Strauss, the serving ambassador to Russia, and Strauss had made phone calls. The word came back that the Russians didn't need the spectrum anyway—it was just the old military establishment trying to protect its prerogatives. The Russian air force might have been able to keep its spectrum if it had simply been fighting the Pentagon, but they had taken on Motorola. Mission accomplished.

Motorola had never seen the opposition coming, though. From the moment the Iridium project was funded, Motorola management was hoping that it would be acclaimed as such a revolutionary system that someone would buy it, and then Motorola would live off the contracts to build and operate the satellites. That's what Motorola normally did: invented new technology, then sold the technology to large operators who would in turn supply it to the consumer. Neither Bob Galvin nor John Mitchell wanted to be the permanent owner of a seventy-seven-satellite outer space constellation, much less a million-plus-subscriber mobile phone company that competed with everyone in the world. Finding another company to run Iridium and then selling multiple generations of equipment to that company was, in fact, the Motorola way. In many parts of the world Motorola loaned 100 percent of the money to build out cellular phone systems, then sold the local operator handsets and switches and other technology as he paid back the loan. But Bob Galvin had always thought "we're leaving money on the table" by not owning telephone systems. He'd always had a hankering for the subscriber money, a revenue stream that would cushion the company against sales downturns if Motorola handsets ever became less popular. Still, if Motorola could realize the full value of Iridium in a onetime sale, it would be silly to risk annoying Motorola's land-based customers. And so the internal debate went back and forth, and as time went on managers became comfortable with both sides of the equation. If they could sell it, great. If they couldn't, they would be even richer. Iridium was obviously the best long-term solution for mobile phones, so there must be dozens of companies eager to buy it.

That's why, as work went forward on Iridium, Inmarsat was assumed to be the ideal Motorola partner—at least at first. Inmarsat was an international treaty organization, set up by the United Nations in 1979 to bring order to communications on the high seas. (The name stood for International Maritime Satellite Organization, and it opened for business in 1982.) Based in London, Inmarsat answered to seventy-nine countries and used huge GEOs positioned over the Equator to cover as much of the ocean surface as possible, but it still fell far short of what Iridium could do. So back in 1990, three months before Iridium was announced to the world, Motorola put out feelers to Inmarsat, with John Mitchell himself making the approach to Olof Lundberg, the Swedish telecom executive who ran the company. Galvin and Mitchell thought the sale would be a cinch—the Inmarsat technology and the Iridium technology had virtually no overlap—and at the very least Inmarsat would want to joint-venture with Motorola. Inmarsat had already shown interest in expanding beyond the maritime industry, but its system was expensive and impractical. The original Inmarsat phone terminal weighed up to a ton and cost between $50,000 and $75,000. Even after ten years in business, it was priced way beyond the reach of small businesses—$30,000 for a clunky fixed terminal that weighed forty pounds and had to be installed aboard ship, then $12 a minute for the actual call. It wasn't portable, it wasn't cheap, it didn't work at all in the extreme latitudes, and it wasn't practical for anyone except multinational corporations, governments, and Royal Suite passengers on the *Queen Elizabeth 2*. Iridium would be a perfect add-on to Inmarsat. No need for the huge 105-foot antennas that homed in on one satellite. People aboard ships could use handheld portable phones for the first time in history, and the ship could never move out of signal range.

Lundberg agreed to bring all his top engineers and marketers to Schaumburg to hear a ninety-minute presentation on Iridium, and after it was over he pronounced himself "dazzled." There were accolades all around for the inventors, exchanges of goodwill, and plans to work together—and as soon as the Inmarsat team returned to London, they started actively working against Iridium ever becoming a reality. Motorola began preparing a private placement memorandum—the formal document used to raise money—and before the ink was dry, Inmarsat had been in touch with most of the potential investors, telling them that "they'll never get the spectrum," that Motorola was an amateur corporation that had never been in the space business, that Motorola would need licenses in every country of the world and wouldn't be able to get them, that the whole scheme was too expensive and too risky, and even that the system was illegal. The seventy-nine Inmarsat countries were prevented by international treaty from investing in a competing system—and Inmarsat regarded this as a competing system. Besides, Inmarsat was going to do the same thing on its own, Lundberg said. Low-earth-orbit satellites were nothing new—the military used them all the time. Within a few months Inmarsat was talking about something called "Project 21," and already making approaches to Hughes Aircraft about how to build and launch its own version of Iridium.

When it became obvious that Inmarsat had taken the business plan and vamoosed, Bob Galvin sent Ted Schaffner down to the Chandler Lab to "build the business case"—to figure out the best way to maximize profit on Iridium. "So I got to Phoenix," recalled Schaffner, "and I found this thing was already inevitable. People were brainstorming various categories of users. The first major meeting I had was with John Mitchell. He had all these documents from Inmarsat, a list of all the countries that were signatories and what kind of business they did. He showed it to me and said, 'What I wanna do is re-create this.' But we had big problems before we could get to that point. We had a technology problem—would it work? We had a spectrum allocation problem. This was partly why they ended up wanting investors from all jurisdictions around the world—they thought it would be easier to get the spectrum allocated that way. But I looked at it and I told Mitchell, 'This is very speculative. What you have here is a start-up venture. I don't think you should raise $3.6 billion'—which was the figure at the time. 'Rather than do that, do a round of financing, build it out, go back for a second round. Do multiple rounds on this, like Silicon Valley.' But he was opposed to that. He wanted committed financing for everything in place from day one. And that wasn't normal. We had an entrepreneurial project here. People were committed to it. People were out marketing it. We weren't pushing back at all. We weren't scrubbing the business plan."

In other words, the Chandler Lab was in a frenzy. Everyone truly believed that "if we build it, they will come." And they had the backing of the ultimate Motorolan, Bob Galvin. It's difficult to imagine any business leader in history who had the kind of charisma Galvin had within Motorola, but he was similar in presence to a Steve Jobs, a Warren Buffett, or, in previous eras, a Henry Ford or Andrew Carnegie. Once a year Motorola brought all its top salesmen from around the world to Chicago, and when Bob Galvin strode onto the stage in front of those Motorola lifers, the thunderous ovation sometimes went on for several minutes until order could be restored.

"The Galvins were big believers in the idea that businesses can reinvent themselves," said Schaffner. "Bob Galvin saw Iridium as a chance to reinvent Motorola. He thought it was time for that. The history of Motorola is full of moments where management decides to back a new and uncertain innovative technology. And that goes all the way back to the car radio. Nobody was clamoring for the car radio either."

The optimism in Chandler was also reinforced by the rampant fear the Iridium announcement had created within the industry. At the time Iridium was announced, the idea of LEOs functioning as cell towers and switching systems in the heavens was all but unknown. By the time the WARC commenced sixteen months later, there were ten other satellite phone systems on the drawing boards, and several of them were already out raising money. Since this had never been done before, the designs for these systems were all over the firmament in various configurations of LEOs, MEOs, and GEOs that would have warmed the hearts of Tsiolkovsky and von Braun because they looked like they belonged on the covers of science fiction novels:

- Project 21, later to be called ICO Global Telecommunications, was Inmarsat's attempt to incorporate handheld portable satellite phones into its existing system. It was called a "Big LEO," like Iridium, but it was actually a constellation of ten medium-earth-orbit satellites flying at about 6,200 miles up in two intersecting planes. (The dull name stood for "Intermediate Circular Orbit.")

- Globalstar was the pet project of New York business titan Bernie Schwartz, gathering international partners for a system of forty-eight LEOs to be designed by his Space Systems/Loral Corporation, a company spun off from Ford Aerospace when the automaker got out of the aeronautics business in the eighties. The LEOs would fly at 1,414 kilometers, almost twice as high as Iridium, and the system would require thirty-eight ground stations. Oddly, Globalstar would leave vast portions of Earth's surface without coverage.

- Odyssey was the highest-flying system of them all. TRW, the defense contractor that built most of America's ICBM systems as well as the first spacecraft to go beyond the solar system (Pioneer 10), was partnering with Teleglobe of Canada on a $3.2 billion constellation of twelve MEOs in 6,434-mile-high orbits managed by seven earth stations.[10]

- Ellipso was a $2.28 billion joint venture by Fairchild, Boeing, and two other partners envisioning twenty-four satellites and twelve ground stations, using elliptical orbits to skew coverage toward the most populous parts of the globe. David Castiel, the CEO of Ellipso, took the lead in sneering at Motorola. "Frankly, my business plan can do without the people on Easter Island," he told *Wired* magazine.

- GONETS was the Russians' answer to Iridium, a thirty-six-satellite system announced shortly before the Soviet Union collapsed. By the time of the WARC they had changed the name of the constellation to the more internationally friendly "Signal," and enlarged it to forty-eight satellites.

- ECCO was a strange system planned by John Higginbotham, a Virginia-based venture capitalist who had cofounded the space insurance firm Intec. Higginbotham signed up Orbital Sciences, Bell Atlantic, and Raytheon to launch eleven GEOs and thirty-five LEOs that would be oriented toward the countries bunched along the Equator—only about 25 percent of the world's population.

- Skycell was the system envisioned by AMSC, the FCC's shotgun-wedding consortium from the eighties. It was already well into the planning stages when Iridium was announced, but the $600 million "briefcase phones" that worked off a single GEO looked positively archaic, and the company was still three years away from launching its first satellite.

There were others, too, as entrepreneurs large and small saw the Motorola announcement as the commencement of a sea change in the way the world thought about mobile phones. There were so many, in fact, that by the mid-nineties the ITU started issuing stern warnings about "paper satellites" and threatening to take frequencies away from people who announced systems and then didn't build them. But the idea of putting all the cell towers in the skies was seen by most as a neon lightbulb moment, a realization that, yes, we'd been doing it all wrong, and surely the cell-phone future would belong to whoever owned the coolest satellites.

What was odd about the latecomers was that, even though they were copying Iridium, they failed to copy the elements that made it unique. All the competing systems used "bent-pipe" technology, meaning the phone bounced a signal off a passive reflector and then down to a ground station, where the calls would fan out through the existing phone system. This meant you still had to position ground stations all over the world. Motorola had learned about cross-linking while developing autonomous systems for Star Wars—specifically the "brains" for exo-atmospheric spacecraft and kinetic energy weapons—so it was already sold on the concept of managing the whole system from space, with virtually no operations on the ground. Everyone else thought of the satellites as signal carriers only.

If outer space switching were to succeed, the stakes could be huge. The promise of Iridium was that the whole world could become wireless, thereby eliminating trillions of dollars in hardware. Every telecom company was burdened by the onerous costs of infrastructure. Cell towers are large, expensive, ugly, and unsafe—fatal construction accidents are so common that there are law firms that specialize in them—and in some geographical regions, including all the oceans and most major lakes, they're completely impractical. Less developed countries will never be able to build towers for rural areas, and even some industrialized regions have never figured out a way to build and maintain cell systems. The national phone company of Poland was so far behind in 1992 that there was a ten-year waiting list for landlines, and they were still using copper wire. If you looked down the list of Inmarsat signatories, many of these companies were investors in a far inferior system—so why shouldn't Iridium be the next step into the future?

By the time of the WARC, Motorola had registered Iridium as a separate corporation and handpicked two top executives. The first was Bob Kinzie, a solemn, poker-faced lawyer who had lived in the COMSAT/INTELSAT world of quasi-governmental satellite companies ever since the 1965 launch of Early Bird. The consummate technocrat, Kinzie had learned about Iridium during one of Motorola's road show presentations, and since Inmarsat was apparently not going to play along, he was hired as chairman and CEO in order to go after the Inmarsat customer list. But Motorola also wanted someone from inside the Motorola culture, so Jerry Adams was hired as President and Kinzie's right hand. Adams was head of Motorola's cell-phone operations in Europe, but prior to that he had run a successful wireless company called Metro One, then worked for two years for Craig McCaw as McCaw Cellular was expanding all over the country. Kinzie and Adams moved into offices on K Street, where their mandate was to sell the private placement memorandum by putting together a consortium of investors and arranging for $3.6 billion in funding, then steal the big European clients away from Inmarsat.

All the top Motorola executives, including Galvin, Weisz, and Mitchell, expected this to be a walk in the park. Mark Gercenstein, Motorola's head marketer for foreign military sales, decided in early 1991 to hold an "Iridium investors conference" and charge $1 million for any company that wanted to hear the presentation. His idea was never implemented—nobody was *that* anxious to hear about Iridium—but it was not laughed out of the room either.

Unfortunately, the result of all the efforts of Kinzie, Adams, Gercenstein, and the Motorola engineers from Chandler was obvious almost from the first day of the first presentation: not just no, but hell no.

"We were dealing with telecommunications companies that in many cases were monopolies and in some cases were government-owned," recalled Kinzie years later. "These were cultures that have no word for 'competition' in their language. It was tough because an American company can't just go out there like the Hudson Bay Company and get complete access. We had to get cooperation."

But cooperation would not come from the powerful postal, telephone, and telegraph monopolies, known in the industry as PTTs. In retrospect, it's not that strange that the five-hundred-year-old PTTs in Europe would dig in their heels against a technological innovation that could erase their borders and make many of their services obsolete. These were companies set up in the Middle Ages to regulate postal rates, and they were the most entrenched bureaucracies in the world. "Governments don't like global things," said Bertiger. "We were stepping on toes." Divestiture and competition, heralded by Margaret Thatcher, praised by Ronald Reagan, had not yet penetrated into companies like British Telecom, which even had a specific part of its charter forbidding it from being in the wireless industry. British Telecom had only recently been formed out of the British Post Office—prior to 1982 it was a division of the postal service—and the British Post Office had a reputation for a full century of obstructionism, beginning when its domination of the oceanic cable business was threatened by Marconi and continuing right up through attempts to stop INTELSAT. An American diplomat in Brussels described discussions with European Union representatives as "vitriolic" when it came to satellites and suspected that their uniform support for Inmarsat was simply anti-Americanism in disguise. "The EU harbors deep suspicions that the U.S. desires to keep [Inmarsat's phones] out of the U.S. market," the diplomat said in a State Department cable.

But it was early in the game, so Motorola didn't recognize right away that they were being stonewalled by Europe. Several Motorola executives flew to Madrid for a presentation to Telefónica, the Spanish communications monopoly, and senior managers redecorated their offices to show off all the Motorola equipment they were using. They oohed and aahed at everything they were told—and then never followed up. In France it was worse. Heading up the Motorola diplomatic effort there was Leo Mondale, nephew of Walter Mondale, the U.S. Senator who had served as Vice President under Jimmy Carter. Mondale was a talented communications lawyer who had worked in the Paris offices of Fairchild Space and for the aeronautics division of defense contractor Mécanique Aviation Traction (better known as Matra), and he was the first hire at Iridium, partly because Motorola thought his connections could bring the big European telecoms aboard. The initial meeting at France Télécom turned out to be an elaborate farce, during which the French executives affected *bonhomie* for their *frères* from across the pond while fishing for competitive information—but that wasn't the worst part of the experience. Someone had managed to place listening devices in the first-class cabin of the Air France flight that carried the Motorolans to the meeting, so the Paris executives knew exactly what Motorola was trying to do and how they were trying to do it. Whether the bugs had been ordered by the French government or the French phone company didn't really matter, since it seems both were opposed to any supranational American phone company. Then, when the Motorola team members returned to their hotel rooms that night, their suitcases had been rifled. Things were getting rough in the world of satellite phones.

The French would continue their campaign against Motorola and Iridium, despite repeated attempts by Mondale, Kinzie, and others to partner with them. At the WARC, France Télécom delegates were telling anyone who would listen that participation in Iridium was a violation of the Inmarsat treaty. Then, after the Iridium system was patented anyway, France launched a complex and expensive legal challenge that resulted in a series of hearings before the European Patent Office in Munich. Frank Bogacz, Motorola's chief patent attorney, took Iridium inventor Ken Peterson with him to the hearings, where proceedings were conducted in French despite all the French lawyers being capable of perfect English. The French legal team presented very weak evidence that the idea of a LEO satellite phone system was invented not by Bertiger, Leopold, and Peterson but by the engineers at a French company called Alcatel Alsthom. Bogacz wasn't really surprised that the French would object to a patent—they objected to everything—but he was shocked when the Iridium patent was indeed revoked. At the conclusion of what he regarded as a "kangaroo court," Peterson went back to his hotel room and placed a six-minute call to Leopold in Chandler to give him the details. When he checked out the next morning, he noted that the six-minute call had cost $106. Of course the Europeans hated Iridium, he thought; they were still selling terrestrial phone service at $18 a minute, which was exorbitant even by the standards of luxury hotels.[11]

Meanwhile, back in Chandler, the dawning realization that Iridium was universally feared had the thrilling effect of convincing everyone that Iridium had improved on sliced bread. All the competing projects, most of them linked to big players like Hughes or Fairchild or Boeing, were taken to be grudging endorsements of Iridium as the new gold standard in telephony. From twelve people in the Strategic Electronics Division, the offices assigned to Iridium quickly swelled to two hundred, and they were adding manpower all the time. There was an infectious enthusiasm abroad—Iridium was the breakthrough that would not only save Motorola's research labs in Phoenix but just might become larger than

Schaumburg. Iridium had now been memorialized in a formal description of system requirements called an A-Spec—thirty-seven pages of text and charts that the inventors needed several months to put together, since the original ideas had been all in their heads—and they proceeded to present the details at every scientific conference and telecommunications gathering in the world. (It was some indication of the value Motorola placed on Iridium that, until the A-Spec was finished, Bertiger, Leopold, and Peterson were not allowed to travel on the same plane.) They were about to launch the largest satellite constellation in history, a fact that made them the big dogs in the systems engineering world, and they had ways of warning would-be opponents that they were willing to use the massive reach of Motorola to crush all comers. The most dramatic public presentation occurred at Caltech, where Bary Bertiger engaged in a head-to-head debate with Andrew Viterbi, the legendary engineer who cofounded Qualcomm and was now backing Globalstar. "Their architecture was just wrong and I proved it," said Bertiger. "We only needed one ground station and they needed dozens. They had no intersatellite links. They were bent-pipe. They couldn't upgrade their satellites once they were flying. They had all the disadvantages of LEOs and none of the advantages of GEOs. The engineering community recognized this, we won all the debates, but Qualcomm was afraid that, if we succeeded, their 'spread spectrum' system would disappear."

Then, four months before the WARC, Motorola sent its entire army to the Palexpo convention center in Geneva for the World Telecommunication Exhibition, a quadrennial event that Ray Leopold called "the most obscene display of opulence I've ever seen in my life." The purpose was to let the world know that Iridium was coming, Iridium was inevitable, nothing could stop Iridium. The Motorola booth featured a huge holographic display showing how the Iridium system worked, and the hologram was so popular that people kept coming back to see it, creating a line that made the booth seem like a museum. On the second level Motorola had installed a dining room and kitchen, and then on the third floor some "very private" meeting rooms, as the effort to persuade the Europeans to join them, not fight them, was ongoing. The Motorola team was housed in $2,000-per-night hotel rooms, and there weren't enough of them in Geneva proper, so the company also rented an old castle in Lausanne, sixty-five kilometers away, where a chauffeured Rolls-Royce was used to take people to and from the train station.

One night Ray Leopold and Jerry Adams, the new Iridium President, got lost in the bowels of the castle and stumbled upon a little Irish-pub-style bar that looked out onto Lake Geneva. They ordered some drinks and tried the local savory pudding, with hunks of rabbit in it, then started telling war stories. Adams had grown up near Chicago and gone to Navy Top Gun school, but his fighter-pilot career ended with a "cold cat launch." He was being catapulted off the deck of a moving carrier, but the mechanism malfunctioned and his plane ended up in the ocean, where he was run over by the ship and sustained injuries that made him unable to fly. It was a hard-luck story similar to Leopold's: after completing his master's degree at North Carolina State, he was three-fourths of the way through pilot training at Williams Air Force Base in Arizona when some of his classmates threw him into the Officers Club swimming pool. The horsing around cost him two herniated discs and made him physically ineligible for combat flight—"and, for the first time, I realized I wasn't indestructible." It was a bonding moment for the two men, who otherwise had very little in common. Adams thought Leopold was a science geek who talked too much. Leopold thought Adams was too serious and too focused on profit and loss. But it was a microcosm of the heady spirits that were driving the Iridium project. People who would never otherwise socialize with one another were pulled together like family by a dream that seemed destined to change civilization itself.

In just a few months the engineers at Chandler and their colleagues had made more than three hundred road show presentations to potential investors around the world. In many places they were received as visionary geniuses. Leopold flew all night to get to a presentation at the national telephone company of New Zealand, but flight delays caused his talk to be postponed until three hours after closing time. Remarkably, almost all the employees chose to either stay late at the office or go home for dinner and return, because when he got there he had an audience of several hundred eager faces. He accepted every invitation to speak at every scientific conference, and one of his favorite tricks was to upset the local agenda by saying, "I think I'll depart from my scheduled topic today—let's talk about Iridium!" He loved the almost audible gasps of anticipation when he would do that. He also loved the question-and-answer sessions, during which skeptics would try to punch holes in the Iridium business case. Everyone at the Chandler Lab would get a surge of energy every time someone said, "It can't be done."

The energy in Chandler was intoxicating. One weekend the Motorola human relations department took the entire senior management team in the Satellite Communications Division out to a ranch in Prescott, Arizona, for "teamwork exercises." It was a typical Motorola technique to create bonding experiences for what could often be large, unwieldy teams of people with disparate personalities. The ranch was outfitted with obstacle courses, some of them twenty feet above the ground, and the team had to figure out how to get through them as a unit. The Iridium guys went through the first five obstacles, but suddenly the Director of Human Relations called off the exercise.

"You guys are dangerous," he said. "You're middle-aged. We're afraid you're gonna have an accident."

In other words, they were too gung ho for mere corporate games.

When Leopold would make presentations—and the team made over a thousand between 1990 and 1995—he would often say, "We're programming miracles."

And Iridium's fame was starting to spread. It was during this period that Edward Teller, cofounder of Lawrence Livermore National Laboratory and the father of the hydrogen bomb, paid an unscheduled visit to the Chandler Lab. Leopold had gone through the Air Force Academy with Teller's colleague Greg Canavan, creator of Brilliant Pebbles, and Canavan was bringing Teller by to discuss Star Wars. But Teller spent most of his time asking questions about Iridium.

"What I like about the system," he said, "is your revisit time." (This was a reference to how often the satellites revisited each point on Earth.)

But Leopold was especially proud when Teller said, "Did you think about doing the same job without using satellites? Did you ever consider high-altitude balloons?"

Leopold replied that indeed they had, and they spent several minutes talking about balloons that would function as cell towers.

And the great man said, "But they would bunch up, right?"

The atmosphere with Teller was collegial and friendly, and the Iridium engineers assumed it was just curiosity on Teller's part, but a few weeks later they got a follow-up e-mail from Teller. He said he had tested the balloon theory at Livermore Lab, and their assumption was correct: solar winds would cause the balloons to bunch up in the stratosphere.[12]

Iridium, in other words, was *known*. It was the favorite topic of the cognoscenti who either ran telephony systems or worked in systems engineering. And perhaps that was why it was so unpopular among executives who were already committed to a certain way of doing things. Because it had now become obvious that the private placement memorandum was a bust. Motorola was asking everyone to buy 5 percent shares in Iridium at $80 million each. But as the rumors swirled, stoked by Inmarsat and the European PTTs, John Mitchell started to realize that the package wasn't going to sell. The objections were many—Iridium's business case was not well thought out, it was "just selling interface" between phone systems, it was a self-serving deal for Motorola—but the main deal-breaker was that Motorola hadn't secured the necessary spectrum. Add to this the fact that Bernie Schwartz had sent his Globalstar salespeople out into the world, and he had managed a very clever reversal of the Motorola selling points. "If you sign up with us," Globalstar was saying, "you can own your own gateway."

By coining the word "gateway"—as opposed to "earth station," "transmitter," "receiver," "terminal," "communications unit," "switchboard," or any of the other traditional terms used to describe the ground facility where calls were processed—Globalstar had created a glamour item. "Gateway" implied a portal into outer space, on the one hand, and a funnel for commerce on the other. A gateway was the place where the door opened into the celestial future, pouring gold into your household coffers. Gateways meant infrastructure and jobs and a strong physical presence. The concept had been especially compelling in Europe, where Schwartz signed up Alcatel, Deutsche Aerospace, Daimler-Benz, and Vodafone as partners. In countries that had only one phone company, it was heady stuff: *You can be the phone company in the future,* Schwartz was saying. And since some of his investors were part of less-than-democratic societies, the gateway concept was attractive from the service-provider end as well: You could sell the phone to some customers and refuse to sell it to others. You were the boss. The gateway made you the boss. You controlled what came in and what went out, like the sultan who controlled the Bosphorus Strait.

Iridium didn't have gateways. Iridium didn't need gateways. Or, to be more precise, Iridium needed only one gateway, a place for calls to be processed when the sender or receiver used an old-fashioned landline or a land-based cell phone.

"We need gateways," Mitchell was telling the troops. "If we had gateways, we could sell the system." And with that, he shut down the road shows and the fund-raising. The team would go out again next year with a new plan. No one was going to invest in Iridium until the company had spectrum anyway. *The only thing I want you to think about for now,* said Mitchell, *is spectrum.*

For those who don't have electrical engineering degrees—and there were a lot of nonengineers at the WARC—the radio frequency spectrum is all but incomprehensible, and the International Table of Frequency

Allocations is impossible to read. It looks like the world's most erratic EKG, on the one hand, but since it's multicolored, it also resembles the world's longest hippie peace quilt. Radio waves are mysterious. They're a tiny part of the radiant energy that was created by the Big Bang fourteen billion years ago and now courses through the invisible universe. This energy can be described in many ways, but the most common is to name radio waves according to their frequency—the number of cycles the wave completes in one second. Hence the table is a twenty-five-hundred-page left-to-right chart using the radio-wave unit of measure, the hertz. Named after Heinrich Hertz, who proved the existence of electromagnetic waves in the 1880s, a single hertz equals one cycle per second. Therefore, one kilohertz equals one thousand cycles per second, one megahertz is a million cycles per second, and one gigahertz is a billion cycles per second. If you were to chart the entire frequency spectrum, it would also include terahertz, which is where microwaves and infrared photography reside, as well as petahertz, exahertz, zettahertz, and yottahertz (10^{24} cycles per second), and there are even higher frequencies that, according to physicists, are "not directly observable" but "may be inferred to exist." In other words, there's a mystical aspect to the radio frequency spectrum, and even people who spend their lifetimes studying it are constantly amazed by its quirks and peculiarities.

Fortunately for nonscientists, the only part of the band that's usable for communications is from 3 KHz to 300 GHz, and even more fortunately, no one has to remember the actual numbers of the frequency bands because over the years the users have given names to the "sweet spots" that everyone competes for. Therefore we have the High Frequency band (3 to 30 MHz), the Very High Frequency band (30 to 300 MHz), and the Ultra High Frequency band (300 MHz to 3GHz). But because of their transmission characteristics—the "cleanness" of reception when you use them—the most highly desirable bands are between 1 and 3 GHz, especially the L-band and the S-band, both of which got their names from Navy counterintelligence units during World War II as a way of disguising what frequencies they were using for radar and electronic jamming.

The 1992 WARC was a turning point in radio spectrum history, when the world realized that all the available sweet spots were going to be overwhelmed if something wasn't done. This realization led in 1993 to the United States spending billions to explore the potential of the rarely used Ka-band (pronounced "Kay-ay band"), between 26.5 and 40 GHz, by launching the Advanced Communications Technology Satellite (ACTS). The ACTS remained active for seven years and was a gift to the world, proving that there were ways to use the Ka-band while avoiding its major problem, which was a tendency to fade or disappear when it rained. In fact, in 1998 ACTS would achieve a transmission rate of 45 megabits per second between the satellite and a Navy ship on Lake Michigan—twenty-two times faster than any previous data transmission from space—making it ideal for the Internet. As all the other sweet spots filled up with users, the Ka-band would become the new frontier.

But all that would come later. In 1992, Motorola wanted a proven, existing sweet spot. They actually wanted three sweet spots, but the first two were easy. The frequencies for the intersatellite links, so that the birds could talk to one another, were noncontroversial, because the band was already used by the Space Shuttle and the International Space Station, and besides, the wavelengths would be absorbed by the atmosphere before there was any possibility of interference. The "feeder links"— communication between the SNOC and the satellites, uploading new software and downloading data through isolated earth stations called T-TACs in the northern latitudes—were equally easy to figure out. Those links used the Ka-band, and Motorola knew all about how to use it from the top secret Milstar program. Iridium would, in fact, be the first commercial firm to use the Ka-band in any significant way.

The problem was the frequency needed for what was called the "subscriber unit," the actual phone. Motorola engineer John Knudsen had been working nonstop on the problem for a year and a half, and he had initially sought two bands in the spectrum, one for uplinks (caller to satellite) and one for downlinks (satellite to caller). Specifically he wanted 1545 to 1559 MHz for downlinks and 1646.5 to 1660.5 MHz for uplinks. But that was spectrum that Inmarsat was using, and no one had the stomach for a head-to-head fight with Olof Lundberg. Motorola's backup choice (1610 to 1626.5 MHz uplink and 2483.5 to 2500 MHz downlink) was spectrum that the European Space Agency (ESA) had gotten specific authorization to use in 1987 for GalileoSat, Europe's version of GPS. But it had never been used, because five years later GalileoSat had yet to launch, and there were rumors that the perpetually underfunded ESA was all but out of business anyway.[13]

But there was another problem. Motorola engineers found a government study showing that if they downlinked at 2483.5, they would be using a crowded band, part of the spectrum used by industries, scientific organizations, hospitals, and even garage door openers, pacemakers, and microwave ovens. Their conclusion was that the frequency was "full of all kinds of crap" and would cause them problems. Much later they found out the government report was wrong—the band would have worked fine—but at the time it caused them to make a fateful decision to seek both uplinking and downlinking in the same band, the L-band, the band everyone wanted.

It was spectrum used by Sweden for radar, used by the military in several dozen countries, used by every radio astronomy listening post in the world, and coveted by the soon-to-be-launched GLONASS system, which was Russia's version of GPS, while coming dangerously close to a band of spectrum used by the Pentagon for weapons testing. It was also the frequency of the Geostar satellites used by Qualcomm for its asset-tracking service. In other words, Motorola was prepared to stampede, elephant-like, through the densest part of the jungle. Fifteen months later, when the Office of Technology Assessment reported back to Congress on the 1992 WARC, the white paper included the following passage:

In the future, government spectrum managers must be watchful that U.S. companies participating in international conferences support U.S. positions and do not promote their own special interests to the detriment of specific U.S. objectives. . . . Sometimes it is unclear who is in charge of formulating U.S. international spectrum policy—the Federal Government or the private sector and its consultants.

And everyone knew who *they* were talking about.

The WARC convened on February 3, 1992, in the Palacio de Congresos y Exposiciones, a white fortresslike complex situated on a landscaped hill in the resort town of Torremolinos, formerly a sleepy fishing village that put the Costa del Sol on the map when discovered by British tourists in the fifties. Here the motley congress of nations fell into chaos almost from the first day and ended up having interminable sessions in which very little was accomplished. There were the usual early fights over the High Frequency spectrum, better known as the "propaganda spectrum," because, at 3 to 30 MHz, it was a virtually unregulated shortwave band that allowed broadcasters like Radio Free Europe, Voice of America, and, increasingly, religious organizations to penetrate into any corner of the planet. Battles over the HF band had been going on for fifty years. Western countries tended to be in favor of expanding it, developing countries in favor of suppressing it, and the usual way to resolve differences was to make new allocations but embargo them for fifteen years. This was what was done, putting off the use of new propaganda frequencies until 2007. It was the last issue of the conference on which Europe would stand with America.

Each region of the world was trying to stake out some area of superiority. Japan was seeking a new allocation for high-definition television, since it had the technology ready long before anyone else. Europe wanted new frequency for something called "Future Public Land Mobile Telecommunications Systems" (FPLMTS), dubbed "Flumpits" by the delegates. Flumpits would provide spectrum for a new digital cell-phone operating system called Groupe Speciale Mobile (GSM) that had the potential to make cell phones compatible all over Europe. This was a problem at the time, as French phones didn't work in Germany or Italy, British phones couldn't be carried into Scandinavia, and the lack of connectivity was frustrating the common goal of a single European market. Thirteen European countries had been working on a pan-European standard since 1982, and they had set aside spectrum at 900 MHz for that purpose, but now that wasn't going to be enough. Of course, one thing that made Iridium seem like such a great idea was that the planet was split up among dozens of competing analog cellular systems. That's why the Motorola engineers thought Flumpits was a direct reaction to Iridium. No one really believed a single global standard for land-based phones was possible—there wasn't enough bandwidth for even the current GSM system—but the Motorolans were suspicious: if all the Europeans wanted was to connect the nations of Europe, then why did they need a *global* allocation?

Meanwhile, the United States wanted new allocations for what was being called Broadcasting Satellite Service-Sound, better known today as satellite radio. The founder of Sirius Radio, a lawyer named Martin Rothblatt, was annoyed that Motorola was coming in at the eleventh hour and disrupting his plans to be the number one American priority, especially since satellite radio enthusiasts had been submitting proposals for twenty-five years and had met all their deadlines. Unfortunately for them, existing radio broadcasters in the United States were so opposed to the new technology that they considered Rothblatt the Antichrist, so the American delegation ended up serving two masters and articulating a mealymouthed position in the debate. Canada and Mexico, by contrast, were gung ho to light up the car radios of North America with Rothblatt's international system, called WorldSpace. This gave Motorola another card to play when Leo Mondale brokered a backroom deal with Rothblatt: *We'll use our influence to get you your spectrum if you'll use your influence to get us ours.* In both battles, Europe was the enemy; Europe was opposed to satellite radio because it didn't own the technology, and they eventually prevailed. The frequencies were approved—but embargoed until 2002.

Then there was the battle over the "Little LEO" spectrum, frequencies desired by four American companies that intended to launch mini-satellites that would be used for short-burst messaging and asset tracking.[14] Once again the United States fought with Europe—and got the frequencies it needed, mainly because they were far from the sweet spots and didn't threaten the PTTs.

All of these were hard-fought battles in which the small Third World countries were used as pawns by the tech giants, and with Europe opposing anything the United States wanted, even when the stakes weren't that high. An example was the fight over Aeronautical Public Correspondence, better known as passenger phones on commercial airlines. This service, which allowed passengers to use seatback phones that could transmit but not receive calls, was in general usage in only one place in the world—the United States—but Europe wanted the worldwide frequency changed for no apparent reason other than to make it difficult for American companies to sell the service in Europe. No wonder the American delegates sometimes looked at their European counterparts as mean-spirited.

But none of the fights compared with that particular slice of radiowave real estate that became known as the "Motorola Band." Like gun control and abortion, you were either for it or against it. Technically it was the Big LEO allocation, and, if approved, it could be used by any low-earth-orbit satellite communications company, based anywhere in the world, but everyone knew that the lobbying effort was not led by Globalstar or Ellipso or Odyssey. Those companies were hanging back and letting Motorola assault the beaches, even as they planned to fight Motorola for that same frequency when it was considered later by the FCC.

When Motorola made it clear just exactly which part of the L-band it wanted, the first howls of protest went up from the European Space Agency. This was a loose consortium of thirteen countries that had its headquarters in Paris but had wisely distributed the Europork with a research center in Noordwijk, Netherlands; a mission control center in Darmstadt, Germany; an astronaut training facility in Cologne, Germany; an astronomy center in Villanueva de la Cañada, Spain; a satellite testing center in Redu, Belgium; and a launch site near the Equator in French Guiana. Ray Leopold and Leo Mondale made a preemptive strike by visiting ESA's research-and-development center in Toulouse to reassure the Europeans that Iridium would work alongside them if they wanted to continue using the spectrum, but they reacted with hostility. Yes, it was true that they never launched the European GPS system, but they had a new system that would be using the spectrum. It was their own satellite phone system, called Archimedes. It would be fourteen satellites in medium earth orbit. Or maybe it would be three satellites in Molniya orbits. Or maybe it would be four satellites. At any rate, *they were going to build it.*

This was the first anyone had heard of Archimedes. There were no specs on it, not even a one-page description. Some of the Motorolans speculated that it had been dreamed up the previous night while the French were consuming a cocktail called the Archimedes at the hotel bar. Things got even weirder when another previously unknown system called LEOCOM was announced by the Italians. Suddenly speculative satellite systems were being invented in Europe almost daily, leading to after-hours barroom joking that they were all invented by Ulf Merbold. Merbold was the West German space traveler whose goofy smiling portrait was ubiquitous in the many offices of the ESA thanks to his being the first European astronaut on a European mission—a fairly irrelevant distinction at the fairly irrelevant date of 1983. Unfortunately for the German physicist, his grin and his name spawned an unlimited harvest of Ulf Merbold jokes as a way to lampoon the arrogant Europeans and their tendency to copy something from North America and then claim it as native. "It's an Ulf Merbold project" became a way of saying, "It's fantasy."

Next Motorola got wind of a campaign by Inmarsat to challenge Iridium on the basis of licensing. If Iridium was unable to get "landing rights" in any given country, argued Inmarsat, then all seventy-seven Iridium satellites would have to be turned off any time they crossed the borders of that country. (This was a legal gray area, since outer space was technically unregulated, and Motorola intended to operate Iridium everywhere, regardless of whether the phone was legal on the ground.) Besides, Inmarsat argued, satellite telephony was not on the formal agenda, and the ITU shouldn't make hasty decisions. Its recommendation? Spend ten years studying the issue before allocating any spectrum. "One advantage Inmarsat had," said Mondale, "is that in 1992 there was no difference between a PTT and a regulator. They were one and the same. The national phone company was also the regulatory authority. And that was the regulatory authority that controlled Iridium licenses."

But the Motorola delegation already had a mole within Inmarsat, so they knew how to fight back. The spy was an Italian named Enrico, and he had shown up unannounced at the end of the first week, asking to speak to the Motorola chief delegate. Enrico said he was frankly "disgusted" by what Inmarsat was doing in the dirty tricks department, and he would like to help. After extensive questioning—vetting him to make sure he wasn't a double agent—Enrico was given a cover story and commissioned as an operative for the remainder of the conference. His cover was that he was dating one of the Motorola secretaries, which gave him ample opportunity to sneak away, pretend to be wooing her, and pass off information to his designated "handler," Ray Leopold.

As the days passed, it seemed that everyone in Europe was more and more determined to shoot down Iridium. Part of the reason was that "one world, one phone" was a concept that threatened Flumpits, since Europe expected GSM to become the worldwide standard. The more visceral reason was that the Europeans simply resented the global dominance of the United States in cellular. And the symbol of that dominance . . . was Motorola.

The Swedes said they needed the Motorola band for radar usage, so Leopold met with them and figured out how to avoid interference. The DGSE—France's version of the CIA—lodged a complaint that the Motorola band would interfere with its counterterrorism operations. There were loud protests from several radio astronomy centers around the world, but especially the ASTRON telescope in Dwingeloo, Netherlands, and the ICRAR telescope in Perth, Australia. There are dozens of radio astronomy installations staffed by physicists listening to radio emissions from distant galaxies. These are the scientists who discovered quasars and pulsars and theorized the Big Bang, and they have chips on their collective shoulders because their needs are constantly pushed aside in the name of commercial interests. The particular reason Iridium annoyed them was that there's a "spectral band" emitted by hydroxyl radicals created by interstellar dust storms coming from the direction of Orion, and that band has a frequency of 1612 MHz—smack in the middle of what Iridium wanted to use for uplinks. The only nonastronomers who fully understand the spectral band are people like Ken Peterson, who was dispatched to the Plains of San Agustin, fifty miles west of Socorro, New Mexico, to meet the astronomers in charge of the National Radio Astronomy Observatory and its twenty-seven antennas. Peterson recalled, "We discovered that the amount of interference is 10 decibels below Boltzmann's constant"—a measurement that allowed Peterson and Leopold to come up with a "picket fence" solution. This was an interface allowing the thirty observatories focused on the spectral band to peek through Iridium transmissions during peak hours. Unfortunately, the Iridium hardware also denigrated the sensitivity of the telescope, thereby requiring twice as much listening time to observe the same number of hydroxide molecules. The astronomers were furious about losing their "quiet skies." "They were dealing with signals from outside the realm of mankind," said Leopold, "and they didn't like sharing with mere humanity."

One by one Motorola created impromptu tiger teams to deal with each objection. If the engineers could develop a compatible usage—a way for both parties to use the same frequency—then they would offer a compromise to the objecting nation. If a compromise was impossible, they would look for something else that country needed and try to get it in exchange for not opposing the allocation. And if all else failed, they would simply agree to a footnote saying that one particular usage in one particular country would be exempted. Footnotes could cause problems later, but it was better to get the allocation approved, even if the uses of it were ambiguous, than to have no frequency at all. During the process the Motorolans learned the art of writing footnotes, which was to make them as technically complex as possible and, more important, as vague as possible.

Week after week the conference dragged on, accomplishing nothing even with eighteen-hour sessions. The turning point finally came when Abderrazak Berrada, a respected scientist and head of the Moroccan delegation, rose for an impassioned speech, castigating the delegates for their inaction and calling it a dereliction of duty. He was speaking primarily to the bloc of thirty-two European nations, dominated by the entrenched postal and telegraph agencies, all refusing action that would threaten their dominance. Berrada then led a two-day walkout of Third World countries. Motorola saw this as an opening and intensified its sales pitch, offering the small countries discounted shares in Iridium, free government phones, and any other incentives they could think of, positioning themselves as the friends of the little guy against the economic bullies in Europe. "I fed the entire continent of Africa in those two days," Jerry Adams told Bertiger.

And the courtship worked. "The Africans ganged up on the Europeans," said Leo Mondale. "Since it was one country, one vote, they had them outnumbered."

Yet the main issues of the day—Flumpits, satellite radio, high-definition TV, and Big LEOs—were all so fraught with emotion that presiding officer José Barrionuevo Peña was afraid to bring them to a vote. A seasoned politician, Barrionuevo was adept at figuring out where there was opposition and passing messages to the players so that everything could be worked out behind closed doors. Three weeks into the conference, he sent a message to the United States: Pakistan will oppose your allocation for Big LEO spectrum.

Pakistan? No one had heard anything about Pakistan, pro or con, during the entire history of Iridium's development. A tiger team was quickly put together to figure out what Pakistan's problem was. The two delegates from Islamabad were found staying in a cheap suburban hotel, eating meals in their rooms to save money. Obviously these were guys Motorola could work with. Soon the Motorola team was buying them dinners, suggesting they move into better digs, and trying to get down to the nitty-gritty: *What is it, exactly, that you don't like about our satellite phones?*

And no one could ever get an answer. Both Pakistani delegates seemed to be lower-level functionaries who were simply afraid to vote for anything that wasn't on the formal agenda. They didn't have reasons, they didn't have instructions from their bosses—they just thought a yes vote would somehow hurt their country. One of the delegates had been a prisoner of war in Kashmir, so there was some speculation that they were voting against it simply because India was voting for it. But Motorola could never extract any information beyond "We are not sure it's good for Pakistan."

And so it stayed, until the last night of the last week of the WARC, March 4, 1992, when the chairman announced that this would be the final session, and if the delegates intended to accomplish anything, now was the time. Everyone panicked—and suddenly Europe and Japan and the United States all came to the table. Europe could have Flumpits if the United States could have the Iridium allocation. Japan could have high-definition TV if it would in turn support Flumpits and Iridium. There was only one problem for Motorola: Pakistan was still determined to veto the satellite phone systems.

As the night wore on past 3:00 A.M., it started to look like Barrionuevo would never bring the Iridium allocation up for a vote. Every time
Motorola sent a messenger to ask him about it, he would say, "We still
have a problem." Finally, a little after 4:00 A.M., Barrionuevo interrupted
the speaker and called for an immediate vote on Big LEOs.

He had noticed that both of the Pakistani delegates were sleeping
at their desks. The allocation passed.

Chapter 6

ROCKET MAN

Trees don't grow on the barren wasteland prairies of Kazakhstan, yet Yuri Gagarin, the first man in space, had planted one anyway. Like everything else at Baikonur Cosmodrome, Gagarin's tree was an alien presence. Attendants assiduously watered and fertilized the sapling until it defied nature and grew into an angry piece of bark, a gnarled trunk with spindly branches, a symbol of the Soviet Union's triumph over an unforgiving, hostile environment. When Dannie Stamp first saw it, he was slightly disoriented. He was strolling down a makeshift "boulevard of the cosmonauts" where later space travelers had planted their own trees next to Gagarin's, and now they all rustled in the incessant wind in a grassless park leading to a bleak observation deck that jutted out over the right bank of the Syr Dar'ya River. The view from the deck was straight out of a Salvador Dalí painting, complete with strange objects poking up out of the landscape, and the river itself was little more than a moist ditch in a semiarid desert where no one except nomadic cattle ranchers and camel traders had dwelt before 1955.

Stamp was surrounded by the largest and oldest space complex in the world, and he was more than a little awestruck. Seven years earlier, United

States Air Force Lieutenant Colonel Daniel Stamp would have been shot on sight if he'd come anywhere near the place. The Cold War was over, but emotional attitudes don't die so easily: at some point in his subconscious this was still ground zero for evil, the womb of every holocaust scenario for the past four decades. The location of "Scientific Research Test Range No. 5"—named after the distant mining town of Baikonur only as a way of disguising it—was chosen precisely because it's in the middle of nowhere, a place where the first stages of ICBMs could fall to Earth and be picked up later by military helicopters, and where scientists and engineers with top secret clearances could be hidden away to do the research of death. This was the place that CIA pilot Francis Gary Powers was planning to photograph from his U-2 spy plane on May 1, 1960, when he took off from Peshawar, Pakistan, ascended to seventy thousand feet, and set a course north by northwest. The KGB had good intelligence that day and an antiaircraft brigade was ready for him at Baikonur, but the Soviet commander ordered the radar turned on too early. Powers detected it, ducked into some convenient cloud cover, and headed farther north. He was shot down later the same day by a missile over Sverdlovsk, setting off a chain of international events that ended the U-2 program forever.

How the world had changed. The one-man spy plane had mostly been replaced by drones and satellites. Sverdlovsk had reverted to its historic name of Ekaterinburg to get rid of associations with Yakov Sverdlov, the Bolshevik who plotted the assassination of the czar. But Baikonur remained. Baikonur was mythical and permanent. For most of his career, Stamp had been charged with inventing ways to destroy anything launched from here or, in a best-case scenario, blow up Baikonur itself. Now, less than four years after the end of the Cold War, Stamp was a guest at the austere Hotel Kometa on Site 95 in the closed Russian city of Leninsk, wandering unmarked streets surrounded by emptiness that made the flatlands of his native Iowa look Edenic by comparison.

Stamp was here because when his bosses at Motorola asked him what launch vehicle he wanted to use for the eighty-eight satellites they were going to put into space, he told them, without hesitation, the Proton. That was the Russian workhorse, one of the largest and most powerful rockets ever built, the one used for that country's Mars and Venus probes, the Mir space station, and, for a while, their one-hundred-megaton ICBMs. The UR-500, as it was officially known, had been designed in the early sixties as a "super ICBM" that could deliver heavy warheads at distances of more than eight thousand miles but could also be used as a launcher for the race to the moon. When an early version was shown to Premier Khrushchev, he noted the monstrous size and said, "What should we build—communism or silos for the UR-500?" Of all the ceremonial events beloved by rocket engineers, none can compare with the moment when a Proton emerges from its assembly plant on a railcar—silver, phallic, and impossibly large—to travel the last two kilometers to the launchpad.

And in the world of rocket science, there is something to be said for size. Stamp had been in the war room in the early eighties, when all the ICBMs in the world were on high alert, and he had done the math, and in the event of an all-out war, he believed that Russian rockets would have prevailed over American ones. In after-hours conversations, the Air Force officers assigned to NORAD would talk about scenarios that resulted in all the American missiles being destroyed and a third of the Russian missiles reaching their targets. The Soviet Union was doing badly in virtually every category of modern civilization, but nuclear warfare was not one of them. Now Stamp had come to Baikonur with Motorola cash, looking to launch satellites alongside the same men who would have once been launching the ICBMs intended to wipe out Cheyenne Mountain and kill him in his bunker. Those men, like Stamp, were still steeped in the stark world of Mutual Assured Destruction, a world that had vanished in an instant on December 25, 1991, when the final act of Soviet President Mikhail Gorbachev was to hand over the nuclear launch codes to Russian President Boris Yeltsin. Overnight Baikonur was hurled into an identity crisis, existing as it did to defend a nation that no longer existed, in a Muslim country newly created, managing nuclear weapons no longer needed. It didn't take Russia long to follow the lead of China and make inquiries about how to get into the commercial space business, but the transition was slow and awkward. This was still a military installation. When you entered Baikonur, you entered a place full of ghosts, especially the ghost of Gagarin, who became a Russian demigod after he died during a fighter-jet training exercise in 1968.

As one of the rare outside visitors to Baikonur, Stamp had expected ritual and history, but he was shocked by the unfinished, haphazard look of the place. Site 45, for example, was a deep earthen crater full of building materials and broken shards of concrete that resembled a messy archaeological dig in progress. It was actually what was left of a Zenit launch that went bad in 1990. The rocket, carrying a Tselina spy satellite, rose about seventy meters into the air, lost thrust, froze, then slowly sank back to the earth, causing an explosion that lifted the entire launch structure twenty meters off the ground, scattering debris three kilometers in every direction, and burning out of control through the night. Since the launchpad was a total loss, the Russians simply left it in place, and with each passing year the blinding dust storms of summer and the mind-numbing blizzards of winter had sculpted the ruins into something that looked ancient, as though it had been the palace of some forgotten Mongol warlord. The Russians were superstitious about things like that. You don't touch it—you leave it. It was like the cosmonaut ritual of pissing on the right rear tire of the bus that took the flight team to the launchpad. Gagarin had done it, and Gagarin had survived. In later years everyone taking off from Baikonur would spend the night before launch at a mandatory screening of *White Sun of the Desert*, a Russian cult film about a Red Army soldier's comic adventures in Central Asia. No one remembered exactly why you had to do this, but it was tradition, it was ritual, it was Baikonur.

Dannie Stamp, silver-rimmed spectacles wrapped around a prematurely bald head, had a folksy way of speaking that made him one of the most likable guys to ever come out of the Air Force space program. More important, he loved vodka, and that meant he got along well with his new Russian partners, all executives at the Moscow-based Khrunichev State Research and Production Space Center, where the Protons were built. Stamp had the eager-to-please manner of a team player—or maybe just the learned behavior of a man who had been chewed out by generals his whole life. "One reason I'm able to do Iridium," he liked to say, "is that I've already confessed every sin"—meaning he'd been responsible for rockets that crashed into the ocean, satellites that tumbled out of orbit, and budgets that ran out of control. He was intimately acquainted with the dreaded word "anomaly," used whenever a satellite stopped functioning for no apparent reason. He had once worked for a two-star general who had a daily 8:00 A.M. operations meeting at which everyone responsible for a space program sat around a conference table and, if called on, was expected to describe every anomaly from the previous day. You weren't allowed to talk about anything that was on schedule or functioning properly—the general only wanted to hear about screwups—and everyone sat in fear of being called to the public confessional. Stamp was so battle-scarred that he didn't trust rockets unless they had a long history of reliability—dozens of successful launches—and he felt pretty much the same way about people. He hired based on reputations, not résumés, because he wanted colleagues who had already felt the agony of a rocket going nose down in the Pacific, or an employee driving a crane through an antenna. Stamp was now charged with building and launching more satellites in a shorter period of time than had ever been attempted in history, managing the largest commercial space project ever conceived. And he needed the Proton to do that.

Stamp was part of the generation of Air Force officers who got locked into a military career as a way of avoiding death in Vietnam. In the late 1960s the Pentagon started running out of draftees to send to Southeast Asia, so the graduate school deferment was eliminated. Since many young men were enrolled in graduate school only as a way of avoiding the draft, their sudden eligibility caused them to sign up for officer training, especially in the Air Force, which was considered safer than the Army, Navy, or Marines. Many of those "ninety-day wonder" officers then remained in the service, reenlisted, and moved up the ranks, and by the 1990s the Pentagon was full of them. One of the reasons the military-industrial complex was so shy about combat—witness the half measures of generals like Colin Powell and Wesley Clark when it came time to clean up the Balkans—was that much of America's officer corps had started their careers as combat evaders. That was something never talked about at the highest levels of the Air Force—unless you were Dannie Stamp. Stamp openly admitted that he only ended up in the service because he was looking for a way to avoid getting shipped to Vietnam.

Stamp came of age in the fifties and early sixties, growing up in a speck on the map called Persia, Iowa, working on his family's two-hundred-acre farm—running cattle, tending hogs, sowing corn, oats, alfalfa, and wheat—and living for the days when he could get away to the outside world—places like Sidney, Nebraska, 423 miles west, where his father bought feeder cattle once a year, or the Omaha Union Stockyards, where his father sold his herd. At the Iowa State Fair in Des Moines, the adolescent Stamp would always enter the poultry competition, since most chicken raisers were female and he had no other way to meet girls. At St. Joseph High School in Neola, ten miles down the road, his graduating class had three boys and six girls, but three of the girls were Stamp's cousins. It was a constricted world, lorded over by the stern codes of a German Lutheran grandfather and Catholic parents still in touch with their traditional European values, but it was a world shaped in Iowa, the state that's produced more aerospace engineers than any other. (It's no accident that Matt Dodson, the eponymous hero of Robert Heinlein's 1948 novel *Space Cadet*, hails from Des Moines.)

Stamp didn't discover space right away, though. First he enrolled at Creighton University, a Jesuit school in Omaha, thinking he would become a lawyer or doctor. But he became fascinated with physics instead and transferred to Iowa State, which offered one of the most prestigious aerospace engineering programs in the nation. It was in his senior year, 1968, that the draft rules were changed, so Stamp decided to be proactive. He went down to the Army recruiting office in Ames and asked what kind of duty he was likely to get if he enlisted instead of waiting to be drafted. The recruiter was honest enough, telling him he would have a two-year commitment and would most likely end up as a lieutenant commanding a platoon. Jungle warfare didn't sound like the answer he wanted, so Stamp went next door to the Navy office. That recruiter told him the Navy wanted a three-year commitment, and he would probably end up as a line officer on a Swift boat. Better than wading through rice paddies, but still active combat. Stamp skipped the Marines altogether and headed for the Air Force, where the recruiter told him he could go through a ninety-day officer training program, followed by a four-year commitment, and he would be placed in a weapons procurement unit based in the United States. Stamp signed on the dotted line.

A few months later the newly commissioned Lieutenant Stamp reported to Norton Air Force Base in San Bernardino, California, where he was assigned to land-based ICBM upgrades and immediately immersed in the scary what-if world of nuclear warfare. By that time the Air Force had phased out the Atlas missile but had two later generations of nuclear missiles on active duty—the Titan and the Minuteman—and there were also various other cruise missiles, decoy missiles, and drones in development. Stamp worked mostly on the Safeguard program, which was a system of lesser-known nuclear missiles—the Spartan and the Sprint—that were designed to destroy incoming Soviet missiles, which were in turn directed at larger Minuteman nukes that had not yet been launched. In other words, the Air Force was developing underground launchers situated near the Minuteman sites in Montana, Wyoming, and North Dakota, so that the Spartans could attack incoming missiles while they were still in outer space and, if that failed, the short-range Sprints could blow them up when they entered the atmosphere. Either way, the

Soviet missiles would be diverted before they could hit the Minuteman sites, and then the Minutemen could be launched unimpeded. (The Americans and Soviets eventually agreed to stop developing antimissile missiles and anti-antimissile missiles, since this kind of thinking is what led to the arms race in the first place.) Stamp's time was mostly spent working on the Safeguard radar used in space to detect incoming missiles, and it was these same phased-array systems that he would one day use for the Iridium satellites. Although they didn't meet each other, Iridium inventor Bary Bertiger was working on the same radars for the same missile systems at around the same time.[15]

Stamp expected his involvement with missile silos and war scenarios to be short-lived, so he used his time at Norton to earn the credentials he would need for a business career. Commuting to campuses in Greater Los Angeles, he completed two master's degrees: one in systems engineering from the University of Southern California and another in business economics from Claremont Graduate School. Claremont was such a center of antiwar activism that Stamp had to keep a change of clothes in his car, lest being seen in uniform would get him pelted with fruit from the nearby citrus grove, but it was there that he first encountered his mentor for life, Peter Drucker, the Austrian known as the inventor of management theory. After his commitment was up, Stamp fully expected to go to work for an aerospace company and start climbing the corporate ladder. But several of his friends had already gone to work for Boeing, and all of them were being laid off at what turned out to be the beginning of a recession. When 1972 rolled around, he took stock of his situation and decided to reenlist in the name of "salary continuity."

Stamp's next tour was at legendary Wright-Patterson Air Force Base in Dayton, Ohio, where engineers developed new aircraft on the site of what was once the Wright brothers' Huffman Prairie Flying Field, long ago converted into a laboratory administering the Air Force's multibillion-dollar science and technology budget. This was a cutting-edge research facility, and one of its defined missions after Sputnik was to "prevent technological surprise." Stamp was assigned to the emerging drone program, which had shifted into high gear after Gary Powers was shot down.[16] Stamp loved the drone work but soon realized he was on the wrong side of history: he was laboring in an organization run by pilots, administering programs designed to get rid of pilots. So he consulted Peter Drucker about the "organizational dynamics" at work and decided, "Dannie, you need some other career in the Air Force." (His observation was correct. The drone technology being developed at Wright-Patterson in the early seventies was not used in any significant way until well into the twenty-first century. The first Commander in Chief to fall in love with drones as a viable military option was Barack Obama. Obama's Secretary of Defense, Robert Gates, went so far as to fire the Air Force Chief of Staff over his reluctance to use the Predator drone.) The immediate solution to Stamp's problem was a request for reassignment to the *other* Air Force brain trust—the Space and Missile Systems Organization (SAMSO)—and that's where he would remain for most of his military career.

Most frequent travelers have flown into Los Angeles International Airport, but very few notice the facility next door in El Segundo. It's the site of the low-key Los Angeles Air Force Base, and its main mission is to run the outer space programs for all branches of the armed services. When Stamp arrived, SAMSO was a ten-thousand-employee team, mostly engineers, that had developed all of America's ICBMs from the Atlas through the Peacekeeper, handled all satellite launch contracts for the Pentagon, pioneered the first polar orbit, and developed a system whereby photographic film could be dropped from a missile or satellite and recovered in midair by a JC-119 aircraft—the main system of photographic spying after the U-2 incident. SAMSO was America's headquarters of rocket science.

Captain Stamp arrived in El Segundo in 1976 and, because he had a background in both engineering and business, was assigned to the contracts for rocket payloads, which meant primarily satellites. SAMSO developed and built six kinds of satellites: spy, communications, weather, GPS, remote sensing, and, at the dawn of the eighties, kill vehicle. Stamp's responsibilities began with million-dollar contracts, then—as he learned from his mistakes—ten-million-dollar, hundred-million-dollar, then hundred-million-dollar payloads on billion-dollar contracts. After additional training at Alabama's Maxwell Air Force Base, better known as Air University, Stamp spent a tour on the front lines, at Buckley Air Force Base in Colorado, where the 460th Space Wing continually tracked, spied on, and targeted the Russians. Ultimately he was returned to El Segundo and put in charge of the top projects in space—multibillion-dollar programs that remain classified to this day, including the early test launches for Star Wars. By the end of his twenty-year Air Force career he was a Lieutenant Colonel, with Captains and Majors working for him, and he was in charge of entire constellations. He would work with the designers of a satellite, then acquire it, develop it, negotiate the rocket contract, launch it, monitor it in orbit, and decide how it would be operated once it was functional. And as part of that job he was a frequent visitor to the Motorola plant on McDowell Road in Scottsdale, where electronics and communications payloads were built for virtually every satellite he put into space.

Stamp's Air Force career, then, consisted principally of learning how the government operated space vehicles—and he knew it was not an efficient model. Everything depended on cost-plus contracts, so that any change made to the original design resulted in huge surcharges, and the cover-your-ass system of testing and retesting new equipment meant it took years to get anything into production and more years to make anything operational. By the time he retired in 1988, Stamp was convinced he could do things better from the corporate side. So he did what many Air Force engineers do and joined some buddies who had formed their own company, Ultra Systems, based in Irvine, California, and designed to live off government engineering assignments. At first the contracts rolled in, but then Congress started cutting all the budgets for war-fighting, and most of those contracts were canceled. In the fall of 1989 Stamp went to his old friend Durrell Hillis at Motorola, letting him know he was available if they had anything. And as it turned out, three of his colleagues—Bertiger, Leopold, and Peterson—were involved with something so top secret they wouldn't tell him what it was at first. In fact, they mercilessly teased him before finally relenting one day and letting him sign a nondisclosure agreement so they could take him into a SCIF and reveal what they were working on. What they were working on was Iridium.

As Leopold recalled later, "Dannie couldn't be still. He was jumping up and down all over the SCIF. He couldn't control his excitement." And within a matter of months, Stamp had left Ultra Systems and joined the Iridium team, charged with managing the contracts to build the satellites and get them launched and functional. He knew it was a historic moment: the first time satellites would be deployed entirely by private industry. The Iridium team would be commercializing space— the very thing that President Kennedy had feared, but now, with the end of the Cold War, the only way outer space technology could continue to advance. Stamp and his boss, Durrell Hillis, were both steeped in the culture of government contracts, and they both hated that culture. "The government system is high cost, low quality," said Stamp. "You make a low bid to get the contract, then you extend the schedule, you have breakage, and you test and test and test. But now, for the first time, we were going to use a fixed-price contract. Everything driven by cost. There's money on the table and that's all you get. So when that happens, things have to work!"

To make sure they did work, Stamp spent three years training the contractors, telling them, "If you want this job, you'll send design teams to Arizona at your own expense and you'll tell your vendors that no one gets any money for two and a half years. If you're still standing at the end of that time, you get a guaranteed contract." He was actually waiting for John Mitchell to finish up the second fund-raising effort after the first one had failed miserably. It finally closed in the summer of 1993, and Stamp was able to hire the two companies that had been working with him. Lockheed Martin won the contract to design the bus. (A bus is a space vehicle without anything on it. When you add the electronics, antennas, and other payloads, you're "loading the bus.") Raytheon—the same company that designed the massive PAVE PAWS system of phased-array radar for the ICBM early-warning system—won the contract for the main mission antenna. Stamp rode herd on both companies, telling Lockheed it wouldn't be allowed to use the facility in Sunnyvale, California, where it normally built space vehicles, because he considered that facility rife with the Pentagon culture of overspending, overtesting, and underperforming. Instead the Lockheed designers were required to open up an entirely new factory in the Old Mill building along the Merrimack River in Nashua, New Hampshire. Likewise, Raytheon was forced to open a dedicated facility for the main mission antenna in Marlborough, Massachusetts. Stamp himself went to Motorola University in Schaumburg to learn the Six Sigma system and all the other manufacturing processes that had made the company so dominant.[17] He kept reminding Lockheed Martin, Raytheon, and everyone else engaged in the project that the final assembly of a satellite built by NASA or the Air Force took anywhere from nine months to one year *per satellite*. But since Motorola needed a hundred satellites, including all the spares and demonstration units, "obviously we don't have a hundred years to build this constellation."

If Iridium had been a government project, for example, Stamp would have had to perform a thermal vacuum test on every satellite, placing it in a chamber that simulates intense radiation and the other brutal elements it would eventually be exposed to in space. The process was time-consuming and expensive, so Stamp made an early decision: they would "thermal-vac" the first one off the assembly line and assume all the rest were sound. Likewise, a decision was made to use commercial parts instead of the usual Class S or Class J radiation-hardened parts used by the military and NASA. Everything that Lockheed Martin and Raytheon built into the satellite had to be available "off the shelf." Motorola at the time was working with Apple Computer on its first-generation PowerBook, so Stamp used that processor. "It was high performance, low power," he said, "exactly what you want in space." Stamp was also innovative about his assembly line, setting up his factory in Chandler so that the satellites were built horizontally instead of vertically. That meant the assemblers could work at waist level, whereas on military satellites, the technicians stood on ladders and leaned over. He trained Lockheed Martin in the new technique, then later bragged about it to his friends at Khrunichev in Moscow, only to be told, "We've done it that way for forty years."

And then there was the matter of what launchers to use. At first everyone wanted to use a Pegasus, the innovative rocket developed by Orbital Sciences Corporation that's shuttled up to forty thousand feet on the back of a B-52 and then launched from the air in order to cut down on all the atmospheric drag that most rocket stages are designed to overcome. The first successful Pegasus launch had been in 1990, and it had since proven more reliable and cheaper than any other rocket, but the problem was that it was too small—no matter how you configured it, the nose cone would be large enough to carry only one satellite—and they didn't have time for eighty-eight launches. Stamp started looking at the specs and launch histories of every rocket in the world and eventually told Motorola, "Look, I feel much safer on a Chinese or a Russian rocket than an American one." All the American rockets were handled by either the Air Force or NASA, all of them had spotty launch histories, few of them had the power he wanted, and he didn't trust the people in charge. Motorola was hesitant at first to launch the entire constellation with rockets built by foreign countries, but Stamp eventually worked out a deal whereby China and Russia got launch contracts in return for a partnership investment, which included a gateway. Russia took a 4.4 percent equity position in Iridium and got a contract for three Proton launches. China took a similar position and got a contract for three launches with its most reliable rocket, the Chang Zheng, or Long March 2C.

Motorola had to use all its lobbying influence to get the Chinese launch contracts authorized. President Reagan had signed an executive order in 1988 that allowed U.S. satellites to be launched by China, but after the Tiananmen Square protests of 1989, you needed a presidential waiver. That proved to be hard to get because Tony Lake, President Clinton's national security advisor, suspected China had sold a Dongfeng 11 ("East Wind") ballistic missile to Pakistan. If true, all Chinese deals could get shut down. There were also rumblings in Congress about the dangers of technology transfers. (This fear made no sense, since the technology transfer was actually in the other direction, with the Chinese letting American engineers see the inner workings of their best launch vehicle.) Clinton eventually decided that the advantages of space cooperation outweighed the dangers of technology transfer and authorized Motorola's contracts. Stamp was ecstatic. He liked the Long March because it had been used since the seventies and proven highly reliable, and it was small but not too small, perfect for launching two satellites at a time. Since Iridium was planned as seven evenly spaced axes of eleven satellites plus one spare per plane, Stamp would need to launch in combinations of seven, five, and two. The huge Proton was perfect for launching seven satellites at a time. The Long March was perfect for two. And the Delta II, built by McDonnell Douglas—regarded by Stamp as inferior to the other two rockets but the best America had to offer—would be used for the five-satellite launches.

Since no Americans had ever visited the Taiyuan Satellite Launch Center, located in a remote mountainous region in China's Shanxi Province, Stamp sent a team of technicians and engineers to check out the facility and prepare it for Iridium. Taiyuan is actually a misnomer designed to disguise the location. The real Taiyuan is the bustling capital of the province with a population of 4.2 million people, but the space complex is 284 kilometers away, across terrain where the skies are frequently blackened by smoke from coal fires. The space complex is hidden in a depression between ranges of the Lüliang Mountains on three sides, with the Yellow River to the west. The Chinese had been operating their ICBM systems out of the site since 1968, so they had given it a series of misleading names, including the Dongfeng 3 test site, the Wuzhai Missile and Space Test Center, and, simply, Base 25. It was a strange and all-but-invisible city full of tunnels, railheads, and underground bunkers where, rumor had it, a rocket had been launched in the eighties and, in the parlance of the profession, "gone horizontal," resulting in the destruction of an entire city. The legend was undoubtedly an exaggeration, since most of the region is comprised of small villages, not cities, but it wouldn't have bothered the Chinese that much, given that they regularly launched vehicles into polar orbit from a place that required rockets to fly either dead north or dead south over highly populous areas.[18] In the United States it's illegal to launch a polar orbit from Cape Canaveral, because the rocket would have to fly over Charleston, Charlotte, Buffalo, and Toronto if directed north, or Havana, Bogotá, and Quito if directed south. This is why all polar orbits from America are launched southward from Vandenberg Air Force Base in California, which sits on a peninsula that juts into the Pacific Ocean, ensuring that the rockets will always be traveling over water during the most hazardous stages. (The only town in danger is nearby Lompoc and the Air Force agreement with the city sets a launch-disaster standard of no more than six dead civilians—seven is too many. This standard is apparently okay with the citizens, or at least all of them except the six dead ones.)

At any rate, the Chinese welcomed Motorola's business, but they were not happy to see Stamp's staff poking around Base 25. One day in 1996 Stamp got an urgent phone call. His team members had retired for the night to their rooms in the Chinese officers' quarters, only to hear clanking sounds as the outer doors were chained shut. Stamp got on the phone with his military partners at China Great Wall Industry Corporation and made it clear: *There will be no imprisonment of Motorolans, and if it ever happens again, forget about commercial space launches in your country.* The Chinese stood down, and there were no further incidents. As Stamp would discover later, the Russians also had a penchant for polite imprisonment. During one visit to Moscow, the Motorolans arranged to visit Dnipropetrovsk, Ukraine, on the way back home, because production officials at a factory called Yuzhmash were anxious to show them the Zenit rocket and make a bid for Iridium business. But on the day they were leaving Russia, the Motorola delegates were held at a meeting, their cars to the airport went missing, and they eventually missed their flight to Kiev. The ploy worked—they lost a day, no longer had time for the extra stop, and returned to Phoenix without visiting Yuzhmash.

The emergence of the "global corporation" was a fashionable idea in the 1990s, as pundits and think tanks predicted that corporations would one day replace nations as the primary way that societies organize themselves. There had always been multinational corporations, like Coca-Cola, and international cities, like Singapore, but Iridium was the first one-world company, offering a one-world service, that was truly blind to nationalities. (Appropriately, it became the first noncountry to receive an international country code. All Iridium phone numbers begin with 8816, code for the "nation" of Iridium.) To enhance that position, as the first truly global corporation, John Mitchell determined that every square inch of the planet would be assigned to one of the Iridium partners. Mitchell had gotten his way, and Motorola had copied the Globalstar sales strategy and started assigning "gateways." A partner buying in could expect to spend about $200 million, including the $40 million to $60 million cost of a physical gateway facility, and for that the partner would receive 4.4 percent equity. The Japanese, the single largest investor group, put up almost $600 million for their gateway and additional equity. The Brazilians and Venezuelans spent $330 million building the Rio de Janeiro gateway and setting up distribution offices in twenty-six countries. Lockheed Martin and Raytheon were pressured to invest as well, and they reluctantly bought small positions. Motorola didn't want to own anything but ended up reserving 19.6 percent for itself—a share later cut to 18.2 as the final partners were brought in—and eventually there were eighteen countries involved, creating a mini–United Nations that included telecommunications firms, government agencies, and businessmen betting on the Next Big Thing. Mitchell's goal was achieved, though: every portion of the planet was assigned to a gateway and "owned" by one of the partners. The problem was that only five of the nineteen partners had experience running phone systems—Kyocera, Vebacom, Telecom Italia, Sprint, and Bell Canada—and together those companies controlled only about 30 percent of the equity. The rest of the partners were either marginally related businesses (an electric company in Taiwan, a power company in Brazil), government entities, or investor groups venturing into telecommunications for the first time. For some reason no one wanted the South Pacific gateway, despite both Australia and New Zealand going through economic boom periods, so that part of the puzzle wasn't locked in until 1998, when the Bakrie brothers of Indonesia took 50 percent and shared the other half with Nippon Iridium and the owners of the Taiwanese gateway. The result was an unwieldy consortium of companies far removed from Mitchell's original vision of "duplicating Inmarsat." Inmarsat was made up of governments and monopoly telephone providers; Iridium had owners who had never been in any kind of communications business. He had ended up with almost the complete opposite.

When he was first hired, Dannie Stamp shared an office with the Iridium inventors and was the only employee of the Space Segment Division. By 1995 he was supervising a hundred people and had executed contracts promising 90 percent of the cash on hand: $850 million to Lockheed Martin, $350 million to Raytheon, and hundreds of millions for rocket launches at Vandenberg, Baikonur, and Taiyuan. At the project's peak, there were five thousand engineers working exclusively on Iridium. By the time his satellite factory was operational, Stamp was spending $1.5 million a day and was supervising more space vehicles and launches than NASA, the military, or any other nation.

Chandler, Arizona, had become the space capital of the world. Every nine days a new Iridium satellite was coming off the assembly line, and during one peak period the rate was one every 4.3 days. Even people far removed from Motorola noticed that something major was brewing. During intermission at the Arizona Opera, people traded stories about what was going on down at the strange building in Chandler and how it would revolutionize telephony. When the first Iridium stock offering was made in 1997, working stiffs in Arizona bought five and ten shares at a time, certain that this was equivalent to the invention of the telegraph or the telephone. It was a thrilling moment in business history and in scientific history, and it probably couldn't have happened anywhere except at Motorola.

Of course, the original scheme for Iridium sketched out in the Chandler parking lot in 1988 had to undergo thousands of changes before it could be declared operational. One question that kept coming from Motorola headquarters was "Why do we need so many satellites?" And so the designers were constantly asked to revisit the issue. The plan was to orbit the satellites at 785 kilometers above Earth, but if you went just a little higher you wouldn't need so many birds. For a time they looked at a scheme that would use forty-eight satellites, but that would entail traveling through the punishing Van Allen Belts, and even though the cost savings were tempting, they eventually decided they couldn't risk it. The other advantage of flying lower was that the lower the orbit, the shorter the distance the actual phone signal had to travel. Even a fraction of a second delay could lead to irritating sound lags in a phone conversation. MCI had purchased forty-eight thousand circuits on the Galaxy GEO satellites built by Hughes in the eighties, only to abandon them when consumer complaints about the half-second delay skyrocketed. Motorola was determined that these would be the first satellite phones that worked just like landlines. (The maximum delay in an Iridium phone signal is 180 milliseconds, which is imperceptible.) But in the unending battle against the cost of the system, Lockheed Martin eventually suggested that designers change the minimum elevation angle of the satellites in space from 10 degrees to 8.2. This would allow them to use six planes instead of seven, reducing the number of working satellites from seventy-seven to sixty-six. The only place this would hurt service was areas near the Equator where there were either tall buildings or large mountains, especially Chimborazo in the Andes and Mount Kenya in Africa. There were a few central business districts in important cities that would be affected—Singapore, Kuala Lumpur, Bogotá, and Nairobi being the primary ones—but the change wouldn't mean loss of service there, just the increased chance of dropped calls. Eventually the decision was made to go ahead and sacrifice the overlapping cell signals at the Equator—some areas would have only one satellite overhead much of the time—in the name of cost. When they broke the news to John Mitchell, his first question was "Wait, what element has sixty-six electrons?" They had already looked it up: dysprosium, a rare earth element used in nuclear reactors. "Dysprosiwhat?" said Mitchell. "We're not calling this thing Dysprosium. It's Iridium." And Iridium it stayed.

Other engineering problems were much more difficult to solve. What was the most efficient way to access the cells? How many cells should each satellite form? How could the cells be turned off at the Poles when there was too much overlap? And, crucially, what should the link margin be? Link margin is the difference between the strength of the signal when it's transmitted and the strength of the same signal after it goes through space, as it weakens because of "path loss" and accumulates random noise before arriving at the handset—so basically a logarithmic unit that amounts to the quality of the signal. The reason cell phones sometimes don't work in certain buildings is that you can't design a signal that will penetrate every surface. The more powerful the Iridium signal, the more power it would need in space, and that meant a bigger satellite, and that meant a much more expensive satellite. Wolf-hard Vogel, a German scientist at the University of Texas, was hired to develop "propagation data"—experiments to see how Iridium would perform in various environments—and the designers used that data to choose a link margin of 16 decibel units, which was 3 decibel units higher than the standard for land-based cell systems at the time. It was strong enough to penetrate tree leaves (the most common obstacle to radio-wave signals) and forty times stronger than needed for operating outdoors, but not powerful enough to penetrate the interiors of steel-frame buildings. In other words, it would work inside some buildings but not all buildings, and especially not inside skyscrapers in central business districts. This was reasonable in 1994, but it would turn out to be a liability as terrestrial customers became accustomed to more powerful signals. At any rate, it was the only possible decision. To create a link margin of, say, 30 decibel units, which would be strong enough to penetrate any building, each satellite would need to generate 1,200 beams instead of 48 beams, and each antenna, instead of being two square meters, would have to be *fifty* square meters, driving up the cost of the system from $3.6 billion to at least $20 billion. Iridium was the most expensive phone system ever built, and Motorola was willing to take that risk, but there were still limits.

And then there were Ken Peterson's math problems. Peterson, the personable coinventor who usually took a backseat to Leopold and Bertiger when it came to selling Iridium or explaining it at scientific conferences, was the go-to guy when it came to anything that required a theorem or an equation. Peterson was yet another Iowan who had attended a one-room country school, in this case near the town of Lake Mills. He then worked at the Farmers Co-op Gasoline Station for a year so he could pay part of his tuition to Iowa State. Once he got there, he heard about the famed electrical engineering program, which had an

85 percent dropout rate because it was so tough, and he knew that was what he wanted to do. He not only survived sophomore physics (the course responsible for half the dropouts) but thrived there and, along the way, fell in love with pure mathematics, especially as applied to radio frequency problems. He would spend a lifetime going in and out of universities, both as student and teacher, even as he was advancing in the corporate world. Peterson's master's thesis at the University of Pittsburgh was titled "Passage of a Sine Wave Immersed in Additive Gaussian Noise Through a Frequency Divider," and it was the beginning of a career at Westinghouse, the Illinois Institute of Technology, Magnavox, and eventually Motorola, where he became known as "the math guy" and one of the leading experts on radar systems analysis at Fort Monmouth, New Jersey, and Fort Huachuca, Arizona. Peterson would always remember the Iridium project as the best job he ever had. "All I did was math problems," he said. "It was wonderful. I would dream about theorems. I would calculate algorithms in my unconscious moments. And it would go on month after month after month."

The first thing Peterson was asked to do was "tessellate the unit sphere." It's a calculation that relates to the process of using and reusing the channels within each cell so that the system never interferes with itself. "There's an algorithm for channel reassignment, defining the sort of handshake agreements between the satellites as to which channels will be used and how often," he explained, "and it's a nasty algorithm." So that was a tough one, but it was nowhere near as tough as the next algorithm he was handed, for "prolate spheroidal wave functions," which is a methodology to identify the most efficient accessing of cells given Iridium's limited bandwidth. He eventually found a twelve-cell reuse concept that had been developed by Bell Labs in 1979, and it was used to maximize the system at 98,586 working channels, so that was the maximum number of Iridium calls that could be active at any time. Every time the system "maxed out" at that level, the plan was to launch eleven more satellites, or one additional plane.

Both of these initial problems took months to solve, but the thorniest math problem of all was figuring out how to shut down the unneeded cells as the satellites converged at the Poles. At any moment, on average, each satellite would have two-thirds of its cells turned on, but that number decreased as it approached the Poles and increased as it approached the Equator. How that gradation functioned was a thorny problem that Peterson worked on for months and never figured out, telling Durrell Hillis, "The algorithm to shut them off is a nightmare." Eventually he handed it over to Yih Jan, a Taiwanese math professor who had joined Motorola from the prestigious Tamkang University, and Yih in turn sought help from Frenchman Jerry Davieau, who worked for the Essex Corporation, a government contractor in Columbia, Maryland, specializing in communications systems. "Each of us would get to a certain place," said Peterson, "and then we couldn't get it any further." Months turned into years. The second private placement memorandum closed. Dannie Stamp started signing contracts for launches. The Chandler factory began production. And still they hadn't resolved the problem. Fortunately a decision was made to launch the satellites devoid of software, then upload it later. "Again Chicago told us not to do that," said Bertiger. "'You cannot upgrade an active system over the air,' they told us. Motorola was in the habit of writing new software for every phone they created. Iridium was the first system to have upgradable software. Again, their advice was wrong." The decision to launch "dumb" allowed daily mission software to be uploaded through earth stations in Iceland and Canada, but its unintended consequences allowed Peterson, Yih, and Davieau to breathe a sigh of relief. They had more time.

From 1993 to 1995 the Iridium engineers engaged in a daily "war on mass," trying to create the highest performance satellite with the lowest possible weight. Mark Borota, the official Motorola general in this endeavor, held meetings at 6:00 A.M. and 6:00 P.M., 365 days a year, in an attempt to decrease the mass. Meanwhile, Motorola had to go through yet another fight to secure the frequency assignment already won in

Torremolinos. During thirteen weeks of FCC rule-making sessions in 1993, Ray Leopold was opposed by five other companies that wanted to share the spectrum using Qualcomm's code division multiple access (CDMA) instead of dividing the system so that Motorola could use its preferred protocol, time division multiple access (TDMA). Globalstar, Odyssey, Ellipso, and ECCO were all willing to use CDMA, which allowed many users to share the entire frequency band, but Philip Malet, a Motorola lawyer, characterized his own company's attitude as "give us the spectrum and then let the others fight over what's left." In other words, *we fought the battle at the WARC—the spoils belong to us.* But this was an area where Motorola was on the wrong side of history. CDMA—better known as frequency hopping, or "spread spectrum"—was a way to use multiple frequencies to send the same signal. This system was invented by Hedy Lamarr, the actress, and would become the industry standard.[19]

Eventually three companies were allowed to use the new spectrum—Iridium, Globalstar, and Odyssey.[20] There was one more potential problem for Motorola, however. The Iridium allocation had been given for a seventy-seven-satellite system, not a sixty-six-satellite system. Afraid that the change would send Iridium to the end of the line and wipe out their hard-fought frequency allocation, Leopold and engineer John Knudsen crafted a filing describing the change as a "minor amendment," and everyone loudly exhaled when it sailed through the FCC approval process without incident.[21]

Meanwhile, the total money raised from partners to build the system was just $1.593 billion—not the $3.6 billion originally projected. It had taken three years and hundreds of sales pitches to find that much money. Bertiger alone had presented fifty road shows in thirty-two countries. Leopold had grown so accustomed to airplanes that sometimes he would awaken from naps grappling for a nonexistent seat belt. But Mitchell was tired of searching the planet for big investors, so in July 1993 he said, "It's over. We'll borrow the rest." Everyone in Chandler jumped on the corporate jet and flew to Buffalo Grove, Illinois, for a celebration dinner where Goldman Sachs handed out symbolic trophies for what was proclaimed as the most successful private placement in history—and it was. "Nobody had ever done anything like that," said Bertiger. "It was unheard of."

A few people inside the company thought the celebration was premature. Ted Schaffner told Mitchell, "No one will lend to this project." But Schaffner was wrong. Discussions had already begun with Chemical Bank and Chase Manhattan, and eventually Mitchell put together a consortium good for $1.55 billion in loans. And besides, there could always be a stock offering—Merrill Lynch was anxious to do one. And if all else failed, there was the junk bond market. People would lend to this. After all, it was Motorola.

One reason Mitchell seemed to be acting with a devil-may-care attitude about funding was that Bob Galvin wanted it that way. It was no secret that several of the Motorola board members were nervous about Iridium, so Galvin set up a dog-and-pony show for the board, with all the inventors present and all the senior managers. As soon as the presentation was over, Galvin took the floor and said, "This is *exactly* the kind of thing this company should be doing." His voice was stern, almost challenging. "The history of Motorola is creating industries and creating markets. And now, if there are no questions . . ."—and he gaveled the meeting to a close. No one on the board had said a word.

It was during this time that Motorola was also trying to secure all the patents involved—and having no luck. Between 1988 and 1995, Motorola submitted fifty-two claims to the U.S. Patent and Trademark Office, and all fifty-two were denied by an assistant examiner with the aptronym of Andrew Faile. There was an apparent reluctance on the part of the government to patent anything involving outer space communications—it was that old JFK prejudice against the commercialization of space. So lawyers were summoned, lobbyists were asked to investigate, and finally, on April 25, 1995, patent number 5,410,728 was issued for the Iridium satellite communications system. Increasingly Iridium was thought of as the future of all cell-phone communication and, more important, the answer to communication in remote parts of the Third World that couldn't afford a traditional system of cell towers. This was a favorite topic of President Clinton's, and in March 1996 he made a special show of support when he visited the New Hampshire factory where Lockheed Martin was building the bus.

By the end of 1995, all the technology for the Iridium system had been "locked," meaning nothing substantive could be changed, and by the fall of 1996 Iridium was ready for launch. As the satellites rolled off the Chandler assembly line, Mark Borota declared final victory in the war on mass: the weight came in at only 1,412 pounds, incredibly light for a satellite and 10 percent lower than Motorola's assigned goal. As Dannie Stamp readied his launch team, Borota informed everyone that, even though each satellite needed only twenty-five gallons of hydrazine for its fuel tank, the actual tank could hold up to two hundred gallons. This was because they were using the only available off-the-shelf tank—there were much smaller ones and much larger ones, but there was nothing commercially available between twenty and two hundred. Stamp wanted to go ahead and fill up the tank, under the theory that "extra fuel in orbit is more precious than life itself and atones for many sins." But the feeling of the launch experts was that they should continue to be ruthless about eliminating every ounce of unneeded weight. They were especially worried about the American rocket, the Delta II, which was already very close to its lift capacity. (The Chinese and Russian rockets, on the other hand, had an excess of power, so the extra fuel wouldn't be an issue.) The fuel wasn't expected to be needed, even at the twenty-five-gallon level, since half of it was intended only for use in case of a de-orbit. A single gallon would be enough for station-keeping. And ten gallons would be sufficient for the occasional outer space maneuver, in which thrusters were used to make tiny corrections in the orbit. The general consensus was that most of the fuel would be deadweight, making the bird less agile, but Stamp coveted the fuel and continued to hammer home his case. Since the rest of the bird had come in so light, the team met and finally gave in. "Fill up the tanks!" said Borota. It would turn out to be one of the most crucial decisions Motorola made.

The next decision involved choosing which rocket to use first. Stamp had developed a standard interface for all the nose cones under contract, so he could put any satellite on any rocket. The Proton was the most reliable, in his opinion, but he didn't want to risk seven satellites on the first launch. So he decided to use a Chinese rocket, the Long March 2C. But when Stamp went to his bosses, they were reluctant, partly because they considered it bad public relations to do the first launch from China. He finally decided to use a Delta II, the McDonnell Douglas rocket, but launch just three satellites on it to be safe. So during the Christmas holidays of 1996 Stamp sent his team to Vandenberg Air Force Base, the ninety-nine-thousand-acre installation that had pioneered polar orbits with Discoverer I in 1959 and was actually a much busier launch site than its more famous civilian counterpart, Cape Canaveral. Discoverer I, like most launches from Vandenberg, had been a classified mission—a fake scientific expedition to cover for the Corona spy-satellite program. Since then Vandenberg had launched every generation of spy satellite—multibillion-dollar programs with code names like Samos, Gambit, Argon, Lanyard, Hexagon, Crystal, and Key Hole—and by the time Stamp's team arrived, the facility had gone through 1,721 rocket launches, most of them ICBM tests hitting targets in the Kwajalein Atoll in the Marshall Islands. Working with the team from McDonnell Douglas, Stamp's men fastened three satellites into the nose cone of the Delta II, but each time they were ready to launch, weather intervened. The Delta II had a five-second launch window, meaning all atmospheric conditions had to be perfect five seconds prior to launch, and the Iridium team was ready to go three times in early January only to have the wind kick up and the launch be aborted.

Then on January 17, 1997, a minor disaster struck. Twenty-seven hundred miles away, at Cape Canaveral, the Air Force was launching a satellite for the next generation of GPS, but 12.5 seconds after liftoff, the launch vehicle blew up and the public address system ordered personnel to "take cover immediately from falling debris." The rocket had traveled only 438 meters before it disintegrated in a blast that could be felt twenty-five miles away, with some parts landing in the ocean while others started brushfires onshore. Unfortunately for Stamp, the malfunctioning launcher was a Delta II. Overnight Stamp's McDonnell Douglas team was called away from Vandenberg and sent to Canaveral to investigate, and the next day the Air Force Accident Investigation Board shut down Iridium until further notice. There would be no Delta II launches until the reasons for the failure were known. Stamp left his rocket on the active pad and went back to Chandler and cooled his heels. Days turned into weeks turned into months, and Stamp could be heard to remark that even now, running the first constellation engineered by private industry, he couldn't escape Air Force bureaucracy.

It was not until May 1997 that the ban on Delta II launches was lifted. For two years Stamp's team had been on schedule. For two years he'd held daily operations meetings at 9:00 A.M. and tracked the progress of the constellation by the hour. He had completed the design process at half the cost per pound of a government satellite, and he had gotten the actual manufacturing process down to forty-eight calendar days per satellite, compared with military satellites that took nine to eighteen months. Now, after all that work, he'd been shut down for four full months, and he was sweating bullets. "We're too far behind now," he told Hillis. "Three satellites are not enough. We have to load her up." So he fastened five birds into the nose cone of a Delta II—only to be, once again, stymied by the Pacific winds. Three more times the countdown was aborted at the last minute, which meant that, for the seventh attempt, most of the Motorola employees who wanted to witness the inauguration of the system had long since given up and gone home.

On Sunday night, May 4, Stamp checked the forecast and again found wind bursts and clouds that were likely to hamper his next try. He got up in the middle of the night to start sending up balloons from the base's weather station so he could see what kind of gusts he might be facing during the initial sixty-five thousand feet. After several balloons were released, he had a jagged track indicating the rocket would get beaten up all the way to the second stage. But then again, that's what computers were for, to calculate the loads and help him lean the nose into the wind when it broke through the sound barrier. At 5:15 A.M. he liked what he was seeing and gave the go-ahead to start pumping RP-1, a highly refined kerosene, into the Delta first stage, and the procedure was completed about the time the sun peeked over the horizon just before 6:00 A.M. At T minus 60 minutes, the wind remained dead calm, so the loading of the liquid oxygen for the main rocket engine began. By 6:35 all the teams—NASA, the Air Force, McDonnell Douglas, Motorola—took up their assigned positions and started the usual fidgeting.

Everyone thought the wind had died too early; the calm couldn't last. The bleachers on the spectator hilltop were empty this day. The last of eight balloons was sent up around 7:45, looking for that sound-barrier place where the gusts were occurring, and the reading came back: *Green. Good to go.*

Stamp held his breath from T minus 10 down, and it finally happened at 7:55 A.M. on Cinco de Mayo 1997. The thrusters burned evenly, the earth trembled, the liftoff was clean, and Stamp watched the reddish-orange burn of the rocket for the full two or three minutes it was visible. Two hundred miles downrange, employees at McDonnell Douglas facilities in Huntington Beach called to say they could see the thick white exhaust tail burning clear and straight until it was a dim red dot ascending through the clouds. Stamp watched the control screens as nine boosters fell away, the main engine cut off, and the second stage separated. Now all he had to worry about was his new dispenser popping the satellites out of the nose cone. According to the telemetry estimates, the rocket would be positioned over a former NASA tracking station in Hartebeesthoek, South Africa, when that was supposed to happen.

There was a countdown monitor calibrated to that moment, and when the monitor reached zero, pyrotechnic charges blew three nuts off the fairing while compression springs pushed the first satellite clear of the dispenser. Everyone froze—until Hartebeesthoek reported, "Separation sighted." A cheer went up in control centers on three continents. One by one the satellites rotated to "top dead center," and thirteen minutes later the last one rolled out into the 395-mile-high parking orbit. Stamp called back to Chandler with the news: "Lo and behold, major normalcy. Booster worked. Third stage worked." All five satellites, which had an expected life of five years, were still flying eighteen years later.

When you're dealing with rocket launches, there's no such thing as perfection—millions of things can and do go wrong—but the Iridium project came as close to perfection as anyone ever had. Over the next twelve months, Iridium would break every record in the history of rocket science, launching 72 out of 72 satellites on 15 out of 15 rockets in 377 days. Nine of the rockets were Deltas launched from California, with three launches from Kazakhstan and three more from China. In one thirteen-day period, Stamp's team ran launches from all three space centers—Vandenberg, Baikonur, and Taiyuan—putting 14 satellites into orbit. After the constellation was operational, there were an additional 23 spares launched without flawing the perfect record—95 satellites loaded, 95 satellites launched. The "infant mortality" rate (satellites that never become operational because they fail before being switched on) was an amazing 2 percent. There were 12 other satellites that never made it into orbit because of "pilot error" during the two-week ascent phase when the space vehicle was being maneuvered toward its "control box" or optimal position in space. Once the Motorola engineers at the SNOC learned how to efficiently maneuver the birds, the rate of failure went to virtually zero. Stamp's team took more or less permanent possession of "Slick Two Ay," Space Launch Complex 2A at Vandenberg, a pad normally reserved for NASA. Iridium had tied up 40 percent of all the launch contracts for the entire planet.

Satellites are not beautiful objects, as a rule, but the Iridium bird had a certain flair that made it attractive to space buffs. Its slender body was tricked out with twenty-eight-foot "wings"—actually gallium arsenide solar panels—and the Teflon main mission antenna was silver coated, reflecting brilliant bursts of light that soon attracted the attention of astronomers. At first these bursts were believed to be meteors, but it soon became apparent that it was a new phenomenon in the skies, called "Iridium flare," as the satellite reflected the sun's rays back to Earth. Brian K. Hunter, a chemistry professor at Queen's University in Ontario, recorded the first flare on August 16, 1997, and at first it was cause for alarm. (Were these UFOs?) As the flares continued, Motorola weathered some criticism for adding yet more "light pollution" and cluttering up an already brightening night sky. But eventually the "perfect hits"—magnitudes of –7 or –8, about fifty times brighter than Venus, which was previously the brightest object in the night skies—became the occasion for organized "flare parties." In October 1997 attendees at the International Astronautical Congress in Turin, Italy, held a flare-spotting soiree on a rooftop. On Memorial Day weekend 1998, a crowd of amateur astronomers at Riverside City College in Riverside, California, cheered so loudly at a 7-magnitude flare that the noise could be heard throughout the city. Astronomy nerds, as they like to call themselves, are fond of pretending to have secret UFO information so they can prank unsuspecting friends into waiting for the precise moment a foreign spacecraft is expected to appear in the sky—thereby setting them up for the flare. It's become increasingly simple to know when the flares will occur, thanks to smartphone applications with alarms outlining the precise time, location in the sky, and intensity of each upcoming event. Ken Peterson, who liked to hunt prairie dogs, was sitting at a campfire one night in the Arizona desert when he claimed to have seen five LEOs in a row. "After millennia of people looking up at the same night sky," he told *Wired* magazine, "we're the first to put up a new constellation since God. It's never going to be the same again." Ray Leopold, a devout

Catholic, went even further in the metaphysical realm. "If you believe in God," he told the *Wall Street Journal* shortly before the constellation was launched, "Iridium is God manifesting himself through us."

Stamp had occasional hiccups, like the time his team was loading seven satellites onto a Federal Express DC-10 for the trip to Kazakhstan, and the operator of the Atlas K Loader "bounced a satellite off the tarmac." Stamp confessed that sin, sent the damaged satellite back to the factory, and made his launch date. In fact, Stamp's operations management had been so successful that he came in $300 million under budget. Everyone in Chandler became a minor celebrity, and the state of Arizona swarmed with the leading space thinkers of the world. All involved would have memories that stuck with them the rest of their lives. Peterson remembered the big party at Nello's Pizza when the manufacturing was declared finished and the Lockheed Martin team was finally sent home, not to mention the day Jerry Davieau called with the final line of the final page of the algorithm for turning off the cells at the Poles. Leopold would remember the day there were 150 guys with Iridium phones walking around on the grounds of the Chandler campus, testing them to see how close together they would work. Leopold went downstairs to ask a senior systems engineer, Norbert Kleiner, what they were doing, and Kleiner confessed that he'd called his mother during an earlier test and she had now become the first customer addicted to an Iridium phone, calling him every day with the same greeting: "Norbert, you can run but you can't hide." But Dannie Stamp would always remember the moment at Baikonur Cosmodrome when he asked to see the launch complex control room.

He didn't get an answer right away from his friends at Khrunichev. "You want to see the control room?" they said.

Yes, of course, he answered, *I have to see where my men will be working.*

The Russians conferred among themselves and said, "We have a problem with that. But we'll show you another control room that's exactly like the one where you'll be working."

Stamp told them that was unacceptable—he had to see the actual room that his men would be using.

The Russians again talked among themselves for a few minutes, then came back to him.

"All right," they said, "we'll take you to the control room. But on the way there, we will go through another room first, and we want you to walk very fast, look down at the ground, and do not raise your head or move it to the left or the right at any time while you are in that room."

All right, said Stamp, feeling the hairs on his neck stand up.

And so they proceeded to the control room. Stamp walked as slowly as possible and made slight movements of his head so he could see as far in both directions as possible.

And what he saw was a carbon copy of the NORAD Combined Operations Center at Cheyenne Mountain, better known as the war room. The only difference was that the giant map had the former Soviet Union at the bottom and the United States and Canada upside down at the top. He noted all the blinking lights, indicating the locations of all the Peacekeepers and Minuteman IIIs, and he suddenly realized that they were still on alert. For whatever paranoid reason, they were still ready to fire.

And then, of course, he toured the civilian control room and shared some vodka with his new friends and acted like he hadn't noticed a thing. When you're starting the first global corporation, the family has to get along.[22]

Chapter 7

FAST EDDIE

OCTOBER 21, 1998
WHITE HOUSE ROSE GARDEN,
WASHINGTON, D.C.

The six and a half months of Ed Staiano's public and private agony had begun with Al Gore fawning over Iridium.

Gore wanted to make the first call. The Vice President made it clear to his staff how wonderful it would be if the brilliant new Iridium satellite phone—the first global phone—were added to his techno-geek credentials. Gore had a reputation going back to the seventies as the proponent of all things shiny and high-tech, associating himself with every cutting-edge innovation from artificial intelligence to fiber optics to early patents in biotechnology. Most men entered politics so they could wield power—Al Gore was apparently in it for the pocket protector. As a young Senator from Tennessee, he was the first member of Congress to appear on C-SPAN in 1979, and it was the Gore Bill that opened up the ARPANET in 1991, leading to the modern Internet. Although he fell short of "inventing the Internet"—as he once famously said, to the glee of his detractors—he was indisputably the leading government spokesman for the "information superhighway," a term that did originate with him. (It was actually a sly reference to his father, who sponsored the Interstate Highway Act in 1955.) He was the first Vice President to hold a live interactive news conference in 1994, and he made a point to lend his name to whatever next-wave technology came down the pike, whether that was robotics or digital earth mapping or magnetic levitation trains. And so it was decreed: Iridium was to be yet another Al Gore "first," and when the idea started filtering down through the White House ranks, the task of arranging the call fell to the only staffer with prior Iridium experience: Dorothy Robyn.

Robyn was a striking brunette economist whose official title was Special Assistant to the President, and she was well known within the corridors of the White House as a dynamic, exuberant intellectual with mastery over dozens of policy areas traditionally dominated by men, everything from shipbuilding to military bases to aerospace to international telecommunications. She first met Leo Mondale and Tom Tuttle, the company lawyer, when the two Iridium executives came by her office to seek help on spectrum and licensing issues. She broached the subject of an Al Gore ceremonial first call and they both thought it would be a great photo op—but who should Gore place the call to? Robyn and Mondale started working up a list of celebrities who would somehow symbolize the "one interconnected planet" theme of the Clinton administration. The person on the other end of the call should be someone far away, in a place previously inaccessible, who was somehow symbolic of freedom, the advancements of science, and all the brotherhood-of-nations ideals the UN holds dear.

One of the early ideas was for Gore to call a progressive leader in the Third World. Unfortunately, that was the kind of thing that would have to go through the State Department, bringing up political issues of which potentate was more progressive than the next, and besides, it didn't really serve the purpose, since the leader of any Third World country already had adequate phone communications. Perhaps a political dissident would be better. How about Daw Aung San Suu Kyi, the Nobel Peace Prize winner who had been under continual house arrest since 1989 as the leader of the opposition party in brutal, isolated Myanmar?

But how would you get the Iridium phone to Suu Kyi? And would such a call be needlessly polarizing? Iridium was already having a hard time convincing the governments of Russia and China that it was not an instrument of the CIA, so using it to bolster the cause of an opposition candidate, no matter how beloved by the outside world, might be the wrong message to send. For a while they ran through the names of famous scientists, and Robyn thought she'd hit on the perfect candidate in eighty-year-old Arthur C. Clarke, now *Sir* Arthur Clarke, science fiction éminence grise and namesake of the Clarke orbit. Not only was he one of the earliest advocates of satellites, but he lived in Sri Lanka, which was sufficiently exotic and distant to symbolize the global reach of Iridium. Mondale squelched the idea, however, and not for the obvious reason that the Clarke orbit was geosynchronous while Iridium used a polar orbit. The reason Clarke lived in Sri Lanka, it was rumored, was that he liked the young boys there. All agreed: *Let's move on from that.* Perhaps the call could be made to one of the many research stations in Antarctica, a continent that itself was a symbol of global cooperation, since some three dozen nations operated there without any serious territorial disputes. This time the idea was ruled out as too esoteric: the Iridium phone was not a gadget to be used by experts, after all, but the first global communications device that could be used by anyone on the street.

As these discussions went on, Iridium executives started to get cold feet, fearing a public relations disaster if their continuing problems with dropped calls weren't solved by the time the Vice President picked up the phone. Perhaps, they suggested, the call should be confined to the "plus-one area," the United States and Canada, to ensure quality control. That limited what could be done, but the Vice President was scheduled to address the International Telecommunication Union at an upcoming meeting in Minneapolis, so wouldn't it be great if he made the first Iridium call that same day and then spoke about it to the delegates gathered from more than a hundred nations? He could call the President, or he could call Iridium CEO Ed Staiano to congratulate him, or maybe he could call his mother in Carthage, Tennessee . . . but none of these ideas was stirring the imagination. How about if someone paged the VP while he was on the plane flying to Minneapolis—because the Iridium pager worked *great*—and then when he got to the tarmac, he could phone the person who paged him. But they had to make sure it was an airport where he wouldn't attract a crowd—just in case the call didn't work!

Finally Robyn said, "Okay, let's just go back to the original phone call and try to re-create some modern version of it." That call had been made in 1876 by Alexander Graham Bell, who wasn't thinking much about posterity when he said, "Mr. Watson, come here, I want to see you." What if they were to find a descendant of Bell's who, on this occasion, would say something more memorable? It turned out there were many Bell descendants, but the most illustrious was Gilbert Melville Grosvenor, a great-grandson of Bell's who had served for many years as head of the National Geographic Society before his retirement in 1996. When his name was offered to the Vice President, Gore said, "Oh yes, I know Gil Grosvenor"—and the choice was made. Grosvenor was an explorer and world traveler, having once donned thermal scuba gear so he could walk *underneath* the ice of the North Pole in order to say that he had "walked beneath the footsteps of Robert Peary." Unfortunately Grosvenor lived nearby, in the bucolic farming community of Hume, Virginia, and, because of the Vice President's schedule, the call would have to be made from Andrews Air Force Base before Gore left for Minneapolis—so the ceremonial first call wouldn't exactly show off the reach and power of the Iridium system. In fact, the call would probably go up to whichever Iridium satellite was nearest and come right back down without going through any of the elaborate switching systems that enabled the satellites to talk to one another. Al Gore and Gil Grosvenor would be so near each other in physical space, social class, nationality, and background that it would be the equivalent of shouting over a neighbor's fence, if not speaking to a relative in the next room. Like everything else surrounding the launch of Iridium, it was a compromise decision that served as a harbinger of future failure. But things were soon to get worse.

On the appointed day, Mondale and Iridium chairman Bob Kinzie arranged for the Iridium equipment to be delivered to Andrews so that, when the ideal photo op arrived, they could hand Gore the phone. But after a dozen people labored to arrange the call, briefed Gore, and worked a reference to Iridium into his ITU speech, they turned on the phone that morning and got an extremely weak signal. Afraid the ceremony might be bungled, they told Gore to go on his way and they would catch him later. Iridium damage-control people got on the phone to Dorothy Robyn and made it clear that the system worked, but they were still turning things on and off during the testing stage. Within a few days they would have full global capability.

So they decided to try again two days later, when Grosvenor would be in his office at the National Geographic Society and Gore would be in the Vice Presidential Residence at 1 Observatory Circle—a distance of 1.9 miles (not exactly "global capability"). At the last minute the schedule was changed so the Iridium call would be Gore's *second* official act of the day, requiring him to be in his West Wing office. Actually he would need to walk out onto West Executive Avenue so that the phone would have clear access to the sky. That meant the distance between Gore and Grosvenor, waiting on the call in his 17th Street office, was now seven hundred yards.

But they needn't have worried about that particular public relations problem either, because on the day assigned for the call, the entire Iridium system went down for maintenance and software fixes. Remarkably, the Office of the Vice President was undeterred and still wanted to do the call. Robyn requested a demonstration phone this time and walked out in front of the Old Executive Office Building to test the voice quality, but she had to try several times before getting a connection. She decided, in the interest of efficiency, that she would be present the next time they attempted the official call—and as luck would have it, they had a space-related event on Gore's schedule. Gore was headed to Ohio to speak at John Glenn Elementary School in Cleveland on the day the seventy-seven-year-old Glenn became the world's oldest space traveler on a Space Shuttle Discovery mission. That would give them a chance to grab the call while Gore was transferring from Marine Two to Air Force Two at Andrews Air Force Base—but that timing turned out to be risky, as Iridium didn't want to make the call during the "seam" period, when the constellation had to pass the signal all the way around the world before it came back down. (Plane 1 and Plane 6 are the only two planes of satellites that didn't communicate with each other.) Gore's staff agreed to switch the timing of the call so that he made it from Andrews on his way back home that night. Dorothy Robyn and Leo Mondale arrived early to test everything in advance. But once again they couldn't get a connection to Gil Grosvenor's location in Hume, Virginia. With launch of service just two days away, they told Gore's staff they would follow him around the next day and just grab it whenever they could.

The next afternoon they were setting up in the White House Rose Garden, where Secret Service agents killed time by throwing tennis balls to Buddy, the President's Labrador retriever. For a half hour they kept the phone activated, testing the signal strength, and were pleased to have a strong connection. The Vice President finally walked out into the garden, and as soon as he did, the signal got weak again. They decided they had no more time, so they handed Gore the phone anyway and he pronounced the official first words to be spoken over the Iridium system: "Mr. Watson, come here, I want to see you." (Wait! Wasn't the Bell descendant supposed to say that? And wasn't the plan to do some modern variation on it? Oh well.)

Gil Grosvenor, fifty-eight miles away, heard Gore's voice over a transmission that went 420 miles up to a satellite, 2,290 miles satellite-to-satellite in space, 420 miles down to an earth station, then 2,238 miles over trunk lines to Hume, Virginia, for a total of 5,368 miles. But Grosvenor wasn't sure what Gore had said.

"Hello, Mr. Vice President," he muttered.

Working from a prepared script, the Vice President then said, "It was 122 years ago that your great-grandfather made the first telephone call. I am calling today on the first system that will allow people to send and receive calls anywhere on earth using a handheld wireless phone. This system completes the telephone coverage of the earth's surface that Alexander Graham Bell began. Thus it seems only appropriate that I should place the call to you. . . . Your great-grandfather was one of the greatest inventors who ever lived. What would he think of today's world of communication if he were still alive?"

Grosvenor mumbled something in response and the two men hung up. A press release went out a few minutes later. The press release failed to report Grosvenor's words as soon as he was off the phone: "I couldn't understand a thing he was saying."

Mark Adams wasn't around for that particular call, but he wouldn't have been surprised by either the distortion of the signal or the distortion of how the story was told later. Adams had closely monitored all the previous Iridium "first calls," starting with the day in 1997 when a technician in Chandler heard the first beep back from the first satellite in space. In future years—long after the signal problems had been solved and the Iridium voice strength proven to be just as reliable as any other cell phone—a half-dozen people would claim to have made the *actual* first call, mainly because call quality was considered by Motorola to be its number one technical issue and various point-to-point test calls were constantly being made as soon as they were possible. Could Motorola engineers eliminate all the interference from space and achieve a sound equivalent to a landline? That was the challenge, and they started doing simulation tapes of what an Iridium call would sound like as early as

1992, using vocoders. "Good luck understanding anything said at that stage," Adams recalled later. "You had Doppler shifts. You had errors. The bandwidth was limited so you didn't have the clarity you could get from a landline. You're trying to squeeze out the best quality possible in a small bandwidth, and it's not happening."

Adams was especially concerned because he was on a mission. It was, like all his missions, not quite secret and not quite public. As an employee of MITRE Corporation, he was commissioned by the Secretary of Defense and the Joint Chiefs of Staff to find out what kind of mobile satellite phones could be made secure enough to be used by the military in hazardous, isolated places. One of the most embarrassing aspects of Operation Desert Storm in 1991 had been the anemic satellite communications in the combat zone. Because of a lack of capacity, orders and messages between the Air Force and Navy had to be printed out and flown by helicopter to waiting commanders. The Pentagon's budget was being cut back after the collapse of the Soviet Union, and commanders knew that private industry was going to be developing outer space systems that, a decade earlier, would have been paid for by the government. So Adams was sent to visit all the American corporations developing satellite phones—Motorola, TRW, Loral, Boeing, McCaw Cellular—and compare them with a view to determining which ones were best. "There were no existing technologies at the time," said Adams. "There was recognition of a substantial investment by the commercial sector. So we were looking at what does it all mean? And it was very promising—a fairly small device that could be used anywhere on the planet—this was something the government didn't have. The government had *big* systems. There was a military system called UHF Follow-On that was global, but not handheld, not nearly as mobile. So I talked to all the engineers at all the companies. And the one that seemed very promising was Iridium, particularly in the area of security. This was the first time you had intersatellite communications—it's a hard thing to do—and we had an acute interest."

Adams was an engineer himself—majoring in physics at the University of Virginia and earning his master's in electrical engineering at Virginia Tech—and he was thrilled by the technology. "It was fundamentally different from anything ever done before—it was not just cutting-edge, it was bleeding-edge. It was networking in the sky in real time. Everything had to be perfectly together." But even as late as 1998, after most of the satellites had been launched and tested, he had his doubts. He kept going back to the Secretary of Defense—first William Perry, then Bill Cohen—and saying, "It has the potential to be a good system, but it's not stable."

And it didn't help that every time Motorola tried to show off the system, the demonstration seemed to fail. Voice quality was the main problem. At a Motorola board meeting in August 1997, when less than half the satellites had been launched, every board member's pager simultaneously beeped just as they were breaking for lunch. "You are now receiving an Iridium message," read the page. It was Bob Galvin's way of telling his potential detractors that the system was amazing. What he was not telling them was that an actual voice call to their cell phones would have been much more problematic.

In the fall of 1997, Adams joined several dozen Iridium and Motorola people for a ceremonial "first call" on the golf course at Paradise Valley Country Club in Arizona. There were red-shirted security guards everywhere. Technicians were scurrying around, making sure the call would be timed carefully. It was to be a handset-to-handset call to some low-level employee at Motorola headquarters in Schaumburg, Illinois. All the Iridium VIPs stood around waiting for the technical issues to be worked out—CEO Ed Staiano, Chairman Bob Kinzie, and their lieutenants—and then, at the proper moment, the phone was dialed by Durrell Hillis, head of Motorola's satellite division. There was a buzz in the crowd, then silence as everyone literally leaned forward—and then some garbled sounds came over the receiver. Again the crowd buzzed, but this time with excitement. Mark Adams was dumbfounded—that was *terrible* sound quality—but the euphoria among the Iridium and Motorola people was palpable. They were celebrating because this time *you could actually hear the person.* Kinzie and Staiano lit up cigars in celebration. Adams excused himself and went back to McLean with yet another downbeat report for the Joint Chiefs. "It's not cutting-edge," he told them, "much less bleeding-edge."

The only person more concerned about call quality than Mark Adams was Ed Staiano. By 1995 the original game plan had failed. Far from being hailed by governments and corporations as the future of telephony, Iridium was being attacked everywhere as a fantasy science project that probably wouldn't work and, besides, was dangerous to the existing telecommunications industry. Engineers liked it, but virtually no one else. John Mitchell decided that the time for consensus-building was over—and Jerry Adams, the original CEO, was too soft for a full-scale war. Bob Kinzie, the chairman, could stay on as a figurehead—he was the kind of guy you needed to get along with government agencies and bureaucrats—but the leader had to be a take-no-prisoners commando. Someone like Ed Staiano. In December 1996 Staiano retired from Motorola after twenty-three years, much of that time spent running the General Systems Sector, which was global and expanding at the speed of light. Staiano had taken cell phones from a $40 million division of Motorola to a $13 billion division, accounting for 36 percent of the company's total sales, so he was considered a five-star general in that world. Mitchell, his boss for all those years of expansion, went to him one night and said, "I'm retired, too, but we can't leave this Iridium mess. We have to win this first." And a month later Staiano reported to K Street as the new CEO of Iridium World Communications Ltd.

The arrival of Staiano in Washington was like a famous football coach being hired by the school that had just suffered ten losing seasons in a row. Staiano was quite literally a rocket scientist—a Ph.D. in mechanical engineering who had studied at the Stanford Institute of Rocket Propulsion—but he also had a reputation for being difficult, imperious, hard-nosed, intimidating, unforgiving, demanding, and downright nasty when he didn't get his way. Staiano stood six-foot-four, had the pumped upper body of a wrestler, and seemed perpetually on the verge of picking a fight. He had grown up in a working-class Italian family on New York's Long Island and been a fraternity-boy screwup at Bucknell University until a professor forced him to take a summer course in engineering—and "it was like throwing a light switch," he said later. He went on to get his Ph.D., but a part of him was still that hard-drinking teenager driving the potato truck for his father's produce business. For better or worse, it was just the type of background that frequently got you to the top in the heavyweight boxing leagues that pervaded Motorola, and Staiano had overcome his abrasive reputation simply by being successful in almost every division of the company, including cell phones, pagers, ground stations, and two-way radios. Staiano was, in short, an alpha male—with a gravelly voice, bulbous nose, and huge, floppy ears—who got up at 6:00 A.M. to run five miles every morning, regardless of whether he was in Washington, Beijing, or Rio. He had contempt for people who were overweight or undisciplined, and it was an unwritten rule that, while on the road, his senior managers should join him at precisely 5:30 P.M. for his daily martini and 11:00 P.M. for his cognac nightcap and cigar. One of his lieutenants was startled on a road trip to find Staiano ironing the creases of his jeans in his hotel room—otherwise his appearance would fall short of perfection.

"Ed was didactic, imperial, rude, tough—he had enormous expectations of everyone," said an Iridium vice president. "A lot of the people at Iridium were young, bright, techie, but they had never been in a high-stress, high-performance environment, and they didn't particularly enjoy it."

In fact, people who had never met Staiano feared him immediately. On his first day he settled into his office, called together the heads of all the departments, and made an edict: the system would launch at 9:00 A.M. Eastern Daylight Time, September 23, 1998—three months *before* Motorola's deadline for making the system operational, thereby triggering a bonus for the Chandler Lab. Until that date, all vacations were canceled. No wonder his nickname at Motorola was "Fast Eddie."

"I was an absolute tyrant on that date," Staiano recalled. "I had to be. Iridium was dysfunctional. It was a big ball of incredible complications. The gateways were a problem. The installations were a problem. How to sell the handsets was a problem. I had no direct authority over the partners, and therefore I had no control over marketing. We had to get licenses to operate from every country in the world, and we had to get special licenses for the gateways. We had a big team of lawyers I had to meet with all the time. We had outside counsel almost everywhere. I had a million miles of flying the first year. I had six meetings with the Chinese Minister of Communications alone. In the middle of it all we had to change the design because of a radio telescope that we were interfering with in France. And then there was the huge pressure of how do we get revenue? That's why I rode hard on everyone. We had to hit that date of September 23, 1998. I couldn't let a single person pull us back from that date or else we would never get the thing launched."

That's the way it looked from inside the mind of the CEO. From the outside Staiano looked like the boss you wanted to tiptoe around and, if possible, avoid speaking to at all. One reason it wasn't a big deal for Staiano to ban vacations was that he didn't like them himself. "Four or five days is all I can take in a relaxed state," he said—and there would be no relaxed states at Iridium. "The work was overwhelming in terms of hard-core engineering," recalled Randy Brouckman, who was frantically trying to make all the gateways operational. "There was six or seven years of work jammed into three. The pacing was outrageous." Yes, Staiano was an impressive executive—he got credit for turning Motorola into the number one cell-phone supplier in the world with a market share of 63 percent—but he had a reputation for overriding minority opinions and getting testy if anything fell even slightly short. "He gives no leeway," said Iridium Vice President Craig Bond. "He shows no mercy."

Meanwhile, in the winter of 1998, the Iridium team started to become aware that the phone would have a limited ability to penetrate the walls of buildings. This would not have been that strange as recently as 1994, when people were accustomed to running outside buildings to get a signal or pulling their car over to the side of the road when they hit a strong reception area, but since then Motorola itself had all but perfected the link-margin technology that allowed users to access cell signals almost anywhere except underground. "We had been testing it for years," said Leo Mondale. "We had a database of user environments, typical use cases, but there's no way to characterize the *actual* user environment with any degree of accuracy until the system is turned on. We knew it would be unpredictable in some environments, but it was unpredictable in buildings *everywhere*. We had a *range* of expectations, depending on how people used the phone and where they used the phone, and as it turned out, the phone performed at the low end of that range. We had hoped for better. Those simulations were optimistic. We should have lowered our sights and called it a line-of-sight service. By *not* calling it a line-of-sight service, we led the public to believe they were getting a different phone than the one they got."

When Staiano heard the early warnings about signal loss inside buildings, his reaction was to order a massive change. Rather than being a stand-alone system, he wanted Iridium to interoperate with every cellular system in the world. This was a capability that had been built into the system as early as 1992, mainly to satisfy small nations with investments in trunk lines and terrestrial cellular. A cartridge could be inserted into the handset so that it would use the land-based cell system when one was available, then use the satellites when out of range—but whereas it was originally envisioned as an add-on or an enhancement, it was now transformed by Staiano into the way the phone would work everywhere. The ground infrastructure had to be altered. The design of the handset had to be changed. The business systems—including the massive billing system—had to be overhauled. The already complicated call-log system now had to accommodate the systems of every service provider in the world.

"When that change was made," said Randy Brouckman, "I looked at my watch, and I told my managers they had ten seconds for fear and anguish. They shouted, 'You're nuts!' I said, 'Nine!' 'I hate this!' 'Eight.' We counted down, then we went back to work."

But most important, the very definition of what the phone was and what it did now became a confused story for the public. In the early nineties, the debate had always been: What will the future of telephony be—land-based wireless or satellite? And Motorola's answer had always been, "Satellite, of course." Because it was more efficient. Because it didn't require massive infrastructure. Because it worked in every country. Because the cost of building cell towers all over the world was so many trillions of dollars that it was virtually impossible. Because, even if you *could* space cell towers all over the world, there were geographical regions where it was impossible to build. But this was all prior to Europe's GSM becoming the worldwide standard, allowing what is popularly known as roaming. The Future Public Land Mobile Personal Telecommunication Systems, or Flumpits, was no longer future; it was now. The GSM standard used in Europe was all but unknown in 1992 but by 1997 had been adopted in more than a hundred countries. (This wouldn't have been noticed so much in the United States, where Qualcomm's CDMA dominated.) Even at the time Iridium technology was locked in 1995, there were only 1.3 million GSM phones in use in the world. Four years later, that number was 150 million. In the year 1999, Iridium's first full year of operation, GSM usage would grow by 330 million. Motorola had won its spectrum allocation at the WARC by letting the Europeans expand GSM—and now it was GSM that was turned against Iridium. Roaming was expensive—roaming charges were, in fact, one of the least popular fees ever introduced to the consumer—but, when you needed it, roaming worked. Iridium would still be the only phone that worked in every country of the world, but land-based cell phones were now able to switch from one provider to another when the user was on the move—meaning they would work in *most* countries of the world. Iridium would have to deal with the "good enough" alternative.

While all these changes were going on, Staiano was relentlessly focused on technical issues. The gateways had to be built, and many of the Iridium partners were dragging their feet. The voice quality had to be perfected. The rate of dropped calls, almost 60 percent in the early testing stages, had to be vastly improved. Then there were the radio astronomers—they were up in arms again, because once the satellites were turned on, the interference was stronger than they had feared. Staiano had to put on his Ph.D. hat—he was "Dr. Edward F. Staiano" when he talked to scientists—and manage these politically explosive situations. The largest radio telescope in the world—so large and ominous-looking that it doubled as a weapon of global destruction in the James Bond movie *GoldenEye*—is in the Arecibo Observatory in Puerto Rico, and the scientists there had to be mollified. But that negotiation was nothing compared with the one in France, where it took six months of fighting with the pissed-off academics in the little commune of Nan-çay who operated a "decimetric radio telescope." The final result was a complicated agreement with the European Science Foundation that involved increasingly strict standards of radio silence for Iridium that would guarantee twenty-four hours a day of "unpolluted" observation time beginning in 2006, so that the astronomers could return to their evaporating comets and death stars.

Staiano had massive political problems as well. Sometime in 1997 the State Department started to make noise about forbidding Iridium service to nations that were under economic sanctions, including Myanmar, Iran, Cuba, Libya, Iraq, Angola, Sudan, and, most alarming, India, which was being punished for nuclear tests. That meant Staiano had to constantly lobby both the State and Commerce Departments to make sure the global phone system, when launched, would be truly global. The Office of Foreign Assets Control held his applications for eighteen months, refusing to say he could do business in a sanctioned country but also refusing to say he could *not* do business there. He eventually took it all the way to Sandy Berger, the President's national security advisor, pointing out that enforcing sanctions against India would not only destroy the Indian gateway but would force the company to fire scores of Indian engineers and possibly make service impossible in thirty other nations.

But it got worse. In early 1998 the FBI suddenly challenged Iridium's FCC license, *threatening the very existence of the company.* The FBI was declaring Iridium to be an outlaw operation since the agency had no way to wiretap calls originating in the United States but going through a foreign gateway. This supposedly created a "safe haven" for terrorists and economic criminals, causing the agency to oppose the license unless every call, both to and from the United States, was routed through the North American gateway in Tempe. Since the whole Iridium system was designed to *avoid* terrestrial systems, this was death to the business plan. The FBI was especially fixated on plans for a gateway in Montreal—a gateway to serve callers in the eastern United States as well as Canada—because officials thought it would be beyond the reach of American court orders. Meanwhile, both Canada and Mexico were asking permission to monitor their own calls through the Arizona gateway, but the FBI was asking to have it both ways: they wanted control over calls that originated from an Iridium phone in the United States, even if the call was processed through a foreign gateway, but they didn't want any foreign countries listening to calls through the American gateway. These issues presaged a future in which new technology destroyed traditional boundaries, making many of the assumptions in the Communications Act of 1934 obsolete.[23]

Eventually Motorola lobbyists complained to the White House, and Dorothy Robyn was assigned to monitor the ongoing debates within the Department of Justice. Once she drilled down into the issue, she found out the controversy was bigger than Iridium. Larry Parkinson, general counsel of the FBI, told her this was "a shot across the bow" to Globalstar, Odyssey, Ellipso, ICO Global, and any other company that intended to sell satellite service, and it came directly from Attorney General Janet Reno, who was fighting mad about telecommunications companies' reluctance to install expensive wiretapping equipment as the world changed from copper wire to digital technology. "Without these restrictions," said Parkinson, "these phones will become the phones of choice for thugs." (The FBI's fears turned out to be well founded. At first the agency was mainly worried about Colombian drug traffickers, but al-Qaeda turned out to be an early adopter, and Iridium quickly became the phone of choice for rebel groups, notably in the Russian breakaway region of Chechnya and on both sides in the Angolan Civil War.) The FBI remained intransigent throughout 1998, to the point that senior FBI officials were warned by other agencies to back off. ("You're trying to cut the heads off a hydra," said a State Department official.) The FCC believed it would be better *not* to force agreements from Iridium but to address national security concerns on an ad hoc basis as the technology changed, and there's evidence that the CIA agreed with this approach. Robyn was ultimately able to jawbone her way through the bureaucracy and work out a compromise between the agency and Iridium's outside counsel, a lawyer who had formally worked at the National Security Agency. That compromise was shrouded in secrecy, but at the eleventh hour, the permit was granted. The important legacy of the FBI battle was that Robyn became known at the White House as "the Iridium expert."

Staiano was a scrapper, toughened up by years of intramural Motorola warfare, so there was no political fight that could scare him. He was also accustomed to having competitors, and as the other Big LEO systems announced plans to go head-to-head against Iridium, he gathered all the intelligence he could find and basically wrote them off as inferior technology. Globalstar would be cheaper, but it wouldn't be available in some of the most lucrative parts of the world, including Asia, and Staiano didn't see any way Bernie Schwartz could build enough earth stations to handle its bent-pipe technology. Inmarsat launched the Mini-M, a notebook-sized phone, in late 1996 and attracted about forty thousand customers, but it still wasn't truly mobile—it had to be pointed at a satellite—and it was considered a bridge product until Inmarsat could go private in late 1999, with plans to launch ICO Global in 2001. ICO would be offering a true handheld phone, but with fifty-five government partners around the world, it would be less than agile in the consumer market. Odyssey, the second most expensive system after Iridium, suddenly announced a stop-work order in August 1997, and the rumors were that its owners—TRW and Teleglobe—had given up trying to raise the money.[24] Fairchild and Boeing were still talking about launching Ellipso, which would now be redesigned to use only seventeen MEOs, but they also seemed stalled. All the other systems were either too small or too far behind to even bother with. So Staiano had two competitors—Globalstar and ICO Global—and neither of them would matter, because neither was truly global and both would be very late to market.

As it turned out, the tragedy of Ed Staiano, like the tragedy of Caesar, was not in any external factor "but in ourselves, that we are underlings." Staiano owed his entire career to Motorola, but when he looked at the contracts with his old company, all of them etched in stone long before he arrived, he saw nothing in his future but debt: Iridium was like a sharecropper who never sees a paycheck because it all goes to renting the land. Starting in July 1993, Iridium had been paying bills to Motorola amounting to about half of what was owed on its "terrestrial network development" and "space system" contracts ($2.85 billion and $3.45 billion, respectively). As soon as service began, Motorola would also receive $140 million per quarter *in perpetuity* for operations and maintenance. (An analyst at Banc of America Securities, reviewing these contracts in 1999, called them "absurdly lucrative for Motorola.") The bottom line for Staiano was that his "churn rate," once he launched in the fall of 1998, was going to be $100 million a month, with 50 percent of that going to

Motorola and 40 percent going to debt service. The initial investment by the eighteen global partners accounted for less than half of the start-up costs. The rest was covered with $1.45 billion in junk bonds and with common stock that was being aggressively marketed by Merrill Lynch.

The unpleasant truth was that Iridium would need a million customers in the first year just to break even, and Staiano wasn't sure how to do that. "The marketing was scary as hell," he recalled. The phone couldn't even be pictured in advertising because it was so enormous. It was bulky. It was a brick after the age of bricks had passed. It was even worse than a brick because it had that thick antenna. It was what one early observer would call "a brick with a baguette sticking out of it." At an Iridium sales meeting in Rome, senior marketing executive John Windolph stood up to say, "This phone is huge! It will scare people!" Staiano hadn't been that concerned about the size at first, since most early adopters know that later designs of any device become smaller even as they become more powerful. But he was being reminded that, for the time being, he had an image problem, so he decided to call in the world's number one cellular marketing guru. That person was an old friend from Rome, the puckishly flamboyant Mauro Sentinelli. Sentinelli was a Telecom Italia executive widely regarded as the man who opened up Europe for cellular, notably by pioneering the prepaid subscription plan. If anybody could sell Iridium, Sentinelli could. When he arrived in Washington, the top Iridium executives breathed a momentary sigh of relief. Staiano was the number one cell-phone executive. Sentinelli was the number one cell-phone marketer. Motorola was the number one designer of telecommunications equipment worldwide. "The reason we all believed in it," said Randy Brouckman, "is that we had all this horsepower." This thing was going to work.

There's a certain type of international corporation that inspires confidence and team spirit simply by its sheer size and ambition. Iridium was the biggest, the first, the best. Everyone wanted a piece of it. After all, an Iridium satellite was being prepared as a permanent exhibit in the Smithsonian's National Air and Space Museum. At Epcot Center in Orlando, eight million tourists a year visited the interactive Iridium exhibit, working touchscreens that simulated Iridium launches and eased the satellites into orbit. A Stanford anthropologist had written a book-length monograph on the future impact of Iridium on the various cultures of the planet. Two hundred opinion surveys distributed through Motorola offices around the world indicated that 93 percent of the world's population wanted a worldwide phone number. Eight of the largest investment banks in the world had studied Iridium and estimated its current value at anywhere from $4.1 billion to $14 billion, with the leader in telecom financing—Salomon Smith Barney—weighing in at $12.4 billion. Iridium was staffing up customer care centers in Sydney, Australia; Maitland, Florida; and Zoetermeer, Netherlands, so that the estimated 12.5 million customers expected by the year 2002 would be able to seek assistance in thirteen languages, twenty-four hours a day, seven days a week. Sure, Ed Staiano was a dictator who barked orders while pumping iron in his Gulfstream jet, running roughshod over anyone who got in his way, but that's how projects this big got done. Sure, everyone was sleeping in the office and straining their family life, but the payoff would come on September 23, 1998, when Iridium became the most famous company in the world, or at least the most famous phone company in the world. Sure, it looked impossible now—the gateways were only half built, the design of the phone was being changed, the financial press was making a big deal out of every failed satellite—but wasn't that what happened during any Broadway tryout? Didn't everything always come together on opening night?

Everyone who worked at Iridium was almost ridiculously proud to be working there. Morale was sky-high, from the new offices in the ASAE Building on I Street (K Street had proven too small) to the computer center in Reston to the operations center in Leesburg to the Motorola Lab in Chandler and the gateway in Tempe. All the hundreds of new Iridium hires had the kind of esprit de corps that management consultants only dream of, and that included contract employees in far-flung places like the gateway in Bangkok and the uplink station in East Iqaluit, Canada, where an aboriginal Inuktitut-speaking native guarded the lonely but crucial outpost. Increasingly Iridium had passed out of the hands of the Chandler engineers and into the hands of sales executives and office workers and technicians around the world. For this was, after all, the world's first global corporation. "This is the first civil application of the global village," said Giuseppe Morganti, CEO of Iridium Italia. "This is a historic event. From the prehistoric period, from creation, it is the first time that mankind can overcome any problem of distance."

Leopold, Peterson, and Bertiger occasionally weighed in on technical issues, but they had already been reassigned to Motorola's big broadband project, Celestri, a $12.9 billion seventy-two-satellite constellation of LEOs and GEOs that was mostly designed by their old friend Greg Vatt from the original Systems Engineering Group.[25] Of the three inventors, Leopold was the most visible. In August 1998, a few weeks prior to launch of service, Leopold was invited to speak alongside one of the inventors of Teledesic at the prestigious Armed Forces Communications and Electronics Association (AFCEA) in St. Louis. Teledesic in those days was always the *other* futuristic satellite system, mainly because Bill Gates had given his support to Craig McCaw's plan to launch hundreds of satellites capable of broadband. Introduced by a three-star general, Leopold regaled eight hundred military officers, politicians, and contractors using video of the successful launches from Baikonur, Taiyuan, and Vandenberg, and by the end of the presentation, the crowd was on its feet. The sheepish scientist from Teledesic got up only to say, "There's no way I can follow that. We've just hired Motorola to build our system, too." That night, Leopold was taken to Busch Memorial Stadium, where Mark McGwire was in pursuit of Roger Maris's home run record, and introduced on the field as one of America's greatest scientists. It was that kind of atmosphere all the time, the feeling that this company was on the verge of changing the world. Meanwhile, back at his office in

Chandler, Leopold was doing Monte Carlo simulations to predict the date that service would launch. "We told Ed Staiano September 23, 1998, was going to be tough," Leopold recalled. "We ran those simulations a hundred different ways and it always came out November or December. We told him not to count on the September date."[26]

But Staiano had to make that date. In his mind it had become the sole standard by which success and failure were judged. Like all the other Iridium executives, he was doing most of his business from planes and hotel rooms. This fact was cited later as evidence for their faith that every international business traveler would need the phone. They were international business travelers, and boy did they need a cheaper and better phone than the one available in the hotel room in Kiev. Every day Staiano was briefed by e-mail, by phone, by teleconference, and every day there was more bad news. The gateways were doing nothing. "Managing the gateways was a cross between herding cats and running the United Nations," said one of the managers. "You had Chinese, Russians, Venezuelans, Italians, Indians, Motorola, Saudis, Brazilians, Thais—you had this unbelievable mix of people. And the politics of that were tremendous. Then, within that, you had situations where the gateway was not the same as the operating entity, and so there was friction between the physical gateway and the people using that gateway. In South America, for example, you had Motorola, the Venezuelans, and the Brazilians, and we were never sure who was in charge. The biggest problems, though, were Africa, the Middle East, and China. We couldn't get our regulatory approvals directly, the gateways had to get them. In Africa you had fifty-four countries and we had five licensing situations for each country, so that's 270 licenses for Africa. In Latin America you had thirty countries. And there was no regulatory framework for any of it. We were really asking for things that had never been asked for before."

None of the gateways cared much about Motorola's Six Sigma gospel, and over time they started regarding both Iridium and Motorola employees as ridiculously overzealous. At an early planning meeting,

Motorola technicians told Iridium South America chairman Alberto Finol that they wanted him to have the Rio de Janeiro earth station fully staffed for two years prior to launch of service so that all the necessary testing could be done, and he looked at them like they were crazy. "You want people pushing a bunch of buttons for two years?" Finol said. "No, we're not doing that."

And there was an even more basic culture clash. Selling phone service is not a simple skill; it's one step less brutal than used cars. Many of the gateways were based in countries that were just a couple of generations beyond the barter system, and others had cultures where high-pressure selling was regarded as impolite. "There is no word in Russian for marketing or public relations," Sentinelli told the *Wall Street Journal*. "I need to get them to sell, but I can't kick their ass because I don't pay them." Yet 100 percent of the company's revenue would be collected through those gateway companies, the gateways would decide what price was charged in each region of the world, and Staiano had zero direct control of them. His only leverage was to appeal to their self-interest—the safety of their investment—and that didn't seem to be much leverage, especially with partners like the Chinese and the Russians who were virtual government agencies. Sentinelli was setting up a $180 million consumer advertising campaign through one of the toniest New York agencies, Ammirati Puris Lintas, but that wouldn't matter if the gateways weren't signing up service providers, doing their own local advertising, and hiring sales forces. Less than a year before launch, very few of the gateways were staffed up, and most of them seemed blasé about timing.

Staiano started calling bimonthly board meetings, with chairman Bob Kinzie working behind the scenes to bring all the gateways in line, but that may have ended up creating more tension. The Iridium board had twenty-eight members from seventeen countries, and its meetings were chaotic and ill-mannered. The partners didn't like one another, partly because of cultural reasons, partly because they felt that certain partners like the Japanese were being shown favoritism, partly because they resented the millions they were forced to pay Motorola to build the physical facilities of their gateway, and partly because they just didn't like Staiano's high-handed tactics. The Saudis, who ran both Africa and the Middle East out of the Jeddah gateway, were especially dissatisfied. They complained about cost, about timing, and especially about pressure—they resented being held to a timetable. Relations with the two Saudi firms got so bad by late 1997 that Iridium was threatening to revoke their franchises, mainly because they weren't building their infrastructure or opening regional offices. "They weren't getting their African licenses," said one executive, "and they weren't going to get them as long as they were thought of as a Saudi company. So I told them, 'If you want licenses in Africa, you've got to move to Africa. You've got to have a presence in Africa.' So they eventually opened up an office in Capetown, but they appointed an Aussie to run it!" The Aussie was, in fact, about as non-African as they come, a privileged investment banker named John Richardson, discovered at London's exclusive Turf Club by Jeremy Soames, Winston Churchill's grandson, who recommended him to Prince Khalid, the brother-in-law of Saudi Arabian King Fahd, majority owner of two Iridium gateways.

All the gateways had entered into the Iridium investment thinking they would be regarded as equals to the giant telecommunications companies of the world, and now they felt as though they were being treated like McDonald's franchisees. Adding to the misery was the cost of frequent board meetings, which were held in Moscow, Kyoto, Rome, New Delhi, Seoul, London, and Rio de Janeiro. Flying owners and their associates all over the world, renting first-class hotel facilities, and dealing with the huge staff required to translate and provide services in exotic locations was running up a bill that was not uncommon in the culture of Motorola but a little scary to the CPAs toiling away in Reston.

Most of the gateway owners were in the telecommunications business for the first time, creating two problems: 1) Staiano didn't value their opinions, and 2) they knew Staiano didn't value their opinions. The Indian gateway owners, for example, had shareholders representing 90 percent of that nation's blue-chip corporations—their sales executive had been the head of Coca-Cola in India before joining Iridium—and they were used to being shown deference, not being told how to run their business. At one board meeting, Staiano was so patronizing that he brought out toy race cars painted green, yellow, and red to show which gateway owners were ahead of schedule and which were behind. It was hard to say what insulted the executives more: the kindergarten-style morality lesson, or the fact that Staiano was surprised when they took offense. "Ed didn't seem to understand that it was a partnership," said Richardson, the head of Iridium Africa. "You had Arabs, Japanese, Koreans—and you just *cannot* approach those cultures like that. Without the gateways he had no business, but he didn't see it that way." Staiano would parachute into Brazil or Hong Kong or Tokyo, try to get what he wanted, then leave so abruptly—on his way to put out the next fire—that sometimes he would strand his fellow executives who weren't able to make it to "Air Ed," their name for Staiano's Gulfstream, before it took off. When John Mitchell decided that the concept of a "gateway" was a glamour item, he apparently forgot that gateways also lead to hell.

Once back in Washington, Staiano recalled, "there was bad news every day." Motorola was refusing to ship handsets to some of the gateways because they claimed the gateway was underfinanced and needed to pay in advance. Kyocera, the Japanese company producing the sleeker of the two handsets, was way behind schedule, and it looked like there would be no phones from that factory on opening day. There were software bugs. There were failed satellites—and each time a satellite failed the stock price went crazy on groundless rumors. Scary stories leaked like a sieve from the SNOC. When some of the momentum wheels on the satellites started to fail, the only people more alarmed than Dannie Stamp in Chandler were the analysts from the major stock brokerages. The momentum wheel was one of the few mechanical assemblies used on Iridium, a device that constantly spun to hold the satellite steady against solar winds. Fortunately Motorola's last-minute decision to "fill up the tanks" meant that anything the momentum wheel couldn't do anymore was now replaced by a little corrective engine thrust. So it turned out to be a minor problem, but no one knew that yet, and the Wall Street gossip mill was causing wild swings in the consumer stock and the price of Iridium bonds.

Then, in the midst of all this volatility, Senate majority leader Trent Lott declared that Motorola was betraying America by helping the Chinese develop ICBM delivery systems! His evidence for this charge: Dannie Stamp's dispenser that popped the satellites out of the nose cone of the Long March 2 rocket and put them into orbit. The reasoning went as follows: Until now China was never able to launch more than one nuclear warhead per rocket, but now, thanks to Motorola, it could launch *two*, and suddenly China had the technology for multiple independently targetable reentry vehicles (MIRVs). The allegations would have been laughable—White House spokesman Michael McCurry called them "flabbergasting"—had they not been so threatening to Iridium's future launches. The story was taken up by the *Washington Times* and the *Drudge Report*, and eventually both the State Department and the national security advisor had to weigh in, assuring the public that Stamp's dispenser was not usable as a missile-delivery system since releasing a satellite into a transition orbit "does not require much accuracy and allows for a wide margin for error," whereas MIRVs are all about precision. At any rate, no MIRV technology was transferred, and the Chinese already had their own MIRV technology, and had had it since 1981. Even with the official explanation, the issue didn't go away, and there were congressional hearings throughout the summer suggesting that Motorola and Iridium were helping the Chinese develop fearsome war-fighting powers.

The year 1998 was one of the most boisterous and volatile in American business history, with the stock market soaring, hundreds of large companies being launched, and alternating moods of euphoria and despair. Was Iridium the next East India Company or the next Dutch

Tulip Bubble? As it turned out, it was both and neither, because the world had entered a new era in which identical technologies could fail one year and succeed the next. The AltaVista search engine was launched in 1996 and was far superior to Google at the time of Google's launch in 1998, but the timing and marketing were bungled. Google survived, using very similar computer coding, while AltaVista was long forgotten. The daily financial chat rooms made the Iridium stock act strangely—on one particular day, the stock apparently moved because of gossip during intermission at a performance of the Arizona Opera—but there were indeed some analysts out there who were asking the right questions, like "Why is Iridium raising so much capital in the junk bond market?" Publications like *Barron's* were also turning extremely bearish on satellite telephony in general, especially since Craig McCaw was rumored to be on the verge of abandoning Teledesic.

In fact, the sheepish Teledesic engineer in St. Louis who refused to debate with Leopold had prematurely released the biggest news in satellite technology for the year. Craig McCaw and Chris Galvin had decided that, rather than get into a bloody battle over who would have the first broadband system in space, they would simply combine their research-and-development efforts. But when the Teledesic engineers flew into Chandler, the Iridium inventors were shocked by the blueprints they brought with them. McCaw's plan was for an 840-satellite constellation in twenty-one orbital planes—there weren't enough rockets in the world. The cost was likely to be $15 billion—Motorola had trouble raising a tenth of that. And more important, in their opinion, it was not fully designed. Quietly and diplomatically, the Motorola engineers replaced elements of Teledesic with their own blueprints for Celestri. They reconfigured the constellation from 840 satellites down to 288—12 planes, 24 satellites per plane—but McCaw drew a line in the sand at that point, refusing to agree to anything less. The compromise system kept most of the features the Iridium inventors had planned for their second generation, including four GEOs that would work with the

LEOs, and the Chandler Lab was able to eliminate most of the project-killer features. "Teledesic became Celestri," said Leopold, "but with the Teledesic name." By the time it was announced, in May 1998, it sounded like the largest, most ambitious confederation of satellite heavyweights ever assembled, with Motorola holding 26 percent of the equity, McCaw and Bill Gates 21 percent each, and 11 percent funded by Prince Alwaleed bin Talal bin Abdulaziz Al Saud, a member of the Saudi royal family well known for his investments in Citibank.[27]

At Iridium headquarters, a block away from the White House North Lawn, there was no time for speculating about macroeconomic forces or the future of the Ka-band. The $6 billion was all spent or pledged. The system had to work. The marketing had to work. Everything had to work. Mauro Sentinelli kept up a steady stream of prepackaged sound bites for the media. On May 19, 1998, two days after a Delta II launch from Vandenberg that completed the constellation, 150 minutes of Iridium testing included a call to Bill Kennard, head of the FCC, who pronounced himself "astounded" by the clarity. The following day Merrill Lynch released an extremely positive report predicting an Iridium stock price of $202 in 2003, a $32 billion satellite phone business by 2007, and $171 billion in space-industry revenues by that same year. Staiano countered press reports about the high costs of the phone (as much as $3,995 at first) by saying it would undoubtedly be discounted, and that he was so far ahead of the game that 170,000 handsets would be in use by the end of the year—in other words, within the first three months of service.

As the launch date approached, Staiano kept hammering home three goals for every Iridium employee: 1) technical perfection, defined as a 5 percent or less dropped-call rate and signal clarity equal to land-based cell phones, 2) sales goals of a million users by October 1999, and 3) launching on time, on September 23, 1998, thereby proving wrong everyone who said, "It can't be done." Staiano relished the sheer size of the assignment. He had a hundred people working for him in the Washington offices, three hundred more at the American gateway and the SNOC in Virginia, and scores of Motorolans permanently assigned to Iridium in Chandler, with the foreign gateways employing thousands in some cases (Japan, Korea), hundreds in others (South America, Europe), and mere dozens in yet others (Russia, China). The morale of the company continued to soar even though his whole team was fatigued and running on adrenaline, and Staiano felt a primal rush every time they conquered one of the technical obstacles.

But how to get the million subscribers? A full decade of market research—extremely expensive market research—had shown that there would be forty-two million "traveling professionals" by the year 2002. These were Iridium's target customers, described as "wireless-addicted" and eager to buy the phone. Motorola's very first marketing study, carried out in 1991 by the Arthur D. Little company, concluded that satellite phone service would be universally adopted by the upscale consumer. This was followed in 1993 by a much more extensive study by Booz, Allen & Hamilton that focused on the technical feasibility of providing satellite service for the world—and finding it entirely feasible. The term "professional traveler" was coined by A.T. Kearney in its optimistic 1996 report *Study of Demand for Mobile Satellite Services by the High Income Professional Traveler*. Leo Mondale then headed up an internal study of twenty-five thousand wealthy business travelers in fifty-four cities and thirty-four countries, showing that they would use Iridium while they were on the move—even the users who currently used Inmarsat. "The market research showed that, for a handheld satellite phone that worked, there was excessive demand," said Mondale. "And there was a willingness to pay for it. The research showed that people would pay direct-dial telephone rates."

The Inmarsat terminals then in use retailed for anywhere from $25,000 to $45,000, so Staiano expected to steal three to four hundred thousand customers from the so-called vertical markets—professionals in specific industries like oil exploration, shipping, forestry, and emergency medical services who frequently went beyond available cell signals. But what Staiano really needed was the horizontal market, the upscale consumer. These were people who already had cell phones but needed them to be more practical. The Strategis Group, a Washington research firm, estimated that fifteen million cell users would switch to satellite within the first year. *Keep your local cell service,* Staiano was telling this customer, *but use it with an Iridium phone. Then, when you go out of range, Iridium will hook up to the nearest satellite and you'll never be without a signal.* When the worldwide advertising campaign was rolled out, part of it billed Iridium as "an international cell phone that works in 600 international cities," thereby emphasizing the places where it was actually least needed. What he was *not* saying in the marketing message was that you had to acquire a special cartridge to make your phone dual-usage, and you had to know when to put it in or take it out. And there was an additional marketing problem: ever since Motorola's introduction of the StarTAC flip phone in 1996, the executive business traveler was accustomed to using a sleek, compact phone. "I could have gotten past that problem," said Staiano. "That was what the media zeroed in on, the size of the phone, but people knew that the size would have come down in future models. That wasn't the biggest problem. The biggest problem was the $50 million a month going to Motorola."

And that indeed turned out to be the elephant in the room: the monthly expense of running the system. The Iridium system was so complex that the software for monitoring each call and then billing it back to the customer had been sold to the gateways for $4.5 million each. This was in addition to the $90 million on average each gateway was already paying Motorola to build out its regional earth station and sales office. All this drove up the price of the service. At the end of the food chain, Iridium was to be paid $1 per minute per call, but because of all the middlemen and the start-up gateway costs, the average call would cost $4. "People were not listening well," said Randy Brouckman, who had spent three years building back offices all over the world and was now watching an approaching train wreck. "The money went to the service provider. The service provider took a cut and paid the gateway.

The gateway took a cut and paid Iridium. Iridium paid Motorola and then took whatever was left. You had the service provider controlling the money collection, the gateway controlling the price, and Motorola demanding money at the end of the chain. The only way to pay all those people was to make the price of the call really, really high." It was the kind of problem that a CEO like Staiano could normally solve—but he lacked the authority. The original plan had been for Iridium to charge a flat fee of $3 per minute everywhere in the world, but the gateways refused to go along. In many cases this was because national telephone companies wouldn't issue licenses unless they were paid for calls they thought were being "robbed" from the national phone system. Their deep-seated fear was that Iridium would displace them entirely, placing even domestic calls beyond their control. All those poor countries that had supported Iridium in 1992 against the Europeans had now turned into beggars with their hands out, creating what one Iridium executive called a "firestorm in the developing world." But that's not to say the rich countries weren't obstructing licenses as well: Spain, Portugal, South Africa, and the United Kingdom were all throwing up roadblocks to efficient use of the spectrum that Iridium had already won.

Staiano couldn't control the gateway owners even when the gateways were wasting their own money. "One day the South Africans called saying they wanted another billing system built from scratch," said Brouckman. "We're just a few months from launch and this is something we hadn't even started on—and these were the people who had all of Africa. So I immediately left for the airport, flew to Budapest, and had a meeting the next day with the only company I knew that could do all the subcontracting for that job. That was a Friday when we had the meeting and made the deal. On the following Monday, four Hungarians flew back to Washington with me to build the system. That's what it was like. It was that kind of breakneck pacing all the time."

As the ribbon-cutting approached, press coverage of Iridium was intense and widely divergent. The lay press didn't know the difference between Iridium and Globalstar, so the narrative was framed as merely a race between two satellite phones to see which could be the first. The savvier financial press questioned the cost of the system. Oddly enough, the media didn't really question the issue of consumer demand. It was assumed that everyone would want one. News reports focused more on price—would it be a toy for the rich or something everyone would be able to afford? A skeptical report on CNN said the system's price tag had risen to $6.6 billion and Staiano would need six hundred thousand subscribers by the end of 1999 to break even. (Both numbers were probably wrong.) The handset cost was still listed at $3,000, and a typical Chicago-to-Paris call would be billed at $1.91 per minute. (These numbers were closer to the truth but missed the point. A person in Chicago calling a person in Paris probably wouldn't use Iridium.) Merrill Lynch, on the other hand, was so wildly optimistic that analysts seemed to be pulling low-ball numbers out of a hat. Retail service would be only $1 per minute, it reported, and phones would sell for $1,000. Staiano blithely told the media that it didn't matter either way: billable minutes would exceed all expectations because of pent-up demand. Merrill Lynch agreed, pointing to its projections of 4.7 million users by the year 2007.

All this speculation was moot anyway, since Iridium by that point was a runaway train. By midsummer, 295 service-provider agreements had been signed in 125 countries, including a contract with PageNet, the largest paging company in the world. Then, in June 1998, Iridium was formally introduced to the public with an advertising campaign that was the first of its kind in marketing history, premiering simultaneously in virtually every country of the world. The campaign was created by Ammirati Puris Lintas, the agency spun off from Young & Rubicam in the early seventies and now known as the number one seller of yuppie luxury brands like BMW, Club Med, Pulsar Time, MasterCard, Schweppes, and Reebok. With 154 offices in eighty countries and annual billings of $7 billion, Ammirati Puris Lintas would supposedly be able to take the global message—the phone that works anywhere—and adapt it to widely varying local campaigns. In Brazil, for example, the gateway wanted to emphasize how handy the phone was in a country where wireless was rare and erratic. In Saudi Arabia, the campaign was based strictly on snob appeal. In Australia, the gateway wanted to go after the large sporting community, so it sponsored an expedition by Peter Hillary, son of Sir Edmund Hillary, called "Iridium IceTrek." It was a hundred-day journey on skis from the Scott Base in Antarctica to the South Pole, using a route never before attempted. The trek had to be abandoned when blizzards, frostbite, illness, and temperatures reaching 58 below sapped the adventurers' resolve, causing them to use their Iridium phones to request helicopter assistance. Nevertheless, they received congratulations from the Queen via their Iridium pagers.

The massive Iridium ad campaign was designed to evoke three emotions: guilt, fear, and snobbery. The key consumer was assumed to be a high-powered executive who felt guilty about missing out on family matters back home, fearful of what was going on in the home office, but proud to be one of the "chosen ones" who hopped all over the globe making important decisions. A guy with that much drama in his life needed to be connected at all times—and Iridium was the answer. John Windolph—the same executive who had said, "This phone is huge! It will scare people!"—had decided to use its hugeness as an asset, suggesting that it was a powerful symbol of masculinity. Sitting with Bob Kinzie at a café in Geneva, he pulled out a prototype phone—"and all these lovelies wanted to talk to us," he told a reporter, implying that women were lining up to be penetrated by its phallic potency. "It's so beautiful, that phone."

Above all, the marketing campaign was based on the universal fear of being out of control—meaning in a place with no service. There were commercials on CNBC featuring Eskimos making Iridium calls, full-page ads in Forbes showing a man making a call from Korea's demilitarized zone, spreads in the Wall Street Journal, Time, and even Variety, all grouped under the tag line "Calling Planet Earth." Typical ads would show dramatic views of the planet accompanied by arch captions such as "A detailed map of your calling area can be obtained from *National Geographic*" or "Welcome to your new office—it measures 197,000,000 square miles." On an ad showing a seaplane flying over a cloud-shrouded mountain, the copy is "If you insist on a life of danger, intrigue and isolation, you can always turn off the ringer." Ads were placed in forty-five countries, direct mail pieces sent out in thirteen languages, sponsored in-flight videos shown on twenty-six airlines, and booths set up in major international airports like London's Heathrow so that people could try out the phone for free. Television commercials featured Alec Baldwin intoning the virtues of Iridium over vistas shot in the Namibian desert ("You're about to get a bigger office"), the Seychelles ("It will impress people—assuming there's anyone around to impress"), Kathmandu ("formerly known as the Middle of Nowhere"), and other generic wilderness areas ("Man has not yet explored everywhere it works"). In some cities, the ad team even used lasers to beam the Iridium logo onto the nighttime cloud cover. In a Yahoo chat room frequented by Iridium investors, there was a rumor that the campaign was so successful that the company already had seven hundred thousand subscribers waiting to be turned on.

As the target date of September 23 drew near, two more satellites failed, causing the stock to slide, and there were rampant rumors that service would never launch on time. Then came word that the phone might not be legal in France on opening day because the government was inexplicably holding up the license. Any license in France is always a diplomatic nightmare, but this one involved what one American embassy official called "real and not-so-real issues" that had to go all the way to the desk of Dominique Strauss-Kahn, the Minister of Economy and Finance, before being approved at the very last minute. The government of Germany was holding up its license as well, insisting that Iridium agree to impossible standards of radio silence to protect the radio telescope at Effelsberg. Ultimately Staiano had to prevail on American ambassadors in five countries to pressure the authorities.

Meanwhile, Staiano was demanding a 95 percent call completion rate, and by August the best he could get was 75. The stock then dropped abruptly when a NASA observer reported that Iridium Satellite 46 was "tumbling." There were, in fact, seven satellites that were not working, but Iridium 46 was not one of them. Iridium had become fair game for the tabloid press. The New York *Daily News* published articles about Iridium's ties to the bin Laden family, since one of their holding companies controlled half of the Saudi gateway. All this random noise created uncertainty around the product. A successful Delta II launch from Vandenberg in September (five spares) alleviated some of the Wall Street nervousness, but that was offset by the official announcement of the retail price of the handset: $3,795.

With various divisions of the company all failing to achieve their goals—there were gateways that hadn't opened for business, the Kyocera handset was still weeks away from delivery—Staiano finally gave in and did what he said he would never do. On September 9, 1998, he announced that the start of service would be delayed. It wouldn't be delayed for long—maybe a month—but it wouldn't be September 23. Immediately Standard & Poor's downgraded Iridium to negative. More ominously, Motorola stock took a hit as well, in a year when it had already lost half its value.

Then, on September 9, news came from Baikonur Cosmodrome that a Zenit 2 rocket had broken up during its second-stage deployment and scattered twelve satellites all over the Altai Mountains of southern Siberia. They were Globalstar satellites, so the disaster had nothing to do with Iridium, but by then all satellite phone systems were being linked together in the public's mind. Globalstar would now need an additional $565 million in financing, and with just eight of its planned forty-eight satellites in orbit, the company had no chance of launching service at any early date. Globalstar stock dropped 40 percent on the news, with high-profile investor George Soros losing $155 million in a single day. But the following day the pessimism rubbed off on Iridium and its stock dropped sharply as well. Satellite phones were starting to look dicey.

Years later, Staiano looked back on the decision to delay the launch as pivotal—and wrong. By missing the announced date—the date he promised Iridium would never miss—he sent a strong signal to an already jittery market and, perhaps more important, to a consumer not certain whether this was a luxury gadget or something he really needed. As the company crept toward the new November 1 launch date, the gateways kept asking, "What's the hurry?" The product didn't work and the sales offices weren't staffed or trained. John Richardson, the CEO for Africa, cornered Staiano after a board meeting—"and I told him, 'Why are you insisting on a worldwide launch? Why do something on that scale? Why does it have to be worldwide? Why not roll it out as each gateway is ready?'" But Staiano didn't want to hear it.

"Staiano wanted the 'dry ice launch,'" said Richardson. "He wanted steam rising from the stage and a fanfare of trumpets."

Up to now everything had been technical—get this built, get that running, get the signal perfected—and the corporate culture was "proctology exams all the time," as Randy Brouckman put it. Overnight the focus changed to marketing. Why weren't the sales offices open? Who was selling the service in Southeast Asia? Why weren't they sending us orders? On October 28, Staiano conducted a conference call with Wall Street analysts and, disguising his anxiety, said everything was rosy and ready to go. "And I believed it," he said. "I still thought we were just working out kinks." Most of his lieutenants were afraid to disagree with him, but not Leo Mondale. He had already challenged the company's conventional wisdom so often that Staiano had fired him and rehired him three times. This time Mondale told Staiano he was launching too early. "The system is not ready," he said. "The system is not nearly ready. You shouldn't be doing this big marketing spend. We have dropped calls. We have instability. We haven't had time to debug it and settle it down. The gateways are not the problem."

But Staiano ignored the advice and pushed on. He was everywhere in the media, especially on CNBC, but much of that time was spent shooting down analysts' reports that now predicted that the so-called international business traveler might have other options. One of those reports, by a consulting firm called Ovum Inc. in Burlington, Massachusetts, came out on Halloween, a week after Al Gore's first "official" Iridium call. It said, in short, that there was no market for the product.

But how could that be possible? Bright and early the next day, Randy Brouckman showed up at his office on I Street and called in four of his fellow executives for the equivalent of breaking a champagne bottle across the bow of the ship. In a few minutes the Iridium phone would be activated worldwide, and they would need to track metrics. So Brouckman showed everyone what he had worked on earlier in the week: an internal "dashboard," so that everyone could call it up on his or her computer screen each day and see, at a glance, how many Iridium calls had been placed, what was the average length of each call, how much was billed on average, etc.

But in the middle of Brouckman's presentation, Ed Staiano walked in and interrupted.

"This isn't anybody's problem," he said brusquely. "We don't need this. Get rid of it and go back to your offices."

Brouckman was stunned. Why wouldn't Staiano want a dashboard? The executives went back to their offices, still excited to have finally arrived at the first day of phone service.

Meanwhile, the official press release went out to media outlets around the world, and it was intentionally epic in tone:

The Global Village just got a whole lot smaller. "After 11 years of hard work, we are proud to announce that we are open for business," said Edward F. Staiano, Iridium LLC Vice Chairman and CEO. "Iridium will open up the world of business, commerce, disaster relief and humanitarian assistance with our first-of-its-kind global

*communications service. . . . The potential uses of Iridium products
is boundless."*

By the time the release went out, Brouckman had settled in at his
desk and, since he was the only guy with a dashboard, started tracking
calls. At first he couldn't believe what he was looking at.

There were no calls.

He rebooted his computer and checked the dashboard again.
Finally, a few figures started to change. He began to see signs of the
occasional call here and there. Then he suddenly realized—those were
all test calls. Those were all either free phones or calls being made by
Iridium employees.

What if a $6 billion company opened for business—and nobody
came?

Chapter 8

MAN OVERBOARD

APRIL 21, 1999
RITZ-CARLTON HOTEL,
PENTAGON CITY, ARLINGTON, VIRGINIA

Ed Staiano didn't want to lie to the banks. He really didn't. But on November 1, 1998, the first day of Iridium service, Ed Staiano knew things that no one else knew. One of those things was that there were no phones available in Europe, mainly because the gateway partners couldn't decide what they were selling—a satellite phone or a roaming agreement—and so they hadn't ordered handsets in time. Another thing Staiano knew was that the distribution system for phones in the United States was so messed up that people who ordered phones as early as June would not have them until late December. But as disturbing as those problems were, they were not uppermost in Staiano's mind, because the third secret was more ominous: failure to launch on November 1 could cost the company $2.1 billion and probably put him out of business before he even got started.

The system still wasn't ready—all the gateways, including North America, were begging for more time to staff up and complaining of dropped-call rates up to 30 percent—but Staiano was in the midst of a complicated financial setup that involved various kinds of bank loans, various kinds of funding from junk bonds, money from a common stock offering, and, of course, the original funding from his gateway partners—money that was spent years ago. His immediate problem was a billion-dollar bridge loan from Chase Manhattan Bank that would be coming due on December 31, 1998. For the average person, it might be hard to understand how a company could raise billions from partners, borrow more billions from banks, issue bonds to raise more billions, issue common stock to raise hundreds of millions, and then end up facing a panicky billion-dollar deadline on opening day. That's why Staiano didn't try to explain it to the average person. He didn't even try to explain it to the average Wall Street analyst. This was the nineties. This was how things were done.

In order to convert that loan into something more permanent—and avoid defaulting on it and causing cross-defaults of another billion—the Iridium system had to be fully operational. Those were the rules. Back in the more collegial days of 1996 and 1997, the bankers at Chase Manhattan had done the math and spoken optimistically of a $1.7 billion "permanent facility"—funds that Staiano could draw down on—but only after the company was open for business. Once Iridium was going, they knew a couple billion more might be needed because cash flow would be erratic in the early days of the ramp-up. Staiano now needed that money. He needed it yesterday. He needed it for marketing, but mostly he needed it to buy time for what looked like a longer-than-usual period of "loading" (the industry term for signing up subscribers).

And now the banking landscape had changed. The same Chase bankers who were begging for his business in previous years were now asking for a few more guarantees. Staiano had already spent $30 million making Chase happy with market research. But now Staiano went into more bank meetings and told them a few difficult truths about how long this was going to take. To his great relief, Chase wasn't that surprised. Motorola already had investments in twenty-six wireless systems around the world with a combined capitalization of $4 billion. Those were in countries where there weren't enough investors to build out the system, so Motorola would take equity and fund the start-up until the local wireless company had enough subscribers to pay its own way. So the banks had been down this road before. In fact, Motorola had taken equity in four of the Iridium gateways—North America, South America, Central America, and India—for many of the same reasons, and the same banks were already involved in those investments. Everybody knew the drill: things go at a snail's pace and then, boom, the floodgates open and the subscribers fall into place. It was a consumer business. Consumers copy other consumers. You can't predict exactly when they'll arrive. What Staiano needed was about a billion and a half in bridge money to get up to speed.

These were the general assumptions, at least, when Staiano's finance team started meeting with their old friends at Chase around Thanksgiving 1998. The plan now was for a syndicate of financial institutions, led by Chase and Barclays, to advance money in the form of "bank facilities," the big-business equivalent of revolving credit card accounts. Once he had this cushion, Staiano was expected to seek long-term financing in other ways. If he needed more money, he would either take it out of his cash flow, issue more high-yield bonds, or go sell more equity on the NASDAQ. The revolving credit account would be secured by the assets of Iridium. But when the bankers asked Staiano how much revolving credit he needed, he said $1.7 billion. That turned out to be a little rich for Chase, so they had a series of meetings that whittled that number down to $1.55 billion, and then they asked Motorola to guarantee at least half of it, since Motorola was going to be receiving the bulk of the early payments that Iridium made. Motorola eventually agreed to guarantee $750 million of the Chase loan, and Chase accepted Iridium assets for the other $800 million.

As Staiano entered the crucial and pivotal year of 1999, he was running the most expensive start-up in the history of American business:

- $1.9 billion raised from Iridium's gateway partners between 1993 and 1997
- $238.4 million in 1996 pledges by the gateways loaned at 14.5 percent after a Goldman Sachs junk bond offering failed

- $223 million from a common stock offering in June 1997, diluting the partners by 8.5 percent
- $446 million from 14 percent eight-year junk bonds issued in July 1997
- $300 million from 13 percent eight-year junk bonds issued in July 1997
- $293 million from 11.25 percent eight-year junk bonds issued in October 1997
- $342 million from 10.775 percent eight-year junk bonds issued in May 1998
- $750 million in a draw-down account from Chase Manhattan of New York, BZW (the investment banking arm of Barclays Bank of London), and sixty other banks
- $750 million in loans from Chase and other banks secured by Motorola
- $800 million in loans from Chase and other banks secured by Iridium assets
- $400 million loaned by Motorola, against money due on the operations-and-maintenance contract, at 12 percent interest due in December 2000

. . . for a grand total of . . .

$6.442 billion, of which $4.081 billion was various kinds of debt.[28]

A softer man—a man less convinced of his mission—might have been chagrined to learn that, after one month in business, the company had made more money selling Iridium T-shirts, watches, mugs, key chains, and cigar humidors than it had earned on the actual phone service. The ever optimistic John Windolph, Executive Director of Marketing Communications, told the media that this was perfectly understandable because "anyone who leads the Iridium lifestyle of international travel"— or even aspires to it—would of course want to be wearing the Big Dipper logo of Iridium as the crest on his sports jacket. He was implying that it was like the Nike logo, conferring instant status. Unfortunately, the thirty banks that had loaned money to Iridium weren't impressed by the rate of early key chain adoption.[29]

When it came time to sign the new credit agreement, in late December, it was the fine print—the part no one was paying attention to—that turned out to be the downfall of Staiano and, indeed, of Iridium itself. In the section outlining the "covenants," or the promises that Iridium made to the bank, Iridium attached its current business plan, which included projections of how many subscribers the company would have each quarter and how much revenue would be accrued. This was a more or less standard part of bank credit arrangements, and the bank almost never held the company to its business plan. In general, you had to tell the bank if you failed to meet 70 percent of your stated projections, and at that point the bank could call the note. In this case, Chase decided Iridium deserved a "haircut" on those standards, meaning there were special circumstances and it was a special company, so the covenants were set at around 22 percent of stated projections. The bank just wanted to be assured that 1) subscribers were being signed up, 2) subscribers were using the system, and 3) subscribers were paying their bills. After all, if you failed to reach 22 percent of your own business projections, something might be seriously wrong.

In later years, Chase would claim that Staiano and his lieutenants lied to the bank—and their evidence was that on December 15, two weeks before these loan agreements were signed, the gateways were forecasting less than 50 percent of the revenues that Iridium put into the business plan. Nevertheless, there was abundant evidence that Staiano truly believed the gateways were underestimating the popularity of their own product. Mauro Sentinelli was especially bullish on sales. "The numbers are coming in, Ed," he told Staiano every morning. In fact, the financial community as a whole was very bullish on Iridium. A report released by Credit Suisse/First Boston on December 23—right before the banking agreements were signed—predicted that Iridium would have $4 billion in revenues by 2002, with 2.9 million paying customers, rising to 10 million by 2008. And Staiano didn't act like a man who was worried. In late December he spent $65 million purchasing Claircom Communications Group, the Seattle-based division of AT&T that provided phone service on more than two thousand airplanes, including the credit-card-activated phones on the seat backs of twenty-two major airlines.[30] Then, in early January, Staiano made an even bolder move. He bought $1 million of Iridium stock with his own money. It was, more than anything, a signal to the market: Iridium was here to stay.

Once his immediate cash problems were solved, Staiano turned to his biggest headache: the gateways. At a board meeting just before Christmas, Staiano scolded the partners for sales figures that were "unacceptably low" and, more important, future projections that were ridiculous. He told them he wanted everyone concentrating on the Iridium cellular roaming service (ICRS), the cartridge that allowed your phone to switch back and forth between land-based cellular and satellite. All the smaller gateways were still struggling—many of them were barely even open for business—but the truth of the matter was that sales were miserable everywhere. Sprint owned 3.5 percent of Iridium but had yet to stock the phone in its stores. Jim Walz, the CEO of Iridium North America, kept pushing back against Staiano's expectations and telling him the sales were going to take longer than he thought. "Those are lofty goals," he said to Staiano, "and we've already done things we never thought we could do, so we'll try. But the handset prices have to go down and the airtime prices have to go down." David Dean, Walz's Market Development Manager, suggested that the company make a deal with the Franklin Mint so that if the current business plan failed, Iridium phones could be sold as collector's items. Fortunately for Dean, this idea was not relayed to Staiano, although the black humor indicates that insiders were more profoundly worried than they admitted.

Some of the overseas partners were more than frustrated—they were livid. John Richardson, CEO of the African gateway, claimed he tested the phone on the streets of downtown Johannesburg and had a measly 35 percent call completion rate. He had also visited a safari park in Ghana, "out in the middle of nowhere," where the guides were using cheap GSM cell phones. Obviously the business plan couldn't compete with that. Richardson said all the sales problems were the fault of the phone itself and of Motorola's unreasonable expectations. "The system was only 80 percent ready," said Richardson, "and I don't think Motorola or Ed Staiano cared that much about the gateways. All they were interested in was selling Motorola product and getting the systems up and running. We asked for time and he said no." But Richardson was known as a complainer and a proxy for the notoriously hard-to-please Saudi partners, and his territory was not considered that important for the overall success of the company. "Richardson was like a character from a beach novel," said one Iridium executive. "He was perceived as the pompous, culturally insensitive, entitled white guy sent in to run all of Africa! He was a complete and total disaster." But Richardson said he was just channeling the disaffection of all the other gateways— "because English was my first language and so I could talk directly to Staiano. They confided in me about things they wouldn't say directly to him, partly because you had a lot of Asians and Arabs who are averse to confrontation."

And one of the things the gateways told Staiano was that the phone didn't work and the billing system was already annoying the few customers they had. A report came in from the European gateway that one of its VIP customers—a prominent European politician—had been billed $19.50 for fifteen seconds' worth of dropped calls, and he was telling the whole world about it. The Italians pointed out that anything longer than three seconds was being billed as a full minute and that this was destroying their reputation.

Staiano, for his part, was resolutely focused on sales goals, and while his gung ho toughness had worked at Motorola, it was considered insufferably insulting by the gateway owners. "Motorola is shipping a thousand handsets a day," he barked at them, "but you're not bringing those handsets to market. The handsets are backing up in warehouses—move them onto the streets." The reaction: the equivalent of blank stares. *We're going as fast as we can,* they seemed to be saying.

The truth of the matter was Staiano cared about only three of the gateways. "While we had twelve gateways and two-hundred-some-odd countries that we had to do business in," Staiano recalled later, "the fact is that from the sale of Iridium handsets and getting that done, the key countries were the European countries, Japan, and the United States, and the other countries were important from the standpoint of being able to use the product there, but not so much from the standpoint of them buying product and selling it in that country. So I kept in personal touch with those key gateways by phone and said, 'I see numbers, but what's really happening?' 'Don't worry, we are going to make our numbers.' Okay. So I had a lot of feedback that said that the key places where we were going to sell product were going to happen."

Staiano still thought most of his problems would be solved by the ICRS, the gadget that made the Iridium phone compatible with land-based cellular. "Cross-protocol roaming," as the industry called it, would eventually become a lucrative part of the cell business, but it was too early. With the Motorola engineers creating sleeves and cassettes that conformed to various cellular systems in various countries, the Iridium "starter kit" began to get bulky with all of its portable antennas, paging attachments, memory cards, and even a "magnetic antenna" that you could slap onto the roof of a car in order to use the phone while driving. A reviewer for *USA Today* said the baffling number of accessories created an effect "somewhere between Transformer action toys and a complete set of Snap-On Tools." Increasingly frustrated, Staiano threw all the devices into a suitcase and stormed into a meeting of engineers in Chandler. Flinging open the suitcase, phones and gadgets scattered all over the table. "You really expect business travelers to carry all this shit?" he raged. He had answered his own question.

Every day Staiano got up and expected it to be the day the orders would start pouring in. Then, around the third week of January 1999, one of the in-house Iridium lawyers went to Roy Grant, the Chief Financial Officer, and said he felt compelled to talk to him. He said he'd been going over the recent Chase loans and had become alarmed that Iridium was in default according to revenue targets. *Which revenue targets?* he was asked. *All of them,* he said. But this didn't make sense—the loan agreements had just been signed around Christmastime and January was not half over—how could they possibly be in default? Well, they weren't technically in default, but based on the daily sales figures coming in, it didn't look like the numbers expected by the bank could possibly be reached. And the reason it mattered was that Bob Growney, President and Chief Operating Officer of Motorola, kept sending e-mails requesting "Iridium loading data," better known as the customer list. What should he do?

The fact was Staiano wasn't giving out the customer list. He wasn't discussing loading data. He wasn't revealing, even to the gateways, what the actual sales figures were. As of December 31, for example, only 35,000 handsets had been produced. To meet the bank covenants, 355,000 handsets had to be available by the end of March, and the rate of production was 1,000 per day at the Motorola plant in Libertyville, Illinois, and zero at the Kyocera plant in Japan because of continuing design flaws—so it was mathematically impossible to meet that particular covenant. Even more ominous, as of January 5 the gateways had ordered just 15,000 ICRS roaming cassettes, of which only 729 had been shipped to customers. The original business plan had been all but abandoned as too ambitious, and the second business plan, pushing back sales goals a full quarter, seemed to be in jeopardy. There was talk around the office now of business plan 3.0, and to enact it, Grant called an emergency meeting of the Iridium banking and finance committee. The next day Staiano issued a warning to the gateways: *Your most optimistic sales forecasts do not meet the covenants of the bank, and if one loan defaults, they all default.* If nothing else would motivate them, maybe fear would work.

Within a period of three weeks, Iridium managers had gone from a feeling of relief—the banks had given them breathing room—to panic. Staiano, almost alone among top Iridium executives, was still not worried. He had been through wars like this before. Your bank is your partner. Your bank is not going to bail on you, because it has just as much to lose as you do. But Motorola was nervous and the gateways still needed a good kick in the ass. So it all came to a head at a huge Iridium board meeting in Washington, D.C., on January 21, 1999.

Staiano designed the meeting as shock therapy. He would start with the horrifying news—the banks could shut us down—and then use it to motivate the troops. The first formal report came from Lars Ernst, the company treasurer, outlining projected earnings for the first, second, and third quarters based on current sales figures. Then Staiano took over.

"Now, what you'll see today in these numbers—and we will talk about this in some detail—is that gateway forecasts that we got in still do not meet the new target plan," he said. "In uncertain conditions, these do not meet the covenants that we have for the banks to be able to have our loans in good shape." Then, to make sure all the partners understood the stakes, he asked CFO Roy Grant to go over a set of charts outlining just how bad the situation was.

"As you can see in this chart in the first quarter based on the gateway forecasts," said Grant, "we're not meeting the bank covenants. . . . Going to the second quarter, again, same analysis. . . . The gateway forecasts do not have us meeting the cumulative accrued revenue and cumulative cash revenue covenants in the credit agreement. . . . The banks have the right to terminate the commitments, they have the ability to declare the loans then outstanding due and payable, essentially accelerating payment under these loans, which will trigger cross-acceleration of all our high-yield debt."

Almost lost in all the verbiage was that Grant was talking about *sales forecasts*, not estimates of earnings. He was starting with numbers that were by their nature *hopeful thinking* and pointing out that they all fell short of what it would take to stay in business. This was the first time everything had been put in such starkly black-and-white terms. For once, the gateway partners were silent. And now it was time to build the morale back up. Lauri Fitz-Pegado, Vice President in charge of gateway management, spoke about licensing issues, regulatory issues, and market-access issues, and that was mostly good news for everyone. The company was making progress toward opening up all the territories that had problems. Then Bruce Dale, Senior Vice President for Network Operations and Engineering, gave a report on the performance of the network itself. That report was an A-plus. All sound quality problems had been solved. All dropped-call problems had been solved. The phone was functioning as it was designed to function. The Iridium satellite communications system was—finally—a world-class operation and a marvel of modern technological achievement.

And then the master of marketing took the stage. Mauro Sentinelli's job on this day was to convince everyone in the room that they could do it. The marketing plan was beginning to work, he said. The advertising campaign had penetrated and would start to show results. There were logjams that no one could have anticipated—technical problems, staffing problems—but now that was all behind them and the company had 137,000 qualified sales leads, and a conservative estimate would be that 25 percent of those would turn into sales. That was more than enough to meet the bank covenants, and—

Suddenly Sentinelli was interrupted.

"We have no orders," said Gordon Comerford.

Comerford was a Motorola vice president.

Sentinelli asked him what he meant.

"We have no orders for handsets beyond the initial order of thirty-five thousand," said Comerford. "We will have to close our production line in Libertyville and the pager factory in Boynton Beach if we don't receive any more orders."

Sentinelli told Comerford to listen to what he was saying. There were 137,000 qualified leads and, of those, 30,000 were from the North American sales team. "We're just waiting on Sprint to push it through their sales network. And we're still waiting on PageNet to roll out the two-way paging service. We expect the pager business to be huge, and they're slow with it."

Comerford sat down, but the stink in the room remained. Comerford was not just any board member. He chaired the Iridium audit committee, among other things, and he had been involved with Iridium since the first fund-raising round. Visitors to Comerford's office in Schaumburg sometimes picked up a leather-bound book he kept on his desk, entitled *Everything I Know About Communications*. Inside, all the pages were blank. Comerford had climbed to the top of Motorola as a pure finance guy.

Sentinelli continued to talk about the roaming markets and the ICRS cassettes that the gateways didn't seem to be interested in selling. Those roaming contracts were now 80 percent of the business plan, especially in the twenty-seven key markets identified as necessary to the success of the company. Unfortunately, Sentinelli said, Iridium had so far signed up partners in only nine of those twenty-seven markets, and of those nine only three had launched the ICRS system. But that explained the low numbers (Sentinelli didn't mention the actual number, but it was indeed low—only 1,555 cellular cassettes shipped). Once the partners were signed up and motivated, everything would change. (By emphasizing the number of sales leads, Sentinelli was avoiding the major issue. Iridium actually had 1.5 million sales leads developed mostly through the Ammirati Puris Lintas ad campaigns, but most of those leads were never followed up on because the gateways didn't have the staff or the expertise to work them.)

The aftermath of the January 21 meeting was somber but not desperate. John Mitchell had been there, and he summarized the situation in a letter to Motorola that was copied to the top thirty executives, including CEO Chris Galvin. These bank covenants were matters of "extreme concern," he said, because any default could be "onerous" for Motorola. It wasn't the first time, but increasingly the goals of Motorola and the goals of Iridium were becoming separated. Staiano had been Galvinized and initiated as a Motorola lifer years earlier, but now he was at the helm of a company that was created by Motorola but likely to be released into the wild at any moment. The monster was about to escape from Dr. Frankenstein's laboratory, and when that happens, the monster has to be killed.

At first Staiano thought he could prevail on his friendships at Motorola to bail Iridium out. He went to Mitchell shortly after the board meeting to say, "Look, John, there's a way Motorola can help. I can't do anything about the $40 million a month in debt service, but I'm paying $50 million a month to Motorola. Maybe you could defer some of that."

Mitchell said absolutely not. Motorola doesn't do that.

"Okay then," said Staiano, "$25 million is for replacement satellites. I'm willing to take that risk. Let's get rid of $25 million a month and make Iridium responsible for replacing satellites. I think they're going to last longer than five years anyway."

Mitchell said absolutely not. Staiano was starting to forget that, when it came to collecting on old bills, Motorola was just a couple of degrees removed from a Mafia loan shark. Bob Kinzie heard about the meeting and told Staiano to be careful. "He was starting to tug on Superman's cape," said Kinzie.

But Kinzie had an old friend on the board who he thought might be able to help. William Schreyer, chairman of Merrill Lynch, had been intimately involved in the ramp-up of Iridium and had underwritten the first public stock offering. So Kinzie and Schreyer came to Staiano with a proposal. Merrill Lynch would do a secondary offering of $3 billion and buy up all the debt, then convert that debt to equity. The result would be cutting the Iridium churn rate from $100 million a month to

$15 million a month, giving everyone a fighting chance. Staiano listened intently to the proposal, thanked Schreyer, and asked him if he had any clout with John Mitchell. Schreyer said he would talk to Mitchell about it right away. That was the last anyone heard of it.

One day after the "hard-truth" board meeting, Merrill Lynch released a report predicting Iridium would reach cash-flow break-even no later than the first quarter of 2000—one year hence—but possibly even by the third quarter of 1999. There was no mention in the report of any problem with the service on $3 billion in debt.

In later years, people involved in the situation wondered why Motorola didn't make more of an effort to relieve the financial pressure on the company. As of the end of 1998, Iridium had already paid Motorola $3.25 billion over a period of five years, and after the new loans were issued in late December, $370 million of the badly needed $1.5 billion immediately went to Motorola. Motorola probably could have increased its equity in the company in return for altering some of the long-term contracts, but then again, that was not the Motorola way. The Motorola way was to decrease equity and increase fees and sales to its partners. Staiano had been a good Motorola general for twenty-three years, but now, in many respects, he had become the enemy. "Ed was brusque with everyone," said Leo Mondale, "and that included the Galvin family. He showed no respect for Chris Galvin. Eventually that affected what could be done."

Still, no one really believed that Chase Manhattan would pull the plug, so the goal going forward was to make the bankers feel good by whatever means. That meant not sharing the fact that, as of January 22, 1999, after two and a half months in business and several months prior to that trying to sell the service, Iridium had precisely 3,637 customers. Even adding a zero to that number wouldn't get the company to the level that would satisfy the bank covenants, which were extremely conservative (52,000 customers by March 31). Staiano suggested to Roy Grant, his CFO, that some kind of soft, sweet whisper of a communication go forth informally to their best friends at Chase that perhaps, just maybe, they might not meet minimum projections on March 31. They *probably* would, but out of an abundance of caution, and to cover their collective asses in case they didn't, they placed a pleasant phone call to make the point. Chase responded amiably, but the bankers asked that Iridium give them some numbers in early February to let them have a general idea of how things were going.

At this point in the history of Iridium, there was no scarier sentence than "Let me see the numbers."

Staiano decided to get through a long-planned secondary stock offering first, so he sent Sentinelli to Schaumburg to make a presentation to a nervous Chris Galvin, then joined two other Iridium executives on a call to Wall Street analysts on January 25. Staiano ran through the list of all Iridium's technical achievements—a 5 percent dropped-call rate, a pager rate of one missed message per 1,500 pages, and a "98 to 99 percent range" for establishing calls with roaming partners—then talked about the 250,000 leads generated by the advertising campaign, leads that would become customers as soon as the handset pipelines were unclogged. (How the number of leads went from 137,000 when Sentinelli was talking about them three days before to 250,000 when Staiano talked to Wall Street was never explained.)

The problem, Staiano said, was Kyocera. Kyocera had been slow to market with its handsets, and a lot of customers wanted that handset because it was fancier, sleeker, and easier to use than the Motorola handset. "I can tell you," said Staiano, "that the number one question coming into our global customer care centers is 'When can I get my phone?'" Later in his prepared remarks, Staiano talked at length about "industrial users"—meaning vertical markets—as well as government users, the military, disaster relief agencies, and companies that were buying phones as a protection against Y2K disruptions.[31] An analyst from Southeast Research asked Staiano whether this represented a change in focus away from the "international business traveler," but Staiano said absolutely not—this had been his plan from day one. He would go after the vertical markets as low-hanging fruit, then get the business travelers later when they signed up for Iridium roaming in concert with their regular cell provider. When an analyst from J.P. Morgan asked the inevitable question—"When will you get to cash-flow break-even?"—Staiano's answer was, "This year. We haven't seen anything that indicates that should be any different." Fortunately, most of the time on the call was taken up with specific questions about Kyocera, as though all Iridium's problems would be solved once the Kyocera handsets started to arrive, and the Merrill Lynch analyst even chimed in with some softball questions about "second-generation systems" and what Staiano might be doing to increase the capacity of Iridium once it maxed out. Among the facts that Staiano failed to mention was that in China, India, and Indonesia—the world's first, second, and fourth most populous nations—the number of handsets sold was zero, because Iridium was still waiting on various government clearances. Two of the three countries had Iridium gateways.

Meanwhile, Roy Grant started drafting the "certification" that Chase was requesting. This was the company's official sworn statement as to whether it would meet its projections for the first quarter of 1999. Normally the company would send a letter to the bank estimating its business progress, but if that letter was certified by the Chief Financial Officer, it became a much more serious document. Chase wanted a certified document. Not because the bankers didn't trust the company but because, you know, we're dealing with billions here. The certification document was especially important to Motorola, because, once it was sent, Motorola intended to ask Chase to release part of the $800 million guaranteed collateral for the December loan.

Grant's first draft of the certificate said Iridium would have $9.2 million in revenues for the first quarter. Over the next three days he would write several more drafts, and by January 29 he had revised that number downward by half, to $4.3 million. That was not enough to satisfy the covenants, but it was assumed that, once Chase saw some progress, the bank would be forgiving. The strategy was to push forward the business plan another quarter—now called business plan 3.5—and Staiano even talked of pushing it forward a full year, but only after the bankers were satisfied.

On January 27 Iridium collected an additional $250 million in what was trumpeted as a successful public stock offering. Public ownership of Iridium was now 13.25 percent.

After the end of business on Friday, January 29, Roy Grant left Washington for a skiing vacation. He was gone for the entire first week of February and didn't call in to the office. Presumably he needed to clear his mind.

On Monday, February 1, Mauro Sentinelli sold all his Iridium stock.

Staiano may or may not have known of Sentinelli's bail-out. He certainly acted like he didn't, announcing to the world on February 5 that Kyocera would now start shipping and all was right with the company. One day around that time Staiano called Bob Kinzie into his office and told him he was about to get an order for $30 million from China and he wanted Kinzie to be on the call with him. The two men sat together waiting for the call . . . which never came.

When Roy Grant got back from skiing on February 7, he had a couple of days to look at the latest numbers he was expected to interpret for their friends at Chase Manhattan. No one had seen those numbers. If they had, they might have been shocked. Entering its fourth month of service, Iridium had 6,009 subscribers and $535,000 in accrued revenue. To put this in perspective, the revenue target in the current business plan, which had been revised downward, was $50.467 million, so the company was at 1 percent of that number. Among "cellular home subscribers"— the international business traveler—actual revenues were $4,000 against target revenues of $8.571 million, or 4/100ths of 1 percent of the goal. In order to satisfy the covenants, Iridium would need to earn $577,000 per day by March 31, which worked out to nine hundred new subscribers per day. Its current growth rate was sixty-five new subscribers and

$23,000 per day—about 7 percent and 4 percent, respectively, of where the company needed to be.

On Wednesday, February 10, Roy Grant issued the certificate. He'd decided to do things the easy way. "Roy Grant was a very competent financial guy," said Leo Mondale, "but he had difficulty standing up to Staiano. Ed Staiano was a very difficult guy to disagree with."

Iridium would meet all its first-quarter targets, Grant told Chase. There would be about $4.3 million in revenues for the first quarter. And the company would have 355,400 voice subscribers by March 31.

Grant had failed to mention several things. He failed to mention that the company had only 6,009 subscribers. He failed to mention that there were only 38,000 handsets available for those projected 355,400 subscribers. And he failed to mention that Iridium's accrued revenue was $535,000, which was far less than the $700,000 that Grant would deposit into his personal bank account the same day he issued the certificate, which was the day he sold all *his* Iridium stock.

And it worked. Motorola's collateral was released two days later. Merrill Lynch congratulated itself on a second successful stock offering. And now added to the bills Iridium was expected to pay was $500,000 for "financial advisor" fees—money that would be going to Ed Staiano's best friend forever, Chase Manhattan Bank.

Over the next month Staiano continued to hide the actual sales numbers from everyone except his closest lieutenants, but Roy Grant couldn't stop talking about them. In meetings with company lawyers and conversations with his fellow financial officers, Grant displayed evidence of a guilty conscience, or at least an awareness that maybe he should revise his certified sworn opinion of February 10. In a conversation with company lawyer Tom Tuttle on February 16, he used the phrase "speculation and fantasy" to describe the projection numbers and said bluntly, "Nobody is buying phones." By February 22, just twelve days after the certification, he was telling Chase informally that "those assumptions no longer apply" and that Iridium probably wouldn't meet any of its targets. But Staiano still wasn't worried. Moving in and taking over the situation, he decided to make it a public relations problem. In an interview with the *Financial Times*, Staiano frankly stated that Iridium would probably not meet the minimum targets set by the bank. At an Iridium board meeting on March 2, Staiano told the board bluntly that sales weren't enough to satisfy the bank and suggested that he take over all marketing efforts for all the gateways. The gateways pushed back against that suggestion, a loud fight broke out, and the meeting ended in a melee of acrimony and finger-pointing. The gateways were starting to turn against Motorola, the company that brought them into the deal in the first place, and they regarded Staiano as a stooge who did Motorola's bidding.

Staiano needed time. He still thought that, given an extra quarter, or two, or three, he would be able to gear up an old-fashioned high-pressure Motorola-style sales effort and the orders would start pouring in. But his immediate problem was that he felt like he was standing on a cliff with Chase ready to push him off at any moment, Motorola acting like it didn't notice he was in danger, and the board of directors assuming he knew how to fly. On March 8 he got on the phone with the entire Chase syndicate in order to ask for help. "We won't meet the covenants," he told the bankers, knowing it was better to say it now than wait for them to find out on March 31. But he still didn't give them the actual numbers. As of February 26—four months after launch—the company had less than a million dollars in revenue.

Again, it worked. Chase agreed to waive the first-quarter covenants in order to help Staiano avoid a cross-default. Staiano now had until June 30 to bring his numbers up, and there was always the chance that he could get the second-quarter covenants waived as well. But with all the gateways still failing to meet their sales projections and complaining that even the revised ones were too onerous, Staiano again went to John Mitchell and said, "I need your sales force." This time Mitchell responded, and overnight Motorola created an "Iridium Action Plan," taking salespeople from its Cellular Resales Group and and turning them over to Iridium. A global sales force of 40 increased to 250 overnight, in addition to the gateway sales operations, and Iridium started offering discounts and incentives to service providers and the salespeople who normally sold land-based service. "Sell the roaming plans," they were told. You didn't need a handset for those. The cartridge would let you use your regular cell phone when you went out of range.

At the monthly Iridium board meeting on March 19, Roy Grant was nowhere to be found. He had left the company, citing "personal reasons," and was replaced by Leo Mondale. Mondale was told to be direct with the gateways. There were tariff problems, he told them—governments were tacking on too many charges and driving up the airtime price. There were problems with handset pricing. All the sales projections were being missed. All the target numbers were going to be pushed back an entire year, but this was probably the last time that could be done. (The gateways still protested, saying that the sales projections that had now been revised for a fifth time were still too high.) The gateways were also briefed on the Motorola Action Plan but were unimpressed. The term "Motorola" itself was becoming a dirty word. The next day Bob Growney, President of Motorola, gave an interview to the *Financial Times* in which he made a strange point: "There is no chance of Iridium going under."

Meanwhile, across the ocean at the largest fairground in the world—the site of the largest computer trade fair in the world—Motorola was introducing a new phone called the Timeport. It was digital, it was small, it used the GSM standard, and it worked on every continent of the world except Antarctica. It had been developed by the Schaumburg guys who had coined the term "Iridiots." As the telecom world gathered at the CeBIT Fair in Hanover, Germany, Richard Midgett, chairman of the GSM Association, representing 323 GSM operators from 129 countries, praised the Timeport's impressive data capacity as well as state-of-the-art voice mail features. "This is what the customer has been waiting for," he said, "a phone that truly enables the GSM dream. A global system for roaming in one simple handset. We challenged the industry to deliver this capability and we congratulate Motorola on being the first to respond." The Motorola gadget guys in attendance called it the first "world phone." It would be sold around the world by the same people who had just been mobilized by John Mitchell to enact the Iridium Action Plan. Flumpits had triumphed. When the Europeans at the 1992 WARC had spoken of "the GSM dream," the Motorolans had scoffed—there was no one powerful enough to get that done. They had forgotten that there was one company powerful enough—their own company.[32]

A week later the Pentagon issued a news release that should have been good news for Iridium—a $219 million contract to buy Iridium phones, pagers, accessories, and minutes. Unfortunately, the fine print indicated that almost all the money would go to Motorola, not Iridium, through a deal with Motorola's wholly owned service-provider subsidiary in Chandler. Later that summer, the State Department would spend another $1.4 million for one thousand handsets, enough to supply America's 260 embassies and consulates—but again, all the money would go to Motorola. All those contracts executed back in 1993 were starting to turn Motorola into the parent that eats its child.

When you have to make public announcements stating you're not going bankrupt, your bankers tend to get queasy. When your Chief Financial Officer disappears like a magician's assistant right before a board meeting, your bankers tend to get panicky. In early April the powers that be at Chase Manhattan Bank decided they might not be getting the precise truth from Ed Staiano, and so they called John Mitchell and told him that his presence was required in New York. Mitchell was told to bring three Motorola vice presidents with him from Schaumburg—including Stephen Earhart, chairman of Iridium North America, and Ed Gams, the Director of Investor Relations who served on the Iridium finance committee—and on April 16 they all arrived at a very high floor of One Chase Manhattan Plaza and settled into a conference room with six Chase vice presidents, including the senior credit officer and the head of the telecom division. Mitchell, speaking for Motorola, eventually admitted the truth—the sales were so low as to be almost nonexistent. At best they were 10 percent of even the lowest forecasts.

The Chase bankers were stunned. They used words like "shocked" and "betrayed" and "lied to." Voices were raised. One of the Chase representatives said, "The credibility of Iridium management has been destroyed." *And by the way,* they told Mitchell, *we want that Motorola loan guarantee restored.* Mitchell was noncommittal on that, pointing out that "that certification didn't come from us, it came from Iridium"—further infuriating the New York bankers.

When Mitchell left the meeting, he knew what he had to do. He didn't do it immediately, which is odd in retrospect. Perhaps he was looking for some way to spare his old friend. It didn't happen until five days later, when Ed Staiano arrived at the Ritz-Carlton Hotel in Pentagon City for the next Iridium board meeting. Staiano was checking in a day early so he could attend a cocktail party with his handpicked candidate to replace Grant as CFO. Her name was Stephanie Cuskley, a banker recruited by Staiano in order to bind the company more closely to Chase. As Managing Director of Chase's High Yield Finance Group, Cuskley was the person who had helped Staiano set up all those junk bond offerings. Mitchell found Staiano at the party and told him he needed to speak to him in private, away from Cuskley. That's when he broke the news: Staiano wouldn't be going to the next day's board meeting. "You have to resign," said Mitchell. "I'm sorry. That's what Chase wants."

Staiano had been clueless. This didn't happen to a Motorola lifer. What happened to his Galvinization? What happened to "Stay past your retirement and help us fix this mess"? Who was the guy who'd bought a million dollars' worth of Iridium stock with his own money? It wasn't Mauro Sentinelli or Roy Grant. It wasn't the gateway owners, who bitched and complained but never pitched in to fix their problems. Staiano was the guy working 24/7 to fix Iridium—who was going to do it now?

The answer, according to the gateway owners, who could now elect anyone they wanted, was John Richardson, the fifty-five-year-old Australian who had been at Barclays Bank in Hong Kong before being hired by the Saudis to run the African gateway in Capetown. Richardson was an adventurer and mountaineer—he had climbed Everest twice, getting within a thousand feet of the summit before the death of two climbers caused his expedition to turn back—but he was best known inside Iridium as the most vocal critic of Motorola. His main advantage at this point was that he was regarded as the opposite of Staiano—elegant, understated, manicured, and handsome in his tailored suits—and, if you could get past his tendency to tell mountain-climbing stories, a capable banker. Many of the board members didn't particularly like him but said, "He represents the Saudis, and the Saudis are the loudest complainers, so let's turn it back on them. 'You think you can do better? It's all yours!'"

Unfortunately, the firing of Staiano—which was pitched to the media as a resignation on principle due to disagreements with the board over strategy—sent Wall Street into a panic. The stock nosedived, going as low as $7 a share after trading at $72 less than a year earlier. The 14 percent junk bonds plummeted to 25 cents on the dollar. Until now, it was possible to think of Iridium as simply having start-up problems—the Kyocera handsets were delayed, the gateways were slow staffing up—and those were problems anyone could live with. The resignation of Staiano implied a fundamental lack of faith in the product itself. Chase Manhattan's anger was about to punish everyone.

The plan was for Richardson, as the new CEO, to work with the banks and the bondholders to restructure Iridium's debt. Wasn't that sort of thing done all the time? After all, Richardson had earned his management stripes at the massive Hong Kong conglomerate Hutchison Whampoa, which had been on the verge of bankruptcy before he whipped it into shape. Iridium needed a reorganization period and then eventually everyone would get paid. Richardson even had specific expertise in managing troubled companies, especially those owned by billionaire investors. His résumé showed him at various times on the payrolls of the richest people in the world, including Sir Douglas "Dougie" Clague of England, Li Ka-shing of Hong Kong, Prince Jefri Bolkiah of Brunei, Khoo Teck Puat of Singapore, Adrian Zecha of Indonesia, Alan Bond of Australia, and Tony Bloom of South Africa. In almost all of those cases, Richardson had been the "workout guy" who came in to turn around failing companies.

Not this time. Perhaps it was some sign of how out of touch Richardson was that he referred to Ed Staiano as a "nice guy"—maybe the first and last person ever to use that appellation for Staiano. Richardson made it clear he didn't like Washington—he preferred London or Hong Kong or almost any other headquarters city—and he seemed a little too blasé for the situation. He concentrated on downsizing personnel— saving nickels and dimes that were not likely to matter much when the company was hemorrhaging $100 million a month—and assumed that he could charm his fellow bankers and businessmen into reaching a gentleman's agreement that would save the company. His first act as CEO was to fire Mauro Sentinelli and the entire marketing department, then bring in a South African, Sue Kennedy, to do the same job with no staff. "The marketing was too arrogant," said Richardson, "and the product package was inappropriate." He then slashed handset prices by 65 percent and minutes by half, but now had to rely on Motorola to sell everything. Next he shut down the "gateway management" department— all those meddlers who reported to Staiano—and settled in for what was primarily a letter-writing campaign: letters of desperation, addressed to Chris Galvin.

Richardson believed, despite all evidence to the contrary, that Iridium could sell its way out of trouble. All he needed was more time. Staiano had been marketing to the wrong customer. Since Globalstar wouldn't be fully operational for another six months, Iridium would still be able to dominate the vertical markets and triumph. "We've already climbed our first ice fall," he told the troops, eliciting blank stares. "The rest are at base camp and haven't even unpacked their gear yet." Remarkably, much of the financial press accepted these statements as evidence of a routine reorganization that would eventually restore the stock price and allow the company to pay its bills. A notable exception was Herb Greenberg, author of the popular Herb on the Street column, who used the term "Iridiots" to relentlessly lampoon the die-hard investors holding on to their rapidly degrading stocks and bonds. Asked in later years whether he really believed he could claw his way to profitability with vertical markets, Richardson paused for a moment and said, "Well, you have to give it a shot."

"I had two priorities," he said. "One was to be absolutely sure where the product was—was it good enough? And the second was the management overlay—I needed managers of sufficient caliber to make the company work. I was dealing with a lot of engineers. They were focused on engineering objectives. I had to refocus them on commercial objectives."

And yet neither of those priorities involved the fundamental issue: debt. After rearranging the deck chairs, Richardson turned his attention to the finance issues, but without much sense of urgency. He brought in a fellow Australian named Tom Alabakis, who had been Richardson's finance executive at the African gateway and was continuing to manage that gateway from Dubai, and assigned him to work alongside Leo Mondale on the tough business of renegotiating all the debt instruments. He very quickly came to the same conclusion Staiano had reached five months earlier: the only possible way out was to get Motorola to defer its monthly bills. "I wrote a personal note to Chris Galvin," recalled Richardson. "I said, 'This could get very nasty if you don't do something. The Sioux Indians are coming up the hill and I'm Custer.' But he never replied."

Meanwhile, Richardson was not exactly inspiring confidence in his inherited lieutenants. On some mornings Richardson would retire to a park bench in Lafayette Square with his newspaper and not return for a couple of hours. "John Richardson would get pissed off when he was reminded of what he'd been hired to do," said Mondale. "He didn't really take the job very seriously. In my opinion he was a straight opportunist looking for a fat check."

"John Richardson was a joke," said Bob Kinzie, who was normally known for his diplomatic word choices. "He was off climbing Machu Picchu most of the time."

"John Richardson was a mystery to me," said Jim Walz, CEO of Iridium North America. "I never could figure out what he was doing, where he was going, or where he had been."

Randy Brouckman, who was essentially running the day-to-day operations at Iridium, was equally uncharitable: "John Richardson had the lifestyle of an upper-class Brit, the casual attitude of an Aussie, and the arrogance of a South African white male."

During the first post-Staiano conference call with Wall Street analysts, Richardson didn't show, leaving Leo Mondale to explain the company's new strategy—"tailoring the Iridium product and service offering," improving distribution, lowering prices—without ever addressing the elephant in the room: $50 million a month in debt service.

Oddly, Richardson focused on the banks, which had secured loans, and on the gateway partners, which had no say in the finance issues, and had very little direct dealings with the players who were most likely to force his hand: the junk bond investors, whose notes were trading at eighteen cents on the dollar by the end of June. Each passing day made the bondholder position stronger, as panicked mutual funds sold off their holdings to so-called vulture funds, companies set up to take advantage of the restructuring, hoping to end up with massive equity positions in Iridium. Richardson didn't seem to notice and instead spent time writing more letters to Galvin, "beseeching him to give us assistance." Eventually Motorola sent in high-level executives from Schaumburg to "hold his hand," as one of them put it. They refocused his attention on the bondholders, telling him that the debt needed to be converted to equity or else bankruptcy was certain. For a couple of weeks, Motorola

President Bob Growney even floated a restructuring plan called "Project Asteroid"—the bondholders would give up their $1.5 billion in notes for 25 percent of the company in return for Motorola investing $400 million. (The $400 million offer was not as impressive as it sounded. Motorola was telling the Japanese and the Germans that they were expected to put up a large chunk of it, and Motorola would not be changing any of the terms of its operations-and-maintenance contract, which meant Motorola would get the $400 million back in a matter of months.) But in the course of the Project Asteroid negotiations, it turned out that Motorola expected to end up owning about 40 percent of Iridium, with the bondholders expecting to keep 44 percent, and neither of the two big dogs would back off and leave anything for the gateway owners who had originally funded the company, much less the holders of the publicly traded stock, who were expected to be grateful for the 4 percent that might be left on the table for them. John Richardson was the representative of the gateway owners, and now he was faced with the prospect of telling all of them—Saudi princes, Indian bankers, Japanese moguls, Russian rocket manufacturers, Indonesian industrialists—that they would have to share 8 percent and they would no longer be in charge of anything. In the middle of that negotiation, Chase Manhattan sent Motorola a letter saying that, by the way, that little charade back in February—the "certification" by Roy Grant—meant that the company owed Chase $300 million. Iridium was increasingly looking like a single plate of pasta that was expected to feed the Roman Legion.

As it turned out, Motorola's fascination with satellites was rapidly evaporating anyway. In late May, six hundred Motorola engineers working on Teledesic at its Kirkland, Washington, headquarters were abruptly sent home to Arizona, despite Motorola's $750 million investment. Executives working for Craig McCaw told the Associated Press that the engineers were "needed for Iridium." Iridium needed many things in May 1999, but six hundred additional engineers was not one of them.

Clinging by his fingernails to the north face of a financial Everest, Richardson decided it was just too hard—and that it was Staiano's fault for accumulating all that debt in the first place. "There was this vital strategic flaw, that all these various debt products had been sold to Staiano with various levels of security," he said. "Staiano was not a finance guy." Publicly Richardson remained blithely optimistic and dismissive of what he considered Motorola's silly predictions of imminent disaster. He told the *Wall Street Journal* in late July, "What [Motorola] said about the alternatives [of liquidation and bankruptcy] was, from an academic perspective, absolutely correct, but from our perspective, it's simply not on the radar screen." *From an academic perspective?* What was this guy talking about? Nothing Richardson ever said did anything to tamp down the helter-skelter break for cover on Wall Street. Mondale tried to step in and do damage control by telling a reporter that the foreigners on the Iridium board would never allow liquidation because, in many foreign countries, the stigma attached to bankruptcy was too enormous.

Bankruptcy may not have been on Richardson's radar screen— perhaps because he was on safari in Tanzania during much of July— and bankruptcy may not have been culturally acceptable to the foreign gateways, but apparently it was pretty okay with the bondholders. Four months after the Staiano firing, the house of cards collapsed. Richardson was offering so little to the bondholders that they grew increasingly hostile. His meetings with Chase were equally frustrating, even though they should have been allies. But it was really Motorola that held the only possible lifeboat, and Chris Galvin was not inclined to deploy it. "Changing the operations-and-maintenance contract could have saved us," said Mondale. "Project Asteroid was too little, too late. We had $400 million in cash. We could have gone two years with that money. Motorola had figured the contract on replacing one satellite per month. It turned out that you needed to replace, at most, one satellite per year. So could we have gotten an eighteen-month deferment on that money? No, they wouldn't do it."

On August 11 Iridium finally defaulted on the Chase loans by failing to meet covenants that had already been deferred twice. Everyone became so spooked, fearing they would lose their place in line, that the bondholders filed a motion in New York to force the company into involuntary bankruptcy—mainly to speed up the negotiating process. Richardson was annoyed by that and sent his own team of lawyers to file voluntary bankruptcy in Delaware on the same day, hoping for a more sympathetic court.

Iridium was now on life support. The *New York Times*, reporting on the teams of lawyers rushing to two different bankruptcy courts, used its notorious "many say" phrase when injecting opinion into a news article. "Iridium now faces its ultimate test," the *Times* reporter wrote. "It must rework a huge debt load, regain the confidence of its investors and find a way to bolster sales of a satellite telephone service that many say is outdated, outmoded and unpopular with consumers." The newspaper of record had delivered a virtual obituary. Asked for a comment about Iridium, Globalstar chairman Bernie Schwartz told *Forbes*, "There is no Iridium."

It was the biggest bankruptcy filing in the history of the United States—and the quickest. The company had been open for business a little over nine months. Thanks to the provisions of Chapter 11, Randy Brouckman and a few other executives stayed on at headquarters, managing the customer list and keeping the company operational in the hopes that someone would take it over. For the first month after the filing, Richardson fully expected to work out a deal that would convert debt into equity, but Motorola was a stumbling block. Motorola wanted someone to ride in on a white horse and bail everyone out—and that someone was Craig McCaw. John Mitchell sent Durrell Hillis to Washington to be Motorola's "acting CEO," even though Motorola had no legal right to do that, and Richardson was slowly eased out. Brouckman started working with McCaw, who said he probably wanted to buy the company out of bankruptcy and combine it with the stillborn ICO Global, which filed for bankruptcy just two weeks after Iridium. McCaw did end up buying ICO Global in November 1999, guaranteeing $1.2 billion in additional investment from his own company and a group of partners that included Inmarsat, Indian media mogul Subhash Chandra, Deutsche Telekom, Ericsson, Hughes, NEC, and TRW, which had already merged its Odyssey system with ICO.

It's odd, in retrospect, that McCaw was so quick to buy the unbuilt ICO system but couldn't make up his mind on Iridium, which he once described as the "technological precursor" to Teledesic. At first he told the Iridium board he was willing to spend $600 million on Iridium, contingent on his due diligence, and he continued that dance until a day in March 2000 when one of McCaw's executives at ICO apparently talked him out of it. Dannie Stamp and his team had relocated to offices in Kirkland, Washington, where for six months they had been preparing for the ownership transition. But on the morning of March 2, a Motorola vice president named Myron Wagner walked into the workroom and said, "Craig has changed his mind. He's no longer interested in buying Iridium." And that was it. The issue seemed to be Iridium's low data speed, but he also might have gotten spooked by a $3.5 billion lawsuit filed by the bondholders against Iridium and Motorola once they saw McCaw's term sheet for the deal. No one knew for sure because McCaw didn't come down from his office. The Motorolans were sent back to Arizona.

Meanwhile, in Washington, Iridium was being dismantled. "I drew the black bean of dealing with the gateways during this time," said Brouckman. "The gateways had to go through the whole anger/grief/denial process. That was real. On the pissed-off scale, India was probably the angriest. Restructuring on the American model was very foreign to them. The Russians wanted to negotiate by holding the license hostage. After McCaw pulls out in March 2000, Motorola is calling, and we sat through about four weeks of bluster about next-gen—in other words, they would sell us the next-generation system—before we decided, 'Go fuck yourself, next-gen is none of your business.'"

Despite the bleakness of the situation, Iridium begged bankruptcy judge Cornelius Blackshear for a few more days to cull back through the twenty-one companies that had expressed interest in Iridium and see if one of them might make a qualified bid. Blackshear agreed to wait until March 15, 2000. A few potential buyers did show up, but none of them had the $10 million deposit. Two days later Motorola shut off service. Brouckman issued a public obituary and apology. "I am deeply saddened by this outcome," he said.

Gloating by Monday-morning quarterbacks was immediate and ubiquitous. Jane Zweig, an analyst at Herschel Shosteck Associates in Wheaton, Maryland, told the *Washington Post*, "It's a very challenging proposition. Today you have wireless pretty much anywhere." (Actually cell phones, at the time, worked on only 8 percent of Earth's surface.) Satellite consultant Jeff Abramson told the *Financial Times* that constellations cost billions, so "even if the satellites last another nine years, [Iridium is] just kicking the can down the road."

"They had the wrong product for the wrong market at the wrong time," said John Logsdon, Director of the Space Policy Institute at George Washington University. "It's not going to be too long before you can take a cellular phone anywhere in the world and have it work by simply throwing a switch."

Logsdon's comments were representative of the conventional wisdom—and especially galling because the universal "switch" he referred to was available only on one phone: Iridium's. Once it was gone, 90 percent of Earth would have no "switch."

Oddly enough, seven months after the bankruptcy, with the system turned off, Iridium common stock continued to trade at price levels that fluctuated between eighty-seven cents and six dollars. To put this in perspective, common stock amounted to only 13.25 percent of the company's equity, Iridium handsets were being sold as nostalgia items on eBay, and the company had issued a public announcement stating that the stock was, in effect, worthless. Yet somewhere, somehow, people were finding the phone, using the phone, falling in love with the phone—and *buying the worthless stock*. Love for the satellites ran so deep that it trumped all available evidence.

The bankruptcy court authorized another $8.3 million for Brouckman and colleagues to wind down the company. A week later, while they were loading things into boxes and shutting down facilities, a tall, mild-mannered guy named Dan Colussy called and asked if he could come by the offices and take a look around.

Sure, come on over, said Brouckman, *but the de-orbit is next Wednesday.*

Chapter 9

THE WONKS SCRAMBLE

MARCH 15, 2000
ROOM 226, EISENHOWER EXECUTIVE OFFICE BUILDING,
WASHINGTON, D.C.

Institutional panic does not resemble human panic. When governments freak out, they do so by stages and in memoranda that begin with phrases like "Unfortunately" and "I am sorry to report" and "Based on the data we now have available." They accumulate studies to justify the alarm, and then they pepper it with meetings at which functionaries gauge which part of the hysteria their agency should be responsible for, thereby to properly express anguish in proportion to the blame. The government panic over Iridium, if charted on an EKG, would resemble the Golden Gate Bridge, with huge spikes at the beginning and end of the year 2000, with periods of calm in between as the less frightened agencies relegated satellites to the cold-case file, and with lots of swaying in the wind as various cabinet secretaries switched sides, overreacted to media reports, and sent conflicting orders to their lieutenants. At the nerve center of all this activity, steady as a rock, was the resident of room 226 of the newly renamed Eisenhower Executive Office Building, directly west of the White House, the same room where Secretary Howard Taft had once run the War Department for Teddy Roosevelt. This was the domain of Dorothy Robyn.

Robyn specialized in sticky situations. She was known as a trouble-shooter and a first-class negotiator who got things done with a combination of personal charm, white papers that "sing," and a level of persistence that sometimes seemed supernatural. When word started circulating around Washington that a whole lot of satellites were about to be crashed into the ocean, she was the first to be notified in the form of a personal letter from a Motorola lawyer. She was puzzled—how did something go from technological marvel to orbital debris *that fast?*—but not overly so. The letter said that even though Craig McCaw decided not to invest, there were other bidders who might show up in a few days to rescue the constellation.

Meanwhile, the rest of the government was being alerted through a chain of events that began with a memo from an obscure bureaucrat in the State Department—Evan Bloom, Director of the Office of Ocean and Polar Affairs in the Bureau of Oceans, Environment and Science. Since Iridium was the only handheld phone that worked in the polar regions, Bloom had been the first to see the official Motorola warning, and he had placed a nervous call to Iridium headquarters in Reston, Virginia. Routed to company lawyer Tom Tuttle, Bloom was told that, yes, the destruction of the constellation had been ordered and would take about a year to complete. Bloom put that news into a memo, and a tsunami of disbelief coursed through the executive branch, eventually reaching the President himself. There were safety issues, there were commercial licensing issues, and, most important to the President, there was the issue of public pandemonium if satellites started falling out of the sky.

By March 15, 2000—the day of the bankruptcy hearing—Robyn was already ahead of the situation. Checking on the outcome of the hearing, she ended up talking to a Motorola systems engineer named Sam Fernandez, and he not only confirmed everything in Bloom's memo but made it worse: the satellites would be coming down in three days! *Surely he couldn't be serious.* Iridium had now turned into an instant crisis, so her instincts told her she had to go to the Pentagon. "When you're looking for a government reason to intervene," she would say later, "it's gotta be a national security reason." Robyn's first job in Washington had been at the Office of Technology Assessment, which sorted out complex issues for Congress, and most of those issues involved the military. She knew the Department of Defense had already spent a lot of money on Iridium, so she dialed her best friend there: Lee Buchanan.

Buchanan was a physicist—a student, in fact, of Edward Teller—who had been head of DARPA's Advanced Materials Office before becoming Assistant Secretary of the Navy. More important, Robyn considered him an intellectual soul mate, especially on the hot issue of "commercial/military integration." The two of them had bonded during a trade mission to Japan to study the ceramics industry, and she knew he would think of Iridium as a golden opportunity to adapt civilian technology to military uses.

She was wrong.

"Don't you guys use these phones?" she asked him.

Yes, said Buchanan, the military did use Iridium phones, but the feeling within the Building was that the system was worthless.

In fact, Buchanan added, the Joint Chiefs of Staff considered buying Iridium two months earlier for the equivalent of two cents on the dollar but turned it down as falling short of Pentagon requirements. The Navy was developing its own handheld satellite phone, called the "Mobile User Objective System."

"Our requirements are that the phone work through a triple-canopy jungle in the rain, and Iridium can't do that," he said.

Triple-canopy jungle? How many U.S. Navy ships ply the waterways of the Amazon rain forest? Robyn wondered to herself but didn't say anything.

"Besides," said Buchanan, "we don't want to depend on a commercial system."[33]

Surprised by Buchanan's pronouncement, Robyn thanked him for his comments and felt a sense of foreboding—this was going to be harder than she thought. Buchanan was no stovepiper—he was one of the good guys—so if *he* didn't want the system . . . She went through her old Iridium phone list, searching for leads, but all her former contacts had left the company during recent bankruptcy proceedings. *What was going on?*

It was a communications system, so surely the FCC would be involved. Robyn placed a cold call there and—bingo!—discovered there were sixteen people working on "the Iridium problem." An ad hoc crisis group had been formed, and now calls and memos were racing through the National Security Council, the FAA, the National Telecommunications and Information Administration, the White House legal office, and, oddly enough, the National Oceanic and Atmospheric Administration. As she sifted through the new information, Robyn noticed that a space policy briefing was scheduled for the next day in room 180. The meeting was for Neal Lane, the President's science advisor, so she called Lane's staff and asked if it would be possible to piggyback on that meeting to get the assembled experts' views on Iridium.

Room 180 was well known to White House staff. This was the "secret Oval Office" outfitted by President Nixon so he could escape the West Wing. It was where the Watergate tapes were transcribed by Rose Mary Woods. It was now technically a working space for the Vice President, but it was more often used for impromptu meetings because it was so spacious. When Robyn arrived there, she discovered that the National Security Council had formed a committee of scientists, lawyers, and policy wonks to monitor Russia's planned de-orbit of the Mir space station, scheduled to occur fourteen months hence. The committee members were now informed, many of them for the first time, that there was another de-orbit scheduled—a de-orbit that would happen in seventy-two hours! To say they were spooked was an understatement. Whereas the Mir would be the largest spacecraft ever to reenter the atmosphere, with a Progress M1-5 space freighter maneuvering it like a tugboat, the Iridium satellites would be tiny objects with nothing but onboard thrusters to lower the orbit before gravity took over. It would be like butterflies flipping around in a hailstorm—the "reentry scenarios" were infinite! Representatives from the Justice Department had already talked to Motorola and were somewhat shocked when the company seemed surprised by the government attention. During a six-hour meeting the Motorola representatives had continually pointed out that the Iridium system was privately owned, which the experts admitted was a unique situation, since none of them had ever been involved with a nongovernmental space system and they weren't sure of the legal implications.

All they could agree on at first was that there was no precedent—and that it was scary. Very quickly they focused on trying to make sure the satellites landed on water, not land, and all agreed that it was desirable to "increase the probability that the satellites would reenter in the Southern Hemisphere." The Southern Hemisphere had two advantages over the Northern Hemisphere: fewer people and less media interest. *Could we hit the Southern Hemisphere?* they asked the Motorola delegation. No one knew right away, so Motorola agreed to get back to the government the next day.[34]

The immediate result of the meeting was a decision simply to ask Motorola for more time. The government wanted to get an "independent hazard analysis" from NASA, and it also needed to check on Motorola's insurance. A 1972 treaty required "launching states" to pay for damages caused by space debris. That meant the United States might be liable for the fifty-five Iridium satellites that had been launched from Vandenberg Air Force Base, even though they hadn't been launched by the government. It was agreed that Motorola should be required to indemnify the government for that possibility prior to getting permission for the de-orbit.

Meanwhile, warnings went out from White House lawyers to White House staff: *Whatever you do, don't get involved with the actual bankruptcy issues. Limit everything to liability, safety, and government exposure.*

Fortunately for the future of Iridium, Dorothy Robyn decided to ignore that particular warning. She felt there must be some way to get the system out of bankruptcy. Something about this didn't smell right. She wanted to see the rest of the iceberg.

But that would take time. Motorola did agree to wait for the independent hazard analysis, and that would give her a month to search for legal loopholes. When that month ran out Motorola would still need two no-action letters—one from the Justice Department and one from the FCC—stating that the government was standing down. Quickly she mobilized a team to come up with reasons to delay those letters.

She soon found an ally in Mike Greenberger, the lead lawyer assigned to Iridium at the Justice Department, and Greenberger agreed to research "public nuisance" and pollution statutes that might be invoked. To do that he called in lawyers from the Environment and Natural Resources Division—the people who handled litigation involving the so-called Superfund sites. After all, wasn't this the same principle? Was there that much difference between leaving PCBs in the soil of downtown Camden, New Jersey, and leaving out-of-control spacecraft in the exosphere? In both cases, you made a mess—and you couldn't walk away, you had to clean up your mess first. More important, you were not the one who decided how your mess got cleaned up—the government was.

Robyn also found a kindred spirit in Kathy Brown, the FCC chief of staff. Soon the two of them were asking questions like "Do we know what foreign government satellites are at theoretical risk from colliding with Iridium satellites during a de-orbit?"—because that would be a hazardous situation that could delay Motorola. Brown conferenced in Chris Wright, the FCC's top lawyer, and he agreed to start researching additional public safety statutes. Unfortunately, he had already found an Aerospace Corporation study that looked at Motorola's de-orbit plan and pronounced it well within Pentagon guidelines—and those guidelines were, in any event, voluntary.

But media interest has a way of trumping the letter of the law, and on April 11, 2000, the *New York Times* business section featured a doomsday headline of the sort everyone had feared:

IRIDIUM, BANKRUPT, IS PLANNING A FIERY END
FOR ITS 88 SATELLITES

The number eighty-eight, like much information about Iridium, was a little misleading. There were sixty-six satellites in the working constellation, eight orbiting spares ready to replace any satellites that wore out, and fourteen dead satellites that had been launched but never made it into orbit, which meant they would return to Earth sometime between now and the twenty-second century. There were also four mass frequency simulators, better known as dummy satellites, that had been launched in Russia and China to test the ability of those countries to handle Iridium payloads. Those would crash to Earth on their own unpredictable schedule. And finally, there were four satellites in the Chandler Lab, where scientists tested software before uploading it to the constellation, and three half-built satellites stalled on the assembly line.

The *Times* article was primarily concerned with the financial woes of Iridium and Motorola, and it didn't contain anything the White House didn't already know, but the nuanced commentary was eclipsed by the words "fiery ending." The article was immediately flagged by Leslie Batchelor, counsel to Attorney General Janet Reno, who sent out a mass e-mail quoting a single sentence:

"They expect to have some parts survive [reentry] because these are pretty big satellites."

This statement was wrong. The Iridium satellites were not big at all. They were among the lightest satellites ever built. But it may have been that very factoid that saved the constellation, because the next day twenty-three people from eight agencies and departments were summoned to a meeting at the Justice Command Center of the Department of Justice Building for yet another "What the heck should we do?" meeting.

What was now being called the Iridium Interagency Working Group—the Mir space station could wait—had gathered to hear Nick Johnson, a scientist at the Johnson Space Center in Houston, present his risk assessment—and it didn't go well. According to what Johnson had cobbled together from data supplied by Motorola and Aerospace, "the probability of someone being struck by surviving Iridium debris is assessed to be 1 in 18,405 per reentry and 1 in 249 for all 74 spacecraft combined." One in 249 doesn't sound like bad odds if your opponent is dealing to an inside straight, but in the world of government liability, it's beyond horrible. (To indicate what an imprecise science this was, a memo sent out two days earlier by the Office of Science and Technology Policy had estimated the odds at 1 in 40 trillion.) So NASA had spoken—and the message was "This is bad."

Based on the questions asked by the assembled experts, the working group started to resemble the city council of Amity Island, the fictional resort town in *Jaws*:

- Could we describe Iridium as "natural" debris, since all the satellites would have eventually become derelict spacecraft anyway?
- In terms of total objects reentering the atmosphere, the constellation amounts to only 25 percent of all orbital debris created in 1999, so could we spread that over two years—2000 and 2001—and say that it's only 12.5 percent of a typical year?
- Why don't we compare it to the years 1988 and 1989 so that the whole Iridium constellation represents only 14 percent of the debris created in either of those two years?
- Should we compare ourselves to the Russians, who have had 1,400 Soyuz and Proton second and third stages reenter the atmosphere, representing 35 percent of all reentries in history?
- What about the whole Southern Hemisphere thing—can we hit the Southern Hemisphere with these? (The answer was no, the de-orbit wouldn't be that accurate.)

All the hedging was designed to avoid their greatest fear: that seventy-four satellites crashing at the same time would scare the hell out of the public. At the end of the day, Lieutenant Colonel Victor Vill-hard from the Office of Science and Technology Policy recommended that they emphasize statistics that didn't sound that bad—for example, that in 1999 there were 433 entries into the atmosphere of man-made space debris, and 84 of those objects had a mass greater than an Iridium satellite. Probably best not to mention the scarier numbers: NASA was estimating that 380 pounds of fragments from each satellite would survive reentry and hit Earth's surface at between 45 and 130 miles per hour.

As all hell broke loose in the wake of the *New York Times* article, Dorothy Robyn missed out on the initial panic because she was in Brussels, preoccupied with one of her periodic fights with "the Euros," whose latest antics involved acting like jerks in a dispute over jet-engine noise standards. The day after the Iridium meeting, she taxied in from Dulles Airport, walked into her office, and saw a note taped to the seat of her office chair: "Call Gene!"

That sounded urgent, so she hurried over to chief economic advisor Gene Sperling's office and listened to a strange story.

"What do you know about Iridium?" said Sperling.

She knew a lot about it, she told him. She was on top of it. Iridium was the first global cell phone. Iridium was such a huge leap forward in technology that it scared the bejeezus out of the FBI and caused the bureau to challenge the company's license. Unfortunately Motorola had bungled the business plan and now they wanted to crash the constellation, but the FCC and Justice Department were already coming up with ways to stop them. Robyn had the equivalent of a graduate degree in Iridium.

"Jesse Jackson wants to buy the satellites," Sperling told her.

Robyn had imagined many things that the President's top economic advisor might say in this meeting, but this was not one of them.

The civil rights leader had just returned from a trade mission to Ghana, Nigeria, and South Africa with twenty-eight American businessmen, and those three African heads of state had apparently asked for help in bringing telephone service to remote villages. At least this is what Jackson said when he managed to get the ear of the President during a recent White House event, telling Clinton he had some African American investors who wanted to take over the bankrupt Iridium system and use it to provide phone service to sub-Saharan Africa. Jackson then sent a Rainbow/PUSH Coalition letter to Bill Kennard, the first black chairman of the FCC, asking him to deny Motorola permission to destroy the satellites: *Check this situation out, get involved, see what can be done. Apparently we don't have much time.*

Robyn returned to her office, her head swimming with the policy implications. A few minutes later she was on the phone with a former FCC commissioner in private practice named Tyrone Brown, and Brown breathlessly informed her of his strange proposal. He represented "the Jesse Jackson group," he said, and he had an amazing scheme to turn commercial bankruptcy into the greatest Third World communications project in history. His backers included Bob Johnson, head of the Black Entertainment Television network, and John Malone, the Colorado media mogul who frequently bankrolled Johnson's ventures. Brown had a business plan, he told her. Brown and his people could save Iridium.

"What do you need from the White House?" she asked him.

"All I need is sixty days to raise $300 million," he said. "Can you get me sixty days?"

Robyn said that yes, she thought she could. What she didn't say was that the whole thing made her nervous, because these people were what were known at the White House as FOBs. The acronym stood for "Friends of Bill," and one thing she didn't need was the appearance of doing favors for the President's cronies. "Ty Brown and Bob Johnson are the last people you want to be close to," she recalled years later.

But this was all she had, so a few days later the FOBs were welcomed into her office. By then Ty Brown had coined the name "WorldTel" for his company, and "the Rainbow Coalition guys," as Robyn was now calling them, were there to show her just how many other FOBs were likely to pile onto this project.

WorldTel's plan for the dispossessed villages of the world was based around the concept of "teleboutiques," and they were very similar to Ed Staiano's plans for solar-powered phone booths in Chinese villages. These were kiosks outfitted with an Iridium phone and run by local "microenterprise" proprietors who would have a system of locating and informing recipients of incoming calls. Price per call would be about twenty cents per minute, which would require subsidies from governments and charities to get the consumer cost down to a dime. WorldTel, said Brown, would bring the Third World out of isolation from the world economy. In Kenya, for example, there were only 2,800 cell-phone subscribers in a nation of 32 million. In the largest African nation, Nigeria, you had 13,000 subscribers for a population of 150 million. New York City alone had more cell users than all of sub-Saharan Africa. In all these cases, the reason for lagging service was the lack of infrastructure.

Brown backed up his case with letters and testimonials from international aid organizations. C. Payne Lucas, president of Africare, had been appalled when Inmarsat charged him full price for a desperately needed phone in the Eritrean war zones of Ethiopia during the famine of 1998, and he was now throwing his full support behind WorldTel. Hamadoun I. Touré, director of the Telecommunication Development Bureau of the ITU in Geneva, had also met with Brown and pronounced himself thoroughly committed to the concept.

What both Robyn and Brown knew was that anything benefiting Africa was likely to make the President extremely happy. "'Can we help poor people in Africa?' was always a question being asked," she said, "and not just on this issue. Clinton cared about Africa. He made many trips to Africa. Africa was always a priority." More specifically, the goal of "universal dial tone" had been trumpeted by international aid organizations for three decades, and Africa—with a "teledensity" of one phone line per one hundred people—was ground zero for that battle. So everything about the WorldTel plan resonated deeply within the Clinton administration. FCC chairman Bill Kennard made regular speeches about the "global information infrastructure" and worked closely with Vice President Al Gore and others in search of worldwide "interconnectivity." Kennard, Gore, and Hillary Clinton were constantly referring to the "global village," using the term coined by Marshall McLuhan in 1960 but twisting it into a sort of social crusade for the dispossessed consumer. And the President himself had made a speech to the United Nations about people "literally disconnected from the global economy" because "more than half the world's people are two days' walk from a telephone." In India there were five hundred million people with no access to a telephone. Iridium seemed, in many ways, like a deus ex machina solution to a long-standing problem. And it wasn't just government agencies that believed this. "A phone in every village" was one of those feel-good ideas that materialized almost immediately after Motorola announced the Iridium de-orbit. Philanthropic and religious organizations got involved, online discussion groups were formed by brainy academics, and Sir Arthur Clarke himself even chimed in with a plan to convert Iridium receivers into "fixed solar powered/wind powered/ clockwork community phones" for "getting the third-world rural poor into contact with the rest of the world."[35]

Brown's expansive presentation also promised a "political victory" for African governments that were too poor to build out cellular systems. This would be "a minority enterprise of epic proportion" that would require the cooperation of the UN, the World Bank, the Export-Import Bank, the Trade and Development Agency, the African Development Bank, and, of course, the Departments of State, Commerce, Treasury, and Transportation. Brown made it clear that Jesse Jackson was especially important to the deal because they would need him to get the monopoly telephone companies in the twenty targeted African nations to make "bulk minute commitments." And it was a good time to be talking about

Africa, since President Clinton was about to sign the Trade and Develop-ment Act, which was directed squarely at strengthening trade with the forty-eight countries of the sub-Sahara. Clinton was also planning a trip to Nigeria, so Brown was pushing for some kind of "deliverable"—an announcement, a deal, an agreement in principle—for the President to take with him.

Robyn promised Brown she would try "friendly persuasion" with Motorola to give the black entrepreneurs time to put their deal together. What she didn't say was that much of what Brown said in her office that day had not only failed to wow her, but was positively frightening. All she saw were red flags, beginning with the constant references to a need for government aid. After the meeting Robyn sought a policy opinion from Will Gillespie, the industrial organization expert on the Council of Economic Advisors and, coincidentally, the resident White House science fiction buff. Gillespie's response was fairly devastating: subsidizing Iridium would cost more than subsidizing terrestrial cell towers, and given the average GNP per capita of $510, "this plan makes no sense economically as a long-term development strategy for Africa."

Still, Robyn thought it was all she had to work with at the time. "Only Iridium can deliver 'universal telephony' to sub-Saharan Africa," she told Duncan Moore of the Office of Science and Technology Policy, trying to win the memo battle. "That's why WorldTel would exist, and Jesse Jackson's involvement would give them an advantage in dealing with African leaders. The President *very much* wants to bring telephony to sub-Saharan Africa. WorldTel would be in a position to do that. This is a unique situation and has to be seen as such."

When Robyn told her boss about the meetings, Sperling responded with a quick note: "Good! Keep me posted. I'm not sure it will work, but it's worth a closer look." Two weeks later Sperling himself took a meeting with Black Entertainment Television owner Bob Johnson at the White House, and Johnson told him big money from cable TV magnate John Malone was waiting in the wings, but only if Malone could be assured of $5 million in monthly revenue. The implication was that Malone wanted the White House to bring in government aid groups that would make the project viable. And that was precisely the problem, Robyn told Sperling—the whole thing depended on handouts, not just from the United States but from a half-dozen international organizations, some of them based in Africa, all of them saddled with approval processes that took months, if not years. They needed a solution that worked *now*. They needed a market-based solution. Besides, doing something good for Africa was the kind of thing that might warm the heart of the President, but it had the potential to completely alienate the Department of Defense. "They're very antagonistic toward anything that seems 'off mission,'" she recalled. "They're constantly asked to do things that are good for the planet but nonmilitary, so there's a culture of 'stick to your knitting.' You can't be portraying this as a project that has nothing to do with defense. They had spent $200 million on an Iridium gateway, indicating *someone* thought it was necessary to the military—*that* is what we had to emphasize. Bring up villages in Africa and you could start alienating people."

For the present, all she could do was stall for time. International law was vague when it came to satellites that were owned by one country but launched by another, and this was causing underwriters longer than normal to write the insurance policy Motorola needed. Good, thought Robyn, the confusion could work to her advantage. Like a football coach preparing her game plan, she now had both a tactical offense and a tactical defense. Offense: Africa. Defense: Safety.

Sometime around the beginning of May, a new name showed up on her radar: Dan Colussy. In the course of talking to her Pentagon sources, she came across a Colonel Rick Skinner, who happened to be at Colussy's "prekindergarten" Pentagon meeting, and he excitedly related that the "Stan Kapalla/Dan Colucci bid" by Polaris Corporation had been endorsed by Iridium management and was now "backed up by a number of former government folks."

Robyn picked up the phone to find out the one thing she wanted to know from Colussy: *How quickly can you put a deal together?* Because by late April her delaying tactics were beginning to fail. "The FCC would have to stand on their heads to get Motorola to delay the de-orbiting," she was told by Mike Greenberger at the Justice Department. "They're spending a fortune to keep the satellites up, and there's not really a legal peg for the FCC or the DOJ to make them delay. There's no controlling legal authority for this. Although Motorola may *think* there is." So Robyn's strategy became: *Okay, let's let them think there is.* That might give the Rainbow Coalition time to raise the $75 million they promised Gene Sperling they could raise, plus at least $200 million more from John Malone—or maybe this new Kaluchi guy would be able to do something.

Fortunately Motorola continued to have difficulties getting its de-orbit insurance in order—in fact, the London underwriters flat out refused to write the additional third-party liability coverage—and that delay had given Colussy time to get a Castle Harlan offer before the bankruptcy court. Help came from the Pentagon as well. Unless Motorola got additional insurance, NORAD was going to refuse to share real-time data during the de-orbit, for fear of government liability. Robyn pressed the military to go one step farther: her position was that there should be no government permission to de-orbit *period* until and unless Motorola did get the insurance.

Meanwhile, Ty Brown continued to call, saying that he was going in another fund-raising direction, this time through a D.C.-based media company called Syncom. The Syncom owners were trying to bring in their powerful friends—friends like Cleveland Christophe, head of Connecticut-based TSG Ventures, an African American investment firm. Brown also pitched Iridium to Navy veteran Maurice Tosé, chairman and founder of TeleCommunications Systems in Annapolis, which had been supplying various types of data services to government agencies since 1987. But Christophe and Tosé both had the same question: Was Bob Johnson investing? Bob Johnson was like a Bill Gates or a Warren

Buffett among minority business owners, especially in the District of Columbia. If Johnson would come in on the deal, so would everyone else in the black community—but that was not to be. After Malone hesitated, Johnson phoned his friend Craig McCaw to ask him what he thought of Iridium, and McCaw told him it was "a Dumpster on a Mercedes engine." It was an odd remark, since McCaw was still hinting that he wanted to get back into the game himself, but based on the dismissive remark, Johnson decided not to invest—and so did the other black millionaires.

One day Ty Brown called Robyn to say, "I don't think it's going well. I think the Dan Colussy group is ahead of us."

Well then, she replied, *maybe you should think about merging with the Dan Colussy group, right?*

She was trying to force a shotgun wedding between the black D.C. liberals and the white Palm Beach conservatives. Like everything else in the convoluted history of the Iridium system, love for the satellites trumped race, gender, politics, conventional wisdom, and the way things had been done for a hundred years.

Chapter 10

THE TOWER

MAY 11, 2000
EXECUTIVE CONFERENCE ROOM, MOTOROLA WORLD HEADQUARTERS,
SCHAUMBURG, ILLINOIS

Scheduling a meeting with the rulers of Motorola in the year 2000 was
not unlike scheduling a meeting with the Vatican, the only difference
being that an audience with the Pope would have been theoretically possible. Chris Galvin, son of Bob Galvin, grandson of Paul Galvin, upholder
of the Galvin legacy, was known for his remoteness, so it was entirely
appropriate that he spent most of his days in that feared and revered
fortress on the Motorola campus called The Tower. He had ascended to
the leadership of Motorola in 1997 more or less by acclamation, as he'd
been groomed for the top job ever since he came out of Northwestern
University's business school in 1973, and even before that, since he'd
worked in the two-way radio division every summer since the age of
seventeen. Two-way radio sales was the equivalent of working in the
stockroom—it was the traditional bedrock of the business—and Chris
was the ultimate Good Son, a serious nose-to-the-grindstone teenager
who spent his spare time shoveling snow for money and peddling home-
made butter door-to-door. Chris Galvin had risen up through the ranks
of sales, manufacturing, and management, earning his bones in the
early eighties when he and Scott Shamlin, the company's Director of

Manufacturing, ran "Operation Bandit" at the company's pager operation in Boynton Beach, Florida, fending off a serious challenge from Japanese competition by cutting the order-to-shipment time from twenty-seven days to two hours.

But now Chris Galvin was fifty years old, the mantle had been passed, and the weight of being CEO and chairman did not sit lightly on his broad shoulders. The immigrant working-class milieu of his predecessors had gradually passed away—the elderly John Mitchell was still around, but he was known as the last of the macho managers—and the Galvin family now owned less than 3 percent of Motorola's stock. The world had changed, the cell phone had become the premier electronics product of the century, and the company had not always positioned itself correctly to take advantage. Galvin was the first nonengineer to run Motorola—and, for that matter, one of the few nonengineers ever to run any division of the company—so his instincts were about as far from creating the next dry battery eliminator or transistor or police radio as any baby boomer could be. He was, in fact, a patrician intellectual who tended to deliberate endlessly, delaying decisions and taking counsel only from a trusted few. He had left his father's management principles in place: *Leave the managers alone! Let them manage!* But now the company was struggling—not just struggling, flailing—and he wasn't sure that the old Galvin way of doing things would suffice. At any rate, he had more pressing matters than cleaning up the Iridium fiasco. When Dan Colussy called Motorola to request a meeting about Iridium, the response was cool. The company, as anyone could have told him, was preoccupied.

That's why Galvin wasn't around on May 11, 2000, when Colussy landed at O'Hare International Airport and drove twenty minutes out the Jane Addams Memorial Tollway to the sprawling Motorola campus, set amid landscaped grounds and duck ponds in the historically German village of Schaumburg. The Germans had largely dispersed by the time of Motorola's arrival in 1976, and Schaumburg was now a polyglot bedroom community full of industrial parks and shopping centers. The Motorola complex appeared placid and manicured on the outside—twelve buildings separated by green space, like a suburban community college—but it was the international nerve center of what had become a profoundly stressed-out corporation. Colussy had seen a hundred other corporate campuses just like this, places where everybody knows who has the best office in the most important building, all the right people have the coveted parking spaces, and the bosses have access to corporate jets and special dining rooms with catered lunches.

Colussy didn't like big companies. He considered them unhealthy. When companies grew larger than about $5 billion in revenue, he found, they lost flexibility and agility. (Motorola was at $31 billion.) Colussy had run big companies himself—Pan Am and what would become Air Canada—and he had spent way too much time dealing with the internal political battles endemic to large groups of ambitious people. It was bad enough in the good times, but when the company got into trouble, that infighting became vicious and self-destructive. Colussy remembered the day in 1980 when Bob David, a classmate from Harvard Business, dropped by the Pan Am Building to visit when Colussy was at the height of his power and influence, with the best executive suite, the catered lunches, and the perks of running one of America's most high-profile megacorporations.

"I guess you're doing pretty good for yourself," David had said that day.

"I'm doing all right," Colussy told him. He told him his salary, what stock options he had, and the rest of his package at Pan Am.

But David smirked. He was one of several Harvard classmates who had become venture capitalists, including Don Burr, who founded the airline People Express.

"What's your net worth?" David said.

It was the moment Colussy decided to resign. He knew David was making an accusation, and he knew the accusation was correct. Colussy had excelled in running other people's businesses, and it didn't make him feel good about himself. From that day forward he vowed never to take a job unless he also had ownership. That was twenty years earlier, and seeing the Motorola complex now, he was reminded of why he bailed out of this world.

As Colussy turned off East Algonquin Road, he wasn't sure whether he was Daniel entering the lion's den or David taking the field against Goliath. He knew that, at the age of sixty-nine, he was about to encounter the biggest challenge of a long career that he thought would have been over by now. It wasn't clear just exactly when the gas jets had been turned on, but the idea of buying Iridium had evolved beyond a what-if and become somewhat of an obsession. Mark Adams, the mysterious man from MITRE, was now calling twice a day, and he had delivered exactly what Colussy expected: a price. "I think the Pentagon will go for $3.5 million a month for three years," he called to say. That was $42 million a year in revenue for the Pentagon contract alone. With a $100 million investment guaranteed by Castle Harlan, his goal of $200 million didn't seem that far away.

Colussy was building a team. John Castle had sent in one of Castle Harlan's new hires, an Israeli named Jonathan Mark, to evaluate Iridium's assets and make an in-house recommendation. Mark had come from Boston's Bain & Company, the asset-management sister to Mitt Romney's Bain Capital, famous for buying failed companies and turning them around, or at least turning them around long enough to harvest their assets. Then there was Stan Kabala, the former Canadian telephone executive, who called almost every day to report on this or that development—new bidders for the satellites, new potential customers, new strategic partnerships—but his business plan for "Polaris" still seemed less than what it first appeared. It was focused partly on the consumer market that had already failed to support the phone, citing a Gartner Group study predicting eight million satellite phone users by the year 2004—a number that seemed crazily optimistic. Colussy was determined to do his own homework, and his leather notebook was rapidly filling up with guesstimates of cost, revenue, asset lists, and all the other data he could glean from his dozens of phone calls and his now almost daily trips to what was left of the Iridium offices in Reston, Virginia.

Foremost among Colussy's new friends was Ed Staiano. Despite losing his million-dollar Iridium investment, Staiano had retired from Motorola with so much money that he now shuttled among four homes—a villa in southern Italy, a ranch in Arizona, a house in the upscale Chicago suburb of Rolling Meadows, and a condo in Alexandria, Virginia—and he made it clear early on that he wanted his Iridium job back. If Colussy and Castle Harlan managed to buy Iridium, he wanted to run the company, and he didn't even want a salary; stock would be fine with him. He believed in the system that much. And Colussy was actually grateful for that offer. He certainly didn't want to run the company himself, and Staiano was the kind of tough guy who often succeeded with a turnaround when more timid souls might fail.

Staiano had already stormed out of a couple of meetings with Colussy and Castle, a pattern that would be repeated several times.

"You don't understand!" Staiano would insist. "I know these guys! Motorola will never go for that!"

"Well, they have to go for it, because we're small and we can't do anything else," Colussy would say.

"You're gonna embarrass us!"

"We can't do that, Ed. Iridium is in bankruptcy. All the rules change in a bankruptcy."

"Okay, that's it, I'm finished," Staiano would bark. "I want no part of this plan. I'm outta here."

And Staiano would scoop up his papers and his laptop and angrily storm out of the room.

The next day Colussy would call him and say, "Ed, we're having a meeting. Do you wanna come?"

"Go ahead and include me," he would say.

So that was Staiano, an almost classic Motorola manager in the confrontational system that was gradually giving way to a kinder, gentler company under Chris Galvin. "The Motorola style," said Randy Brouckman, "was to punch the other guy in the nose and see if he got up."

That's another reason Colussy thought this was going to be a tough meeting. Privately he thought Motorola was a little out of control. He was suspicious of companies that loaded up their executives with expensive perks and stock options and country-club memberships and fancy offices, even though in his private life Colussy himself loved expensive cars and fine food and wine and belonged to several private clubs. The difference between corporate perks and private ones was that you couldn't make good business decisions when you didn't really know how much things cost. Motorola was one of those companies that added everything to the bottom line and charged it back to clients. All those trips on the corporate plane eventually showed up in the cost of the microprocessor, or the transponder, or the satellite. In his briefcase Colussy had documents showing that Motorola carried the cost of its Chandler engineers on the books at $375,000 per person per year. Yet he knew that the average engineer's salary was just $70,000. The other 80 percent consisted of Motorola peripheral charges—embarrassing overhead even by the standards of a Hollywood movie studio, much less what was being ballyhooed as one of the most efficient corporations in American business history.

When Colussy drove onto the Motorola campus, he was less than twenty-four hours away from what the creditors said was the final deadline. He had one day to come up with a solution to Iridium or else the banks would move in, the satellites would come down, and the lawsuits would begin. Arrayed on Motorola's side of the conference table were Rick Severns, a bespectacled accountant type who was the Chief Financial Officer for the telephone division of Motorola; Ted Schaffner, the pale Germanic Executive Vice President regarded as the company's troubleshooter or "special-purpose guy"; Sidney Cruz, an administrative assistant who was there to record the proceedings; and Ron Taylor, a tall, athletic military type in his forties who headed Motorola's government facilities in Arizona, which were developing the "secure module" the Pentagon wanted so badly. The ghost presence at the meeting was Keith Bane, the Executive Vice President for Global Strategy and Corporate Development. Colussy had been told that Bane was the gatekeeper; he was the one who would report back to Chris Galvin and tell him whether Colussy passed muster or not.

Arrayed on the Colussy side of the conference table was his usual contingent: himself.

Colussy started with some brief background—people always seemed impressed by the fact that he had run Pan Am, even though his years at UNC were much more pertinent to the situation—and then he emphasized the financial strength of Castle Harlan and John Castle's personal assurances. He told them a little about his plans for running the company as well, but he could see by their reaction that nothing was happening. They were not being dazzled. They were actually downright skeptical, especially Ron Taylor. *Was Colussy aware that the system had no broadband capacity? Was Colussy aware of what Motorola was spending per month to simply keep the satellites in orbit? Was Colussy aware of the eighteen other partners who had a stake in Iridium beyond Motorola?*

"Surely you're not thinking of using this for *data!*" sneered Taylor.

And yes, Colussy was aware of all this. He didn't say so at the time, but he thought the data capacity was fast enough for asset-tracking applications. And he was certainly aware of Iridium's vast "gateway" system, which he considered a mess in thirteen languages. Since Motorola was a minority partner—albeit the most important minority partner—there was only one reason it had any control at all over a company that had already declared bankruptcy: Motorola operated the SNOC. The Satellite Network Operations Center, nestled in a fifty-thousand-square-foot building on ten bucolic acres of the Lansdowne Office Park in Leesburg, Virginia, was where a hundred technicians kept an around-the-clock watch on the constellation, nudging birds back into place when they strayed off orbit, sending up software fixes when a communications hiccup occurred. Iridium had a contract to pay Motorola $45 million a month to operate the SNOC, but since those payments had ceased nine months earlier, Motorola executives felt they had the legal right to shut everything down, which would mean, of course, destroying the satellites. They had withheld that decision as long as Craig McCaw was talking about buying the company, but they were not willing to do the same for Colussy. How could Motorola be sure that, if they turned the SNOC over to Colussy, he would even be able to afford to keep it running?

Colussy didn't say so, because he didn't want a Staiano-type reaction, but he had already done research indicating he could run the SNOC for $4 million a month, not the $45 million Motorola claimed to be spending. It was such a ridiculously low figure that he was sure Motorola would scoff at it. Staiano already had. But as he parried the various objections of the Motorola executives, he started to feel better. They weren't kicking him out of the building. Their resistance was diminishing. Someone broached the idea that Colussy's investment group should pay $5 million a month in return for Motorola's agreement not to shut down the SNOC during the due diligence period—and Colussy did the same thing he always did when someone mentioned a cost that he never intended to pay. He said, "Yes, I understand," in order to change the subject.

There were other things that Motorola wanted—$100 million for this, $50 million for that—and Colussy dutifully wrote down every single number, knowing that any one of them could derail his company completely. After listening to everything the executives had to say, he asked his own questions: Could they deliver the security module the Pentagon wanted so badly? Yes, they could, in about five months. Could they transfer the government contract to him? Yes, they could. Could they give him the handsets and pagers they had already manufactured? Yes, they could, although those would be the last pagers he would ever receive, since the manufacturing facility in Boynton Beach, Florida, had been shut down and all the toolmaking equipment had been destroyed. How many replacement satellites were ready for launch? Four. How many rockets to launch the satellites had been built and paid for? Most of them. Could he have the intellectual property—the patents? He could license it for a small fee, they said. (A small fee to Motorola always began at seven figures.) Could he have the operating facilities in Virginia and Arizona? Yes, although Motorola wanted to continue to operate the Pentagon facility in Hawaii. Would they support all his appeals to the FCC? Yes, they would.

The only time the meeting came to an impasse was when they talked about the de-orbit. Colussy said there would be no de-orbit because his company would continue to operate the constellation. *Yes, yes,* they said, *we understand, but let's suppose your company fails anyway, and the satellites have to come down. Motorola doesn't want to get sued for space junk injuries later.* Colussy said that by that time the company would belong to a different corporation so he would absorb all the lawsuits. *Yes, yes,* they said, *but people don't always sue the person they should sue, and they might come after Motorola for building the satellites in the first place.* Colussy said he understood that, and he was already on top of it. He was willing to buy $2 billion worth of insurance against that possibility. They instantly said that wasn't enough, indicating they had calculated the number themselves. They weren't specific about who might sue them for more than $2 billion, but they said the deal was impossible without "complete indemnification." Insurance wasn't enough. They wanted to know that, once they turned over the keys to the satellites, they would never have to worry about them again.

And that's how both sides left it. Colussy thought it was a good meeting. If their main concern was insurance, things could be worked out. And from outright skepticism at the beginning of the meeting, they had closed with a little warmth, or as much warmth as Motorolans were capable of. Colussy left the building and went back to O'Hare and suddenly wondered: What happened to the deadline? Supposedly he had to have a deal by the next day or else everything was going away. What went away . . . was the deadline. It was never even mentioned.

Colussy would have been even more amazed by that "little warmth" he felt in The Tower that day if he had known the pressure those executives were under. It is said that civilizations look most solid—their rituals in place, their hierarchy established, their borders well defined—right before they collapse, and perhaps that's what was happening at Motorola in May 2000. The company was at the end of three decades of world domination. After a fivefold increase in market value in the 1980s, tripling its earnings and sales, Bob Galvin had handed over the chairmanship to George Fisher in 1990, and Fisher had doubled the size of the company yet again. By the end of 1994, Motorola controlled 60 percent of the cellphone market worldwide—not just the handsets, but the construction of cell towers and wireless ground stations as well. But that's when Fisher abruptly left to run the troubled Eastman Kodak Corporation. Fisher told friends at the time that he saw no reason to wait around to be replaced by Chris Galvin, and that seemed to be where all indicators were pointing.

The board didn't act immediately, though. Lifelong Motorola executive Gary Tooker, very loyal to the Galvin family, was brought in from the semiconductor division to serve as CEO, and he let all the division heads run the show, with pretty good results. Revenues were up 31 percent during Tooker's three-year tenure, 1994 to 1997, and profits were up a remarkable 53 percent. Tooker knew he was a placeholder executive, though, and the inevitable finally happened in January 1997, when the board named the forty-seven-year-old Chris Galvin to the CEO post, the culmination of a thirty-year grooming process. Galvin's first message to the troops was that Motorola was entering a new phase. The company was too engineering-driven. There was too much internal rivalry. One of his first actions was to change compensation schemes so that everyone was rewarded according to company-wide performance instead of performance by division. (Ironically, he was going against some of the principles of Six Sigma when he did this. By the mid-nineties, the leading cheerleader for Six Sigma was not anyone at Motorola, but Jack Welch, head of General Electric.) *We will all get along*, Galvin was saying. *We will not punch each other in our respective noses any longer.*

The downfall of Motorola, and Chris Galvin, began with the most popular phone the company ever created: the StarTAC. Modeled after the communicator carried by Captain Kirk in *Star Trek*, it was the world's first flip phone, the lightest phone ever mass-produced at 3.1 ounces, and a product that flew off the shelves as soon as it hit the market in 1996, eventually selling sixty million handsets despite being initially priced at $1,500. The StarTAC was a status symbol, a frequent prop in big-budget Hollywood movies, and seemed to be a continuation of the Motorola dominance of the market that had begun with the DynaTAC in 1983. The StarTAC had only one problem: it was analog. Even that wouldn't have mattered were it not for the fact that the Finnish electronics company Nokia introduced the world's first digital phone within months of StarTAC hitting the street. In 1997 it was far from certain that the whole world would go digital, even though there were three features that digital phones promised that analog phones could never match. Those were call forwarding, voice mail, and texting (eliminating the need for pagers). Robert N. Weisshappel, head of Motorola's cellular phone division, thought those were pretty lame reasons to prefer digital over analog, especially since Nokia couldn't match Motorola's miniaturization. What people wanted, Weisshappel told the troops, was a small phone—the smaller the better—and a phone that was "cool." The StarTAC was smaller than a pack of cigarettes and so cool that every teenage girl in America was begging her parents to have one.

Weisshappel, in fact, was so confident the StarTAC would soon dominate the world that he started squeezing service providers. His "Signature Program" was created to put distributors on notice that they wouldn't be allowed to carry the StarTAC in their stores unless at least 75 percent of their product line came from Motorola, and they would also have to support the $1,500 StarTAC price point with stand-alone displays. At a famous meeting in Bedminster, New Jersey, an executive at Bell Atlantic listened to Weisshappel's terms and sarcastically remarked, "Do you mean to tell me that you don't want to sell the StarTAC in Manhattan?" Executives at GTE and BellSouth also refused Motorola's terms and increasingly turned to other companies. The result: digital took off, without Motorola. Weisshappel, with his explosive temper and his dramatic presentations, famously yelled, "Forty-three million analog customers can't be wrong!" at a Motorola sales meeting. But his intractable position eventually doomed the company's cell-phone business. From 60 percent of the wireless market in 1994, Motorola's share had fallen to 34 percent in 1998, the first full year of Galvin's chairmanship. The company tried to counterattack Nokia in Europe with the Shark phone in 1999, but that initiative failed miserably as Motorola lost even more market share to both the Finns and the Germans, in the form of a new digital phone introduced by Siemens.

The wireless phone division wasn't the only part of Motorola that was suffering. In the summer of 1997, one of Motorola's oldest friends— Illinois Bell, which now did business as Ameritech—told the company it couldn't wait any longer for Motorola to catch up. Ameritech was switching all its wireless phone business from Motorola to Qualcomm. Then, in March 1998, Motorola lost a $500 million contract with Bell Atlantic's wireless division, PrimeCo, this time over a quality issue. Meanwhile, Motorola's equipment business, building earth stations and cell towers, rapidly lost ground to Lucent Technologies and Nortel Networks, mainly because Motorola never managed to develop an integrated switch of its own. After 54 percent growth under Fisher and 27 percent growth under Tooker, Motorola shrank under Chris Galvin to the point that Nokia passed the company as the wireless leader in 1998 and by 2000 had reduced Motorola's market share to only 14 percent. Motorola eventually made the switch to digital—long after Weisshappel had parted ways with the company—but it had lost almost three years to Nokia. On May 10, 2000, the day before the Colussy meeting, Motorola stock dropped 16 percent, completing a 50 percent free fall that had begun in March. The reasons: it didn't have enough cell parts to keep up with Nokia, and it had lost yet another contract, this time with longtime customer British Telecom, which switched to Nortel Networks. No wonder Chris Galvin wasn't at the meeting. His world was falling apart.

The day after the meeting in The Tower, Keith Bane called Colussy. He was the executive reporting directly to Galvin, and obviously he had a favorable briefing from Severns and Schaffner. "There are pockets of love for Iridium at Motorola," said Bane, "as well as pockets of hate." *It's like this,* he said: *make sure I don't get sued for falling satellites, and we can work this thing out.* Of course, he said, he would want "compensatory pricing."

"Yes, I understand," said Colussy.

Bane wanted Motorola to get some cash from the deal and to get some equity in the new company, and he said, "Ed Staiano is a close personal friend." He wanted Colussy to know that Staiano's involvement would be looked upon favorably by Motorola. Which was fine with Colussy—he liked Staiano and would be happy to use him, no matter how many times he quit.

Colussy had never met Bane and didn't know much about him other than that he was a die-hard Motorola lifer ("The sun never sets on Motorola," he was once heard to say), but he would have been intrigued by two items on Bane's résumé. Besides Staiano, Bane had another close personal friend: Craig McCaw. The two of them had put together a joint venture to rescue the wireless company Nextel Communications in 1995, and that was one of Motorola's few successful initiatives of the

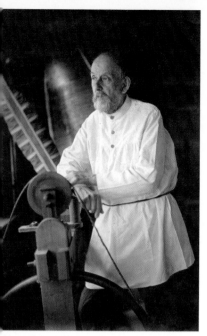

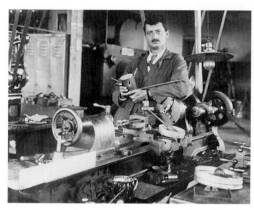

Transylvanian Hermann Oberth was rejected by the University of Göttingen as a "romantic futurist" when he proposed the multistage rocket, but found his calling on the set of the first outer-space movie.

e original satellite geek was Konstantin olkovsky, a half-deaf high school math cher regarded as a crackpot in his little n on the Russian steppe: his calculations de orbital vehicles possible.

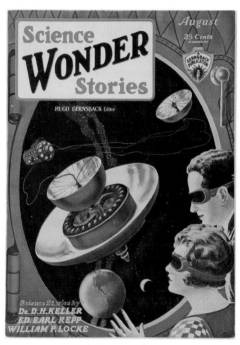

man Potočnik, a Slovene engineer, wrote *Problem of Space Travel: The Rocket Motor* couldn't get it published in Europe.

Potočnik's study was serialized under a pseudonym in the American pulp magazine *Science Wonder Stories*.

Nazi engineers painted the logo of the first outer-space movie on the fuselage of the feared V-2 "Vengeance" missile that terrorized London from the Peenemünde Army Research Center in the Baltic.

On May 2, 1945, the Nazi officer corps of Peenemünde—chief among them Wernh von Braun (with broken arm) and Genera Walter Dornberger (to his right)—posed f this picture after surrendering to General Patton's Army in the Alpine village of Oberjoch. On that day the history of satell was forever altered.

The first rockets launched at the White Sands and Atlantic missile ranges in the 1940s and 1950s were captured V-2s, the prototypes for America's ICBM system.

Arthur C. Clarke, proud member of the British Interplanetary Society, first calculated the height of the geosynchronous orbit, known today as the Clarke orbit, but was vetoed as the recipient of the first Iridium call even though he lived in exotic Sri Lanka.

As head of the Army's Redstone Arsenal in Alabama, von Braun schemed throughout the fifties to launch intergalactic space vehicles, befriending Walt Disney in his quest, and was disgusted when the satellite program was awarded to civilians.

ССР
ПРОЛЕТАРИИ ВСЕХ СТРАН, СОЕДИНЯЙТЕСЬ!
КОМСОМОЛЬСКАЯ
ПРАВДА
н Центрального Комитета ВЛКСМ

...vda gloated about Sputnik only after the ...viets realized they had scored a propaganda ...up in the west.

Vanguard, the first American satellite, traveled four feet before exploding, prompting the Soviets to offer the U.S. participation in its program of technical assistance for backward nations.

...international press was brutal after the failed ...ch on December 6, 1957.

Von Braun (right) and the Army triumphantly celebrate the launch of Explorer I, the first American satellite to reach space, in January 1958. With von Braun are William H. Pickering (left), Director of the Jet Propulsion Laboratory, and designer James Van Allen of the University of Iowa.

Children were encouraged to build scale models of America's first civilian satellite, Vanguard, but sales were less than brisk.

Communications satellites were promoted as ecumenical and peaceful to make sure nations didn't try to claim ownership of outer-space air space.

The first true communications satellite was Echo, a voluminous "satelloon" launched by NASA and Bell Labs in 1960.

When President Reagan announced the Strateg Defense Initiative in March 1983, it was deride as science fantasy.

President Kennedy rammed the Communications Satellite Act of 1962 through Congress as a way of keeping AT&T from monopolizing outer space.

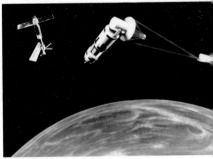

An Air Force rendering of the lethal Star War kill vehicles in 1984 looks eerily similar to the Iridium satellites that would be built a decade later by the Motorola engineers who worked unmentionable" programs within the Strate Defense Initiative.

ston Churchill used a Handie-Talkie at
t Jackson, South Carolina, during a secret
sion to meet with President Roosevelt in
e 1942. Later in the war Motorola would
oduce the Walkie-Talkie, which revolutionized
lefield communications and led to decades of
tagon contracts.

Motorola always took the lead in miniaturization,
beginning with the mass production of
transistors and portable radios.

e ancestor of the cell phone was the car
ne, which Motorola introduced in 1946,
it had a very limited customer base.

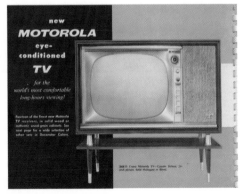

Motorola was late to the television game but
barged into the wild markets of the 1950s and
1960s with price points $100 cheaper than RCA.

John Mitchell, the powerful Motorola executive who drove research for the car phone, cell phone, and Iridium, tested "the Brick" on the streets of New York as early as 1973.

Bill Weisz, Bob Galvin, and John Mitchell comprised the three-man team that forged Motorola into the "American Samurai," the only U.S. electronics company that could take on the Japanese and win.

In 1996 Motorola introduced the StarTAC, the first flip phone and the lightest phone ever mass-produced. Ironically, it was the beginning of Motorola's decline—because it was analog.

Dan Colussy, the eventual savior of Iridium, was married to his wife Helene at the Coast Guard Academy chapel the day after his graduation in June 1953.

Lieutenant Junior Grade Dan Colussy and a shipmate at Yokosuka Naval Base before returning to their Korean War patrol in the loneliest, coldest, stormiest region of the Pacific Ocean.

...ussy started his civilian career testing ...raft engines at General Electric's ...ndary River Works plant in Lynn, ...ssachusetts.

Colussy was expected to meet with heads of state angling for the prestige of Pan American World Airways service, including Italian Prime Minister Francesco Cossiga (second from right), who bestowed honorary knighthood on the Pan Am President in 1978.

Three engineers at the Chandler Lab—Barry Bertiger, Ray Leopold, and Ken Peterson—worked out the final algorithms for the Iridium system on the whiteboard in the Motorola parking lot.

The herringbone arrangement of six planes of satellites that race up one side of the planet and down the other, acting as interconnected switchboards in space, was acclaimed as one of the most elegant engineering systems ever conceived.

The strange phenomenon of "Iridium flar was believed to be either a meteor or a UF when first observed in 1997 but is actually the sun's rays reflecting off the silver-coate antenna, creating a streak in the sky fifty times brighter than Venus.

polar launches in the United
[Sta]tes must be from California's
[Van]denberg Air Force Base, for
[saf]ety reasons, which is how
[Do]nnie Stamp ended up there,
[eve]n though he considered the
[Del]ta II inferior to Russian and
[Ch]inese rockets.

The humongous Russian-made
Proton, Stamp's first choice,
was the only launch vehicle
large enough to put seven
Iridium satellites at a time into
orbit.

The final five spacecraft,
completing the constellation of
sixty-six working satellites plus
six spares in a "storage orbit,"
lifted off from Vandenberg on
May 17, 1998, breaking every
record in rocket science—the
successful launch of seventy-two
satellites on fifteen rockets in
377 days.

[Th]e actual launch pad is pretty much the only part of China's Taiyuan Satellite Launch Center visible
[to t]he outside world, and Stamp's employees were the first Americans to stay there overnight—
[som]ething they realized only after the doors to their sleeping quarters were chained shut.

When the momentum wheels on the satellites started to fail, mission control technicians at the SNOC used the hydrazine thruster to fight the solar winds.

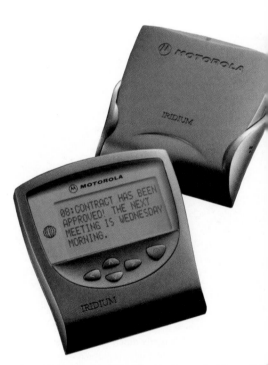

The Motorola handset for Iridium was so thick and heavy that one observer called it "a brick with a baguette sticking out of it."

The Iridium worldwide paging device worked beautifull for the one year before consumer pagers became obsole

The globe-hopping Ed Staiano ran Iridium from a Gulfstream jet, dubbed "Air Ed" by employees, who lived in fear of his "shock and awe" management style.

flamboyant Mauro Sentinelli was hired ndle Iridium's worldwide marketing, dumped his stock and fled back to ie in 1999, where he was knighted, red at Telecom Italia, and awarded the est pension in Italian history.

Cornelius Blackshear was the tough New York bankruptcy judge who jawboned Motorola into postponing the destruction of the satellites.

le billionaire Craig McCaw, pioneer e cell phone revolution, played ie with Motorola, the Pentagon, and ssy for more than a year, suggesting anted to buy Iridium and fold it into uristic Internet-in-the-sky scheme as developing with Bill Gates.

DOCTOR FUN   12 May 2000

Copyright © 2000 David Farley, d-farley@metalab.unc.edu
http://metalab.unc.edu/Dave/drfun.html
This cartoon is made available on the Internet for personal viewing only. Opinions expressed herein are solely those of the author.

"I bought one of them I-ridium Satellites on E-bay!"

Editorial cartoonists had a field day during the summer and fall of 1999, as the government prepared for the simultaneous crashing to Earth of ninety-two satellites.

Merchant banker John Castle promised $100 million for the Iridium rescue effort, then got cold feet, leading to public embarrassment for Colussy and panic at the White House as Motorola's "de-orbit" plans commenced.

The Iridium rescue plan began in the room where President Kennedy wrote *Profiles in Courage*, in a Palm Beach vacation home that Castle had restored to look exactly as it had in the 1960s.

Prince Khalid, one of the most successful businessmen Saudi Arabia, such a private man that none of the Iridium partners ever him, much les met him.

Civil rights leader Jesse Jackson (right) worked with his confidant, former FCC Commissioner Ty Brown, to float a plan to use the satellites for phone service in sub-Saharan Africa. The scheme didn't work out, but the belief in it may have been what saved the system.

Terry Jones (left) an Herb Wilkins were principals in Sync the firm that ende up sponsoring the sub-Saharan plan, knowing that help Africa was a cause dear to President Clinton's heart. W Colussy lost his funding, Wilkins t colleagues, "You're gonna believe this for the first time i life, I'm gonna ba white man."

... Dan Colussy's home overlooking the Severn River, near the Naval Academy in Maryland ...

ere were three war rooms during the tle over Iridium ... The Tower—red and revered centerpiece of torola's headquarters campus in aumburg, Illinois ...

Chris Galvin, the third-generation Motorola CEO, tried to convince the White House that the satellites should be put out of their misery.

, and the venerable Old Executive Office Building, ere presidential advisor Dorothy Robyn encouraged e Rainbow Coalition group backed by Jesse Jackson and rican American media mogul Bob Johnson.

Dorothy Robyn, here briefing President Clinton aboard Air Force One, decided to throw the full weight of the White House against Motorola's single-minded determination to crash the satellites.

Rudy de Leon was the wily Deputy Secretary who "took the temperature of The Building" before orchestrating a Pentagon showdown with Motorola.

Dave Oliver, number five on the civilian side at the Pentagon, was known as the master of the dirty job. Secretary Cohen told him to fix the Iridium problem.

Secretary of Defense Bill Cohen always listened to his wife, Janet Langhart Cohen, a television diva who told him to meet with her boss—Bob Johnson, owner of the Black Entertainment Television network.

By 2006 Senator Stevens had become a staunch supporter of Iridium, appearing with Colussy at the ribbon cutting for a new Alaskan ground station, along with Alaska Governor Frank Murkowski (second from left) and Air Force General Joe Ralston (far right).

...en Alaskan Senator Ted Stevens opposed ...ussy's deal to save Iridium, he soon felt ...wrath of bush pilots, wilderness guides, ...imos, highway patrolmen, and stevedores in ...Chukchi Sea.

...fierce opposition of California Congressman ...y Lewis resulted in what one Pentagon official ...d "a Barnum and Bailey show" on Capitol Hill.

By 2015 the eighty-four-year-old Dan Colussy was "just a shareholder" and had retired to a life of golf and dinner parties in Jupiter, Florida—when he wasn't flying to the board meetings of three "green" technology corporations he continued to invest in and work with.

Matt Desch took the helm of Iridium in 2006 and not only made the next generation of satellites possible but formed a consortium that would revolutionize air traffic control.

Thanks to the enthusiasm of the French aerospace industry—including some of the same men who once tried to sabotage Iridium—the second-generation "Iridium NEXT" system will be built in Toulouse, then launched on Falcon 9 rockets provided by Elon Musk's SpaceX. The satellites will fly into perpetuity.

Anyone who finds himself alone at the ends of the earth, like U.S. Marine Corporal Sean McMullen, still has one friend—an Iridium phone.

late nineties, mainly because it was digital. The other résumé item was more metaphorical: Keith Bane had started the Vietnam War.

As the gunnery officer on USS *Maddox*, Bane engaged three North Vietnamese torpedo boats in the Gulf of Tonkin on August 2, 1964, firing his five-inch guns 280 times, much to the chagrin of Lyndon Johnson. As Colussy would soon find out, Keith Bane wasn't a shy man, and he was not afraid to pull the trigger.

Chapter 11

ET TU?

JUNE 8, 2000
DRY BED OF LAKE FUCINO,
AVEZZANO, ITALY

Dan Colussy was a man who rarely raised his voice—when criticizing someone he tended to use a quizzical tone, as in "Ed, that can't be right"—but throughout the summer of 2000 he would spend many of his days getting yelled at as he encountered all the angry, skeptical, pessimistic souls who believed Iridium was a scrap heap of broken promises and outdated technology. A few days after his meeting in The Tower, Colussy arranged for a teleconference to tell the Iridium board of directors—most of them gateway owners—what his plans were. He spoke for about twenty minutes, outlining his preliminary conversations with Motorola, the Pentagon, and Castle Harlan, and then he put on his crash helmet and bunkered in for the sharp questioning and outraged rants he knew would be coming. He was putting himself directly in the line of fire, but he was doing it for a reason. Colussy prided himself on being a peacemaker, the kind of guy who could calm nerves and make everyone come together for the common good. He would offer them stock—maybe 1 percent of the company for each of the thirteen gateways. The gateways would come around—he was sure of it.

Colussy thought that his toughest meetings would be at the Pentagon and Motorola, but now he knew those were just the most obvious ones, and he would be facing a bottomless chasm of aggrieved parties held over from the days of Ed Staiano. There were six dozen more entities that had to be appeased, enticed, or strong-armed if this deal was ever going to get done, and chief among them were the business leaders around the world who had paid Motorola an average of $100 million each to build call-processing centers that now stood idle, like shiny new refrigerators in a village with no electricity. Back in the early nineties John Mitchell had used the gateway system to sell the investment, ending up with fifteen gateway companies and eleven physical earth stations located in Rio de Janeiro, Moscow, Tempe, Beijing, Mumbai, Fucino, Seoul, Bangkok, Nagano, Taipei, and Jeddah, with construction plans for ten more gateways in Montreal, Mexico City, Düsseldorf, Jakarta, Agüimes, Alice Springs, Buenos Aires, Santiago, Caracas, and Novosibirsk. Colussy didn't really need any of these partners to operate the system, but he considered them potential investors—perhaps they would double down instead of simply writing off their gamble—and in many cases they controlled landing rights, which were the licenses allowing you to operate the phone in those countries. Theoretically you could carry the phone into a country and operate it anyway, but without landing rights you wouldn't be able to sell it there.

One advantage of having Ed Staiano on his team was that Staiano had personally met with every partner in every gateway, and he knew all their idiosyncrasies. This is not to say he had influence over them. A group of Indian banks and insurance companies, known collectively as Iridium India Telecom Ltd., were so incensed that they would sue Motorola on three continents for the next decade and end up arguing before the Indian Supreme Court that Motorola, though a corporation, should be classified under criminal law as "a person with a guilty mind"—and the court agreed! Then there was Aburizal Bakrie, a floppy-eared Indonesian industrialist in monk's glasses who happened to be President of the Chamber of Commerce of the world's fourth most populous nation and a personal friend of President Suharto's. Bakrie had bought licenses for half the South Pacific, including Australia and New Zealand, and now considered himself fleeced by the Iridium deal—and he was powerful enough to turn the Indonesian government against the system, leading to denunciations of Motorola in the national parliament. In Moscow, Taipei, Bangkok, and Hong Kong, the story was much the same, so that by the time Colussy got an audience before the raucous Iridium board, his remarks about how he intended to buy the company out of bankruptcy and turn it around were all but drowned out by largely irrelevant rants by the twenty-eight members, which included everyone from sophisticated telecom executives like Herbert Brenke, chairman of the German wireless company E-Plus Mobilfunk, to bureaucrats like Anthony Kiselev, director of the Khrunichev rocket-building plant in Moscow, and, on the back bench of the board, industry neophytes like Alberto Finol, a Venezuelan dairy magnate. Everyone who had owned a gateway was now fuming, threatening, demanding explanations. The Indians were madder than the Thais but not so mad as the Koreans, who were in turn only slightly more upset than the Australians and Brazilians. All over the world prominent businessmen felt humiliated by the bankruptcy of Iridium.

Thanks to John Castle, who had a long history of dealing with Arab investors, the Saudis were quickly identified as the most likely to invest in a new version of Iridium, and for a strange reason—Castle thought they were embarrassed by the bankruptcy. There's no term in Saudi Arabian culture for a "reorganization bankruptcy," and in fact the penalty for bankruptcy in Saudi Arabia is imprisonment for fifteen years or until all debts are paid, whichever comes first. Saudis, in other words, don't go bankrupt, ever.

Two Saudi companies had run the Africa and Middle East franchises out of a gateway facility in Jeddah, Saudi Arabia. The principal investor in that operation was Mawarid Overseas Company Ltd., owned by Prince

Khalid, best known for the colorful pink-and-green racing silks used by Juddmonte Farms, his Thoroughbred racing operation in England, Ireland, and the United States. "The Prince"—brother-in-law of King Fahd and himself descended from one of the most revered family trees in Saudi Arabia—was sixty-three years old and had technically turned his Mawarid Holding Company over to his four sons to manage, but his vast empire of companies dealing in cement, television, chemicals, industrial pipe, catering, and insurance was cordoned off by a small army of lawyers, consultants, and aides. Prince Khalid had made his fortune by partnering with big American companies like Browning-Ferris and American Express, companies he trusted implicitly because of their size and prestige. That's why the Iridium investment was something of a sore point. In 1993 John Mitchell and Motorola CEO George Fisher had made a personal presentation to the Prince's eldest son, Prince Fahd, and the family had gotten into Iridium on the equivalent of a handshake offered by two highly respected businessmen. Now John Castle wanted to convince Fahd to reinvest despite the family's heavy losses—but there was bound to be more scrutiny this time. In early June Castle invited the Prince's chief of staff to a meeting at Castle Harlan headquarters in New York and, in his usual theatrical manner, took the Saudi representatives to an ornate conference room lined with paintings of Castle's own Thoroughbred horses. He then ceremoniously dialed the captain of the *Marianne* on an Iridium phone as his yacht approached the Pitcairn Islands, the archipelago famously inhabited by the survivors of HMS *Bounty,* and told them that, in his opinion, Iridium was a technological marvel that would someday find its market. The Saudis were noncommittal about the investment, but Castle could tell they were intrigued—he told Colussy to stand by.

The other part of the Iridium franchise—the gateway serving the Middle East—would be a tougher nut to crack. It was owned by the Saudi Binladin Group (SBG), founded by Mohammed bin Laden, the one-eyed bricklayer who befriended King Ibn Sa'ud during the Great Depression and became the "royal builder." During the forties, fifties, and sixties, bin

Laden married twenty-two times and sired fifty-four children, including the famous terrorist, while building the nation's largest construction company, responsible for all mosques and holy sites including Mecca and Medina. After Mohammed bin Laden's death in an airplane accident in 1967, SBG had diversified away from construction and now had international investments in manufacturing, engineering, urban development, and telecommunications, including chunks of stock in Microsoft and Boeing. Colussy considered both Saudi groups strong prospects for an investment, and for a fundamental reason. With Saudis, losing money is not the worst sin; losing face is. Years earlier, Colussy had spent several months working out an agreement for Pan Am 747 service between New York City and Dhahran, and during that process he had learned some of the peculiarities of the Saudi business system. He knew that, for Saudis, reputation is everything.

Colussy would travel thousands of miles during the summer of 2000, trying to solve the "gateway problem" along with every other smoldering issue left by the absconding Motorolans, and in many cases the former partners and clients of Iridium attacked Colussy simply because they'd not yet had a chance to vent against *anyone*. Colussy was not unsympathetic to their plight. He'd once had a huge public business failure himself when his start-up regional airline, Columbia Air, announced short-haul service on the East Coast, then was abruptly put out of business when President Reagan fired all the air traffic controllers in August 1981, causing the FAA to revoke the new airline's assigned "slots." Columbia had been backed by prominent politicians in Baltimore, who expected it to become the anchor airline at Baltimore-Washington International Airport, so when it was shut down on the brink of its first flight, it became such an embarrassment to the city that Colussy was shunned by old friends and ridiculed by public officials. So he knew what it felt like to announce big plans and see them collapse, but he also knew that you couldn't start building anything new until you'd fully realized the losses of the past and counted them as losses. The more reasonable gateway owners could still be potential sources of capital.

Unfortunately, that was the kind of problem usually worked out over time, and Colussy had no time. Once he realized the complexity of Iridium's business structure, he apologized to his wife, telling her that, yes, he was going to come out of retirement because this deal was going to be all-consuming, but he *promised* he would find someone else to run the company. He wasn't going to be a CEO at his age. Next, he told his friend John Castle that he sensed the banks and gateway owners were all sick of Iridium, and he thought they could get the sale price down to something more manageable than what they originally thought, maybe something as low as $60 million. Castle had promised $100 million from his own bank, but that number now seemed to be in flux. At one meeting in New York, Colussy heard someone from Castle Harlan throw out the number $75 million, and at a later meeting it had gone down to $50 million. It was a little troubling to hear your partner scaling back his commitment, but it probably didn't matter, since potential investors seemed to be coming out of the woodwork anyway. Stan Kabala was bringing people to the table. Ed Staiano said he had investment sources he could draw on, including a group of families in Boston who might put in as much as $300 million through a company called the Bollard Group that specialized in making rich people richer. Even Randy Brouckman was talking about big Iridium contracts. He was in contact with the Secretary-General's office at the United Nations, which had already been using Iridium phones at its Balkan refugee camp in Neprosteno, Macedonia. "There's a real need," said Brouckman, "but in terms of bureaucracy they make the U.S. government look like a start-up." He also had promising talks with SpaceHab, a Virginia company that serviced the International Space Station. Money wasn't going to be an issue. The problem was going to be putting a company back together that had fractured into ten thousand pieces. Colussy's main weapon became his deep, soothing voice and his impenetrable calm in the midst of chaos.

Hence, on a humid 82-degree afternoon, Colussy's quest for the Iridium satellite system brought him to a dry lakebed seventy-four miles east of Rome, where three effusive Italian executives spoke exuberantly and waved their arms toward the ninety antennas and earth stations that stretch across the plain where Lake Fucino used to be. The Italians were probably the least angry of all the gateway owners, but that was a moot point since the European gateway at Fucino was jointly owned by Turin Telephone Finance (part of Telecom Italia) and a Düsseldorf wireless company called Vebacom, and Vebacom had majority control. The Italians were desperate to stay open, but the Germans owned two-thirds and were already suffering billion-dollar losses fighting Deutsche Telekom for control of the German cellular market, so they had no time for Iridium. If a board vote were taken, the gateway would be shut down immediately and sold for scrap. So the Italians had come to Colussy with hat in hand: if Colussy's new company could somehow guarantee some continuing income, maybe they could talk the Germans into leaving the gateway open a little while longer. After all, Fucino was also a backup control center—it could replace the SNOC in case of emergency—and it was the premier gateway after Tempe.

It helped, of course, that Colussy was Italian American, as was Ed Staiano, who had been well liked by his Italian counterparts while running Iridium. Since Staiano had a vacation home in Sorrento, he agreed to make the three-hour drive from the Amalfi Coast to be at the meeting with Fulvio Ananasso and Antonio Marzoli, the Iridium Italia executives who were most eager to get out from under the control of their German overlords. After much food and even more drink, all converged on the Italian satellite center of Fucino, an odd place that was at once a symbol of national pride and national folly and an apt metaphor for Iridium itself.

It was Emperor Claudius who first tried to drain Lake Fucino in A.D. 50 by digging a tunnel that would supposedly create more arable land and make the River Garigliano navigable. Unfortunately the collecting canal his workers dug through a mountain was too crooked and small, so it backed up during the grand opening, wiped out an entire gladiatorial exhibition, and caused Claudius to run for his life from the flooding waters. During the next century Emperor Trajan reopened the tunnel but failed to make it work, then Emperor Hadrian gave it a shot but couldn't solve the engineering problems. After defeating three emperors, the malaria-ridden lake was regarded as a devilish force that defied human intervention, as one local ruler after another proposed and then abandoned various schemes to enlarge and straighten the cursed tunnel of Claudius and make the countryside bloom. Meanwhile, sediment started to build up, earthquakes further weakened the ineffectual collecting channel, and Lake Fucino remained stagnant and unwanted for more than a thousand years—but not long enough to make an Italian forget. Holy Roman Emperor Frederick II once again launched the ambitious public works effort in the early 1200s—and failed. Finally, in 1863, a Roman nobleman named Alessandro Torlonia wagered all of his tobacco-and-salt fortune on yet another attempt to drain Fucino, so that he could use the land for large-scale farming. It took him twelve years, but this time it worked, mainly because Torlonia used Swiss engineers. King Victor Emmanuel II gave Torlonia the title Prince of Fucino for his trouble, and history was (finally) made as the last drop of water receded from the lakebed.

As Colussy looked out over the open plain, he saw a surreal blending of an antenna farm pointed at outer space against a backdrop of ancient ruins, like something out of a Stanley Kubrick film. Looming from one of the surrounding peaks was the gray medieval fortress of Piccolomini, which had menaced all the adjoining countryside for five hundred years as various counts, dukes, popes, rebels, French regents, and Spanish usurpers occupied it briefly, then moved on into the bloody history of the Abruzzi region. The soil itself was still rich and fertile, but the area was mostly famous for being the satellite operations center for the European Space Agency, many other satellite companies, and for the European gateway of Iridium. It turned out that the draining of the lake had resulted in perfect terrain for satellite communications—a flat open basin surrounded by protective mountain ranges.

Colussy was more than willing to bargain with the Italians, but he realized just how far apart the two sides were as he was wined and dined by Staiano's old friend and colleague Mauro Sentinelli, an animated caricature of the Happy Italian, now working with Telespazio, the satellite arm of the national phone company. That evening Sentinelli invited everyone to his villa, part of a planned community designed by Mussolini, and the Italians continued their festive sales presentation at an elegant restaurant overlooking a beautiful artificial lake crafted by the finest fascist designers. Colussy told the Italians that their gateway *was* attractive to him, mainly because they didn't care about who listened in on calls that went through Fucino. The United States had a law forbidding foreign nations from wiretapping calls from locations inside the United States. If Iridium were to use just one gateway, then a call that originated in, say, Australia and was received in, say, Russia would be untraceable by both Australian police and Russian police, but visible to American law enforcement. This meant neither country was likely to grant licenses for Iridium to sell the phone. The solution might be Fucino, since the Italians apparently didn't care that much about who listened to its telecommunications traffic. It would be possible for foreign security services to station employees inside the Fucino gateway specifically for call-intercept purposes. And there was also the simple fact that Iridium needed a backup gateway in case anything happened to the one in Tempe. Where better than in western Europe?

Ultimately the Fucino gateway was the only one to get a deal offer from Colussy. It was a complicated formula of discounted Iridium minutes, stock in the new company, and the rights to assets that it could sell for anywhere from $6 million to $12 million, but no cash up front and no ongoing payments. Of course, all those terms could become better, Colussy suggested, if the Italians wanted to invest in the new Iridium. When that possibility was mentioned, he could tell by their hesitation that no cash would be forthcoming from Italy. In fact, the Italians were expecting to be paid several million a year by the new company in order to keep the gateway operating. The problem with taking over a $6 billion company that had been running through upward of $100 million per month was that no one could quite get his brain around the concept of "zero." Everything in Colussy's strategy for the company involved bartering airtime, existing assets, and equity in exchange for a chance to make cash later. For some gateway owners, who had first invested in 1992, "later" had come and gone long ago. Colussy eventually flew back to Washington with everything still up in the air. The Italians never said yes and they never said no. All they said was that they loved Iridium and they wanted to see it survive.

One thing Colussy knew when he got back home on June 10 was that he didn't have time to go on any more jaunts to foreign gateways. The honeymoon with the guys in The Tower had been short-lived. Yes, Motorola was willing to listen to Colussy, but the top executives would rather have been doing almost anything else. Colussy's daily phone call with Ted Schaffner was suffused with so many threats, veiled and otherwise, that Colussy wondered whether Motorola even realized that it was the company with the bankrupt asset. Schaffner wanted Colussy to "prove up the business case" for his new company—something that was really better handled by the bankruptcy court—and that took day after day of number crunching based on thousands of pages of documents supplied by Motorola and Iridium. At some point the Pentagon decided that "everyone would be more comfortable if a major company operated the SNOC"—meaning someone other than Castle Harlan. It might have made the Pentagon comfortable, but it didn't make Colussy happy at all, since it would add another layer of cost. Colussy easily set up an agreement with Boeing to run the SNOC—that company being the obvious choice since it had recently merged with McDonnell Douglas, which operated the Delta II rocket and had been working in the SNOC since day one—and he thought that perhaps Boeing's involvement would lead it to invest in the new company as well. But the insistence by the government that Colussy use a major defense contractor was an example of everyone seeking the typically expensive solutions of the military-industrial complex instead of doing what Colussy's frugal investment group would have done—simply rehire the same employees who were already running the constellation.

And then there was the one Motorola problem that would never go away: insurance. It's not uncommon for people in business to use the word "indemnification"—meaning they want someone else to pay the bill for any potential lawsuits—but usually it's the kind of thing that gets negotiated into an agreement stating that each party will pay for its own mistakes. In this case that wasn't good enough for Motorola. They really did seem to believe that the satellites might eventually fall out of the sky and kill a busload of nuns on their way to work at an orphanage. There's no other way to explain their obsession with a) getting billions in de-orbit insurance, b) having someone else pay the premiums, c) getting billions in product liability insurance, d) getting someone to pay those premiums, too, e) making sure whoever paid the premiums was a Fortune 500 giant that could afford to pay more billions when the insurance company billions ran out, and f) making the new owners of Iridium sign an agreement to cover all Motorola's liability costs even if those costs were caused by events prior to the year 2000. Schaffner used the words "complete and total indemnification" so many times that Colussy stopped writing it down. Colussy had several phone sessions with Jean-Michel Eid, an agent with the French insurance company Aon Space, and Eid eventually agreed to insure the constellation for $500 million per satellite and $1 billion for a "one fell swoop" de-orbit, plus $45 million for "political insurance." None of this was good enough for Motorola, so Colussy arranged for nine people representing Motorola, Boeing, Iridium, two insurance companies, a brokerage firm, and Colussy's investor group to meet in London to speak directly with British and French underwriters, and the deal they worked out—the largest and most extensive in satellite insurance history—was, alas, still not good enough.[36]

The fact was the Iridium satellites were not going to kill anybody. For starters, there was already an Iridium insurance policy in place, written by Aon in May 1999, and Aon obviously didn't think it was going out on a limb as far as liability was concerned. The premium was just $283,500 a year, and that was good for $500 million *per satellite* up to a maximum of eight satellites per year. Risk assessments carried out by aviation insurance experts hired by Boeing estimated the probability of personal injury from de-orbiting the Iridium satellite system at 1 in 1.063 trillion. Since the population of the world at the time was only 6.07 billion, you would need to add five more populated continents to the planet in order to make the odds of getting your arm broken by a falling satellite 370,424 times worse than the odds of winning the Powerball lottery, which has the worst odds of any lottery in America. Looked at another way, the odds of being injured by a de-orbit were six million times less than the odds of being struck by lightning. In the year 1999, 425,000 pounds of space junk had fallen out of orbit, according to Aerospace Corporation, and of that amount, about 84,000 pounds had survived reentry and 21,000 pounds had struck land. Total casualties: zero. Total property damage: zero. It's actually more likely you would be hit by parts falling off an airplane than by an errant satellite component—to be precise, 106,300 times more likely. According to the best estimates by Aerospace, the only part of the lightweight Iridium satellite that might not burn up on reentry was the eighty-pound titanium fuel tank, although it would undoubtedly weigh much less than eighty pounds by the time it struck the earth or the ocean. Furthermore, it's impossible to control a satellite once it's knocked out of orbit, and some of them have such strange crash patterns that it takes them a hundred years to reach Earth. It's possible, of course, for them to come straight down, but the average de-orbit would be around twenty-five years. What did Motorola propose? That the Colussy group buy a hundred-year insurance policy for, say, $5 billion per satellite, just to guard against the possibility that falling debris would clobber a highly visible, high-net-worth individual like Warren Buffett? In the world of risk assessment, there was a word for the potential liability of a satellite system. That word was "safe."

And yet the anxiety over falling satellites that would bedevil Colussy for months would turn out, ironically, to be the fear that saved the constellation. The quick-and-dirty NASA study ordered by the Iridium Interagency Working Group said the chances of being hit by falling debris would be 249 to 1, with the chance of an actual death at 40,000 to 1. The NASA analyst believed that, in addition to the titanium fuel tank, five other parts would survive reentry: the battery, three structural brackets, and the electronic control panel. Since no one has ever been killed by a falling satellite part, it's not exactly a precise science, but the difference between 40,000 and 1.063 trillion is a little hard to explain. The latter number was done by an aerospace contractor that had everything to lose if the numbers were wrong. The former number was done by a government employee working on a deadline for the use of a panicked White House. Fortunately for history, it was most likely that very panic that motivated intense interest by the Clinton administration. And it wasn't so much the odds of injury that scared the administration as the timetable for the de-orbit: Motorola said it would take fourteen months to complete, raising the specter of untold thousands of news reports every time four more satellites were released into crash orbits.

In the meantime, Colussy would hire brokers and lawyers to work on the indemnification issue, assuming it would turn out for the best, and concentrate on getting his deal done. But what looked like a fairly straightforward deal when he was chatting with John Castle from the deck of the *Marianne* had been transformed by May into a spindle of interlocking toxic assets and, by June, into one of those puzzle boxes in Clive Barker horror movies that unleash demonic forces when opened. The board was a mess. The ownership structure was a nightmare. The pricing system for the actual Iridium phone calls would challenge the comprehension of an MIT math professor. The arrangements for future rocket launches would require negotiations with both Russia and China. Every time he met with one of the service providers—companies that actually sold the Iridium subscription plans—he had to allow extra time for the first hour of the meeting, which was completely taken up with yelling about how irresponsible and infuriating Motorola had been.

And then there were the lawyers. The legal profession was very happy to assist in the demise, restructuring, and rebirth of Iridium, and by one estimate the lawyers were burning through $4 million a month in fees approved by the bankruptcy court. You had lawyers for Iridium, lawyers for the creditors, lawyers for Motorola, lawyers for Boeing, and lawyers for every gateway. You had lawyers on staff and outside lawyers for specialized matters like licensing and insurance. The White House counsel was involved, as were the employees of the biggest lawyer of them all, Attorney General Janet Reno. There were lawyers from the Departments of Commerce, Defense, State, and Treasury, including one—David Cohen—whose title was Senior Counsel to General Counsel in the Office of the General Counsel. Colussy hired lawyers to deal with the FCC, the various aviation insurance companies in Paris and London, and the International Telecommunication Union in Geneva. He also decided to bring in his own personal lawyer, Isaac Neuberger, an Orthodox Jewish rabbi from Baltimore who had been Colussy's close friend and advisor for twenty years. Neuberger was, by most standards of American business, eccentric in both manner and dress, sporting wool berets of the type favored by the Beat generation and keeping four speakerphones active on his desk at all times, punching and stabbing at them like a perpetual game of Whac-A-Mole. But Colussy trusted him implicitly, having used him to navigate the shoals and hazards of dozens of business deals while running UNC. Neuberger constantly amazed his friend with his network of fellow rabbis that seemed to stretch to every corporation and government agency around the world, but Neuberger himself was so devout that business had to be planned around the six days in the week when he was available, since he would never agree to work on the Sabbath.[37] The Saudis were wary of Neuberger—bringing a personal lawyer into the middle of a business deal struck them as a conflict of interest—but they agreed to work with him anyway. Some of the Motorola lawyers, on the other hand, couldn't stand him. Sometimes it seemed that everyone had at least one lawyer that some other lawyer couldn't stand.

And the chief lawyer of them all, when it came to matters Iridium, was an imposing sixty-one-year-old jurist named Cornelius Blackshear. Blackshear was the presiding judge in the Iridium bankruptcy, and he was exerting every ounce of influence he had to prevent Motorola from destroying the satellites. In many ways Neil Blackshear was an unlikely candidate to unravel one of the largest bankruptcies in American business history, but then again, the musty old courtrooms of the Southern District of New York, just steps from Wall Street, were accustomed to every kind of business failure, and Blackshear had already presided over the demise of such notable American corporations as the department store chain Alexander's, the direct-marketing catalog Spiegel, the electronics giant the Wiz, the beloved New York icon 47th Street Photo, and Pan Am (many years after Colussy's departure), as well as the bankruptcy of financier Eli Jacobs, which resulted in an auction of the Baltimore Orioles.

Blackshear was not the type to suffer fools. Born poor in Sanford, Florida, he left for New York City the day he graduated high school in 1957, then scrounged around pushing dress racks in the garment district, working as a shipping clerk at Macy's and Gimbels. Seeking broader horizons, he joined the Navy in 1959 and, after his commitment was up, attended the New York Police Academy, rising through the ranks to become a member of the SWAT team, a detective on the 79th Precinct squad, and, ultimately, an internal affairs investigator during some of the worst corruption scandals of the NYPD. Meanwhile, he went to night school at the John Jay College of Criminal Justice, graduating at the age of thirty-two, then on to Fordham Law School to enter the bar at the age of thirty-eight. In 1979 he became the first black bankruptcy trustee in history, and it was President Reagan who promoted him to a judgeship in 1985. Since then he'd attained a reputation for inventive decisions that preserved assets. He knew the mean streets of Brooklyn as well as the cutthroat business world of the garment district, and he brought up five sons. There wasn't much that got past him.

Blackshear had already sat through seven months of false hope by the time Colussy got involved with Iridium, and there were days when he was under intense pressure to go ahead and let Motorola destroy the constellation. He had sorted through the zany filings of twenty-one bidders, including a brothel owner in Nevada and an Israeli Internet publisher called HotJump that described itself as "the leading random prize-portal provider," whatever that was. Big Flower Holdings, the largest printer of advertising inserts in the United States, proposed a plan to buy Iridium, destroy the satellites, and use the frequency spectrum for other ventures. A Minneapolis "church growth consultant" named Carl George had put together a board comprising himself and "the world's foremost satellite experts and business restructuring talent," including Air Force General Tony McPeak and former employees of Teledesic and Rockwell, and he was offering a $61 million purchase price. An eccentric theologian, William Welty of Southern California, was listed on George's "team" even though he also announced his own separate bid through a company called Leading Edge Technologies that would have used the satellites to evangelize the world. Welty's principal partner was Robert d'Ausilio, a Utah businessman running a company called Intraspace that planned to launch "space tugs," nuclear-powered vehicles that would clean up space garbage, pull satellites into their proper orbits, and provide power and propulsion for the International Space Station. (How Iridium fit into his plans was never clearly defined.) Then there was Michael John, a Las Vegas software developer who was proposing Iridium as a delivery vehicle for his "WormHole Technology," a system to transfer compressed data. John offered $100 million but made no deposit. Three Internet architects even set up a crowdfunding system called "Save Our Sats," attempting to get three million people to sign up for a NextCard credit card, thereby triggering an automatic fifty-dollar contribution to a fund that would buy Iridium out of bankruptcy. The common denominator of all these companies seemed to be that they had even less money than Colussy and didn't seem to realize that it took tens of millions just to keep the satellites flying.

Chase Manhattan Bank was the managing partner for the only secured creditors—a consortium of thirty banks from around the world—but its 7 percent of the deal was actually smaller than Barclays', at 11 percent, and equal to Citibank's and Kemper Insurance's. The banks had been unimpressed by all the bidders so far, and now sent the consulting firm of Alvarez & Marsal to Iridium headquarters in Reston in order to identify anything that could be sold or turned into revenue. Joe Bondi, the restructuring lawyer Colussy spoke to most often, reported back to Chase that he thought there was about $413 million in assets he could save. That included the $150 million cash on hand plus a "capital call" of $243 million—in other words, money that the gateway operators had legally pledged back in the early nineties but never paid. Colussy, who had spoken to almost all the gateway operators, said, "Good luck with that one, Joe."

Fortunately, the advent of Castle Harlan caused Judge Blackshear to issue one last extension. Colussy and Neuberger zipped up to New York on Amtrak in late June 2000 and told Blackshear that Castle Harlan was prepared to put up $50 million, and Blackshear told them they could have one more month of due diligence to get it done—until July 28. This development had the unexpected side effect of enraging a company called Venture Partners that had already tried to buy Iridium in March and failed. Gene Curcio, chairman of Venture Partners, had offered $600 million for the company—the same price McCaw was toying with—but failed to post the required $10 million refundable deposit. Blackshear had seen this sort of behavior before—trying to use company assets to make the down payment—so he dismissed the offer out of hand. Venture Partners now filed a sneering motion opposing Castle Harlan's request for a due diligence period. Why was Castle Harlan allowed to submit a proposal based on all kinds of conditions—government contracts and the like—when Venture Partners was willing to pay $51 million on the spot and be done with it? (Judge Blackshear was unimpressed by the motion's arguments, since he wasn't concerned so much with the purchase price as with the ability of the bidder to run the system.) Curcio was now offering a million more than Castle Harlan—but the fine print showed that his offer was only $2 million cash and $49 million in Iridium resources.

From Motorola's point of view, all these shenanigans were just muddying the waters and causing more delays. Motorola lawyers told the judge it was "highly unlikely" that Castle Harlan would be able to come up with a deal and, unless someone gave them $9 million for their expenses, they intended to de-orbit the constellation anyway. Blackshear called the lawyers into his chambers and calmed Motorola down, but after the hearing Motorola demanded that Castle Harlan spend a week in Arizona proving it could run the constellation. Ted Schaffner bombarded Colussy daily with requests for information and demands for decisions. Then, no sooner had Colussy suffered through the June 22 de-orbit deadline than Motorola issued another one—July 1. What happened to the judge's instruction to wait another month? That only applied to Iridium, answered Motorola, *not to us*. Obviously all these latecomers like Castle Harlan weren't going to be able to run these satellites, they kept telling the court.

Adding to Colussy's general level of daily stress was that, after avoiding it for most of his life, he had become a Beltway commuter. To get to the Iridium offices in Reston, Virginia, he had to spend up to an hour and forty minutes on the infuriating stop-and-go freeway that surrounds the capital. Eventually he hired a driver so he could spend the time doing work in the backseat. The Iridium office was one of those boxy, modernistic buildings surrounded by a few puny trees that looked like they were the last holdouts against a vast, empty parking lot, and it was a gloomy place even before the bankruptcy. Now it was the site of an autopsy, with most of the cubicles empty and the ones in use on the outer edges of each floor occupied by attorneys. Most of the building contained computers—room after room of expensive servers that processed the mind-numbingly complex pricing system of the various gateways. Each country billed its Iridium phone service in local currency, and each customer had a choice of four different service plans. That meant that a call could originate in a country licensed to use Zambian kwachas, go through a gateway priced in Brazilian reals, and terminate over landlines priced in Peruvian nuevo soles. If the phone had been purchased in, say, Oslo, then all of this would have to be fed into the Reston computers, where it was converted into South African rands for the originating gateway and Saudi riyals for the gateway owners, translated into American dollars for Iridium, and summarized in a monthly bill denominated in Norwegian kroner. The diciest part of the whole system was the terminating leg of the call—the part that continued on land after the satellites passed it down. These "tail charges" could be as low as 4 cents a minute or as high as $2.50 a minute, depending on where the call ended up, and that was on top of the already exorbitant Iridium per-minute charges—anywhere from $3 to $7. Iridium had hired Andersen Consulting to program ten million lines of computer code for the billing system, but the system was unstable, functioning so erratically that Leo Mondale eventually developed an in-house Excel system that worked just as well. Naturally the customers were annoyed by the unpredictability of it all, and one of the first decisions Colussy made was to change all billing to one price and one plan. Iridium would get 80 American cents per minute, period. Service providers could charge whatever they wanted—most charged around $1.50—but the wholesale price was 80 cents. (Ten years later, that was still the Iridium price.)

Periodically Colussy would gather his little makeshift brain trust in Reston or, as time went on, out by Dulles Airport at the SNOC. From the air the SNOC looked like a Japanese fan, with its half-moon parking lot and tilted satellite dishes, and from the ground it looked like the student union of a college in the Midwest. It was an unassuming place, but it was where the hard-core satellite men worked, and over time Colussy preferred to be in a cubbyhole out there rather than stuck in a dreary Reston cubicle. Wherever Colussy went, Ed Staiano would follow, and Mark Adams, the techno-spook, tended to pop up at every meeting as well. Colussy came to rely on both men for getting him up to speed on satellites, although some of Staiano's old colleagues at Motorola didn't like being on the opposite side of the negotiating table from him.

"Please control Ed Staiano," said Motorola vice president Rick Severns at one point. "We would appreciate it at our upcoming meetings if you would hold any emotional displays to a minimum."

In late June Colussy held a formal teleconference with all Iridium's creditors, telling them he was working rapidly and that he expected the whole thing to be wrapped up shortly, but every briefing was like a graduate-level seminar. He was in one of those total-immersion summer-school programs, trying to become an expert on the satellite business in the time you would normally spend learning how to drive.

Colussy's search for customers and investors was aided by Jonathan Mark, the Israeli analyst sent over by John Castle, but Mark turned out to be a mixed blessing. On the one hand, he asked the right question—"Where is the revenue?"—but on the other he had an extremely short-term view. He would run off on tangents, like touring the plant in Libertyville, Illinois, where Motorola built the Iridium handsets, or spending weeks talking to Rupert Murdoch's News Corporation about buying the Iridium spectrum for a pilot program to bring the Internet to airplanes. Why talk about selling the spectrum? That would amount to putting Iridium out of business, and as far as Colussy could tell, News Corporation had no other reasonable uses for Iridium. Mark also started getting into memo wars with Ed Staiano, mostly chiding him for wasting Castle Harlan's money on unneeded staff. That was an easy thing to do, but who had time for it? A typical Mark e-mail was a mixture of veiled threats and restatements of the obvious: "If we cannot get our cost structure to match our revenue stream out of the gate (with some ramp-up possible) we will not do the deal." Mark, in other words, kept repeating the mantra "Castle Harlan wants its money back quickly." Everyone wanted money back quickly. It was a statement of the obvious that didn't do much to advance due diligence.

Still, Colussy had to admit that Mark was right about one thing. When you're taking over a company that failed to attract customers the first time around, you have to find new customers. Colussy spent a lot of time on the phone with Denver billionaire Pat Broe, owner of Omni-TRAX, the largest privately held short-line railroad system in North America, comprising seventeen rail lines and two thousand miles of track. Broe was fascinated with Iridium as a way to keep tabs on his fleet of rail equipment. Colussy had several conversations, trying to convince him to become a customer and possibly an investor, but Broe would never totally commit. Then there was ParView, a Naples, Florida, company known as the pioneer in GPS systems used in golf carts to tell players how far from the green they were. The CEO said he was ready to expand in the form of a two-way asset-tracking system of his own and either invest $10 million or provide $1 million a month in revenue. Colussy also spent quite a bit of time chasing down Paul Maruani, a strange Frenchman who was paranoid about how easy it was to hack e-mail accounts. Maruani wanted to launch a company called EZ Bank that would be billed as the first truly secure e-commerce funds-transfer mechanism, and to do that he thought he would need Iridium. It was not a bad idea—and he was correct that Iridium could handle secure e-mail—but he, too, turned out to be maddeningly indecisive.[38]

Sometimes Colussy had a hard time discerning who was real and who wasn't. A California company called GlobalTrak was selling several devices that used GPS to track the movements of everything from small children to Greyhound buses. Colussy had numerous conversations with GlobalTrak CEO Eric Benson, unaware at the time that Benson was under investigation for running a pyramid scheme in Orange County, California, where his "office" was a Mail Boxes Etc. He also found out about "the Jesse Jackson group" and invited all the members to dinner at the Metropolitan Club, but when they told him about the phones-for-Africa movement, he didn't put much faith in their ability to help. Although Colussy didn't know it at the time, and for reasons that had nothing to do with Africa, it turned out to be the only lead that mattered.

Meanwhile, Colussy hustled around the District of Columbia talking to some of the most unexciting people ever to refuse a martini. The life of a satellite tycoon, he was discovering, was an endless succession of dim fluorescent-lit waiting rooms arrayed around hallways where the clickety-clack of the secretary's high heels echoed across marble floors. The Washington bureaucracy fully deserved its reputation, and any agency with the word "license" or "permit" in its charter was bound to be staffed by an army of people whose deliberately measured passive-aggressive voices threatened to put you to sleep at any moment. The goal of the process seemed to be the collection of bulky brown folders full of data. At the headquarters of the FCC, a gray stone building with a porte cochere that resembled a Native American bingo hall designed by Mussolini, Colussy and company had numerous meetings about transferring the 165 licenses held by Motorola. At the Pentagon, which had perfected the art of the incomprehensible purchasing process, Mark Adams took Colussy to obscure offices specializing in things like the Defense Working Capital Fund (it should have had "slush" somewhere in the name). Over on K Street, Colussy hired Carter Phillips, a pink-faced sandy-haired senior partner at the vast Sidley Austin firm, to carry his case before the FCC, the insurance companies, and any other regulatory agencies that cropped up and to ensure that all the licenses were transferred. In the rarefied world of "regulatory enforcement law," Carter Phillips was king, having been an Assistant Solicitor General before going into private practice, and currently holding second place on the all-time list of cases argued before the Supreme Court (in excess of fifty)—cases touching such obscure and complex parts of the federal bureaucracy that, if placed end to end in a bound volume, would render the most dedicated law student comatose. Unfortunately, Carter Phillips would be hired by Bill Gates less than a month after Colussy put him on retainer, and Phillips would become preoccupied with Microsoft's massive antitrust case. Still, Colussy wasn't worried. From the looks of Sidley Austin's massive McPherson Square office building, there were plenty of soldiers to fill the breach.

When Colussy was not shuffling around Washington, he was shuttling around the country—Schaumburg one day, Tempe the next, then to New York City, where he would check into the Pierre hotel on the Fifth Avenue side of Central Park, just a five-minute walk from the 58th Street skyscraper where Castle Harlan was based. There the team members would discuss everything from the purely technical—the link margins, the handsets—to the expected life span of the satellites, the projected purchase price, the spare satellites that still needed to be launched, and their Achilles' heel: marketing. Iridium North America, the main gateway in Tempe, was another thorny problem because it was not wholly owned by Motorola or Iridium and it had a separate management team. Getting Motorola's 53 percent would be no problem, but they would also have to negotiate with Sprint and Bell Canada in order to acquire total control.

By late June John Castle started to drift away from the Iridium project, partly because he was once again joining the *Marianne* on her around-the-world voyage, partly because he had delegated the due diligence jobs to Jonathan Mark. Mark's incessant refrain was "break-even within the first year"—a goal Colussy considered unreasonable, if not fantastical—but he agreed that advance contracts would be attractive to Castle Harlan's co-investors. Sometime in June Colussy found a gold mine of data in the form of Ted O'Brien, an executive at the old Iridium who had been regarded as so marginal to the company's success that he didn't even show up in organizational charts. O'Brien had headed the division that sold phone service to industrial and maritime users, closing boring service contracts with businesses while the "real" marketing people like Mauro Sentinelli sold the phone to the "international business traveler." O'Brien was happy to be found and happy to be heard, and after a single meeting Colussy hired him. The number one target customer, O'Brien said, should be shipowners. Not cruise ships—those were a waste of time, taking huge markups on calls and never going out of range of the competing Inmarsat system. But smaller ships would flock to Iridium, mainly because of the phone's portability and its much lower cost. The only problem was that EuroCom Industries, the Danish company that manufactured the Iridium maritime units, had shut down production earlier in the year. O'Brien had assurances EuroCom could get back up to speed within thirty days, though, and he said fishing fleets would immediately become prime customers. He felt Iridium could be competitive globally by sticking to a few core business categories: maritime, corporate aircraft, oil and gas exploration, forestry operations, and mining companies in remote locations. Any company that needed global coverage—whether in ships, planes, or operations that moved around like police and medical services—would eventually choose Iridium over any alternative.

This was sweet music to Colussy's battered ears—someone who saw the future of Iridium instead of cursing its past—and it confirmed his earlier suspicion that there was nothing wrong with the Iridium handset. Yes, it was a brick, but an oil field operator wanted a brick. This was not a product you slipped into the pocket of your Armani suit, but a device you strapped to your belt or left on the wooden table in a radio shack. It was a tool, not a toy, and it was made to be used in the harshest of working conditions. The Iridium market was any place in the world where your phone was the *only* phone. The Royal Canadian Mounted Police had tested the Iridium system in small planes it used for the Yukon Territory and pronounced it "a possible solution for the northern latitudes." Ed Staiano had initiated talks with Raytheon to offer the Iridium Ka-band to the Antarctica Research Station, where scientists generated so much paperwork that it had to be airlifted out. If it were possible to somehow provide limited broadband service over the Iridium system, all those documents could be transmitted via e-mail, at a bargain cost to the National Science Foundation of $3 million per year.

One week in June the Colussy team checked into the Pointe South Mountain Resort in Phoenix to have intensive due diligence meetings at the Chandler Lab and the gateway in Tempe. Staiano showed up, only to get into a tiff with Jim Walz, the Iridium North America CEO, over Walz's refusal to release his thirty-thousand-name customer list. But the real fight came later, when the service providers arrived. Colussy had asked Ted O'Brien to invite them, since, in the world of industrial communications, the service providers own the business. Globalstar had finally launched "full service" four months earlier, and now Iridium's bent-pipe competitor was stealing customers left and right, especially in the Caribbean, where angry Iridium users were being offered $500 rebates to switch. Not to be outdone, Skycell started offering $400 rebates on its briefcase phones for anyone stuck with Iridium service, and some of Inmarsat's service providers matched Globalstar's offer for anyone switching to the Mini-M, which was now being marketed as the TracPhone. In short, anybody selling a satellite phone was picking apart what was now regarded as the dead carcass of Iridium.

Even though he knew the service providers would be loaded for bear, Colussy was taken aback by how angry they were. Honeywell Inc., which had burned through $40 million developing the AirSat device so that Iridium could be used in aircraft, spent a full two hours yelling at Colussy about things that he couldn't have possibly been responsible for. But that was just the beginning. They were all mad—Stratos Global, BearCom, Seven Seas, WorldCom, Station 12, Earthwinds, S.P. Radio, Infosat—and for good reason. These were companies with very specialized sales forces—more in the nature of outfitters or communications consultants—and they were used to being rulers of their domains. SAIT Radio Holland, based in Brussels, had actually invented ship-to-shore communications in 1901 through a partnership with Guglielmo Marconi, and was still servicing merchant ships when the last marine radio officer was phased out in 1999 to make way for satellites. Each of the service providers had invested up to $1 million in setting up their Iridium sales operation. They then spent huge amounts of time and money on every single sale, because their expertise was in knowing how to deal with finicky customers like oil-rig operators, mining companies, and jet leasing firms. They were the only companies in the world that had recognized the genius of the Iridium system from the beginning, and they had been big losers when the plug was pulled.

Colussy kept his mouth shut during their rants against Motorola, figuring they needed to vent before he could get anywhere, and then patiently explained his plans for the future. He knew that ultimately his superior product would prevail, and in the meantime he said all the right things about billing systems, territories, discounts, and the rest. None of the sophisticated users trusted the company anymore, but at least they left Tempe partially mollified.

Once they were gone, Colussy turned his attention to facilities. From his years of acquiring companies at UNC, he was a believer in keeping people and getting rid of real estate, and he followed that formula here. The only building he wanted was the Tempe gateway, which resembled a state park nature center flanked by three giant globes designed like Rubik's Cube sculptures. (Called "radomes," these were the actual earth-station antennas that processed the calls.) Motorola had donated the land for the gateway to Arizona State University, then leased it for eighty years, so Colussy would simply take over that lease. The Chandler Lab, on the other hand, wouldn't be needed. Colussy wanted Motorola to give him everything for free in return for taking the system off its hands, but Ted Schaffner had come down from Chicago and was computing how much each transfer would cost. Motorola was the kind of company that would, on the one hand, say that something was too small to bother with (it was shutting down two plants that could have been of value) and, on the other hand, demand fees and transfer payments and percentages of future revenue for tiny assets that were the equivalent of the albatross computer building in Reston. There turned out to be seven spare satellites, not four, that needed to be launched. (It's hard to say how you misplace satellites, but apparently Motorola had done so.) And the original idea of changing the name of Iridium to "Polaris" fell by the wayside when Colussy discovered vast warehouses full of promotional materials—brochures, user guides, advertising campaigns—with the name Iridium all over them. He was deep enough into this by now to know that he was going to have to save every paper clip. Iridium it was and Iridium it would remain.

At first the Arizona employees were in a "Who the flip is this guy?" mood, especially since they'd all suffered through the Craig McCaw due diligence meetings, but after a week Colussy started to gain their confidence. The trip to Tempe was a feel-good milestone. Everyone was energized—everyone, that is, except Jonathan Mark. Mark had become Colussy's sole contact with Castle Harlan, and Mark kept reminding everyone that "this is not the kind of deal that Castle Harlan normally gets involved in." Mark said this not just to Colussy, but loudly and often to whoever was around. Since a lot of Colussy's credibility was based on the backing of Castle Harlan, this was troubling behavior. John Castle himself rarely weighed in on any of the business problems. The most excited he'd gotten in the past month was when he sent an e-mail full of lame ideas for Iridium advertising slogans ("The way 21st Century spies stay in touch with Whitehall—you can, too!"). Now Colussy was losing confidence in Castle's lieutenant. One day Mark called with sales leads from Stan Kabala, even though Castle and Colussy had decided not to work with Kabala. In May they'd discovered that Kabala had put out feelers for a private placement—attempting to raise $20 million he didn't tell his "partners" about—and after that they no longer trusted him. But Mark, in his zeal to prove himself the king of due diligence, was meeting not just with Kabala but with Kabala's partners. Another time he called to say he'd talked to the Saudis and that they didn't like Motorola. This was like telling everyone it's hot in Miami in August. But the fact that Mark would make a point of it made Colussy nervous.

Mark wanted Colussy to know that the Iridium signal quality was bad. He'd gotten this information from Eagle River, the company owned by Craig McCaw, and apparently it was based on McCaw's personal experience aboard one of his many yachts. (McCaw was such an avid yachtsman that his 118-footer, *Extra Beat*, was used as his "day sailer," while for longer trips he used either the *Tatoosh* or *Le Grand Bleu*, both longer than a football field.) McCaw had tested an Iridium phone while at sea and couldn't make it work. The story was not that outlandish. Free Iridium phones had been given to generals and admirals who tried to use them upside down because they didn't read the instruction booklet, and there were congressmen who tried to use them inside the Capitol Rotunda without success, not realizing that a satellite phone required access to the sky, since the "cell tower" was 485 miles away. McCaw was dyslexic, which meant he never read anything if he could help it—he formed opinions through the verbal reports of his lieutenants—but it was still hard to believe that one of the original cell-phone tycoons didn't know how to extend the antenna and stand on the open deck with the phone. Because of the rapidly moving Iridium constellation, it was hard *not* to get a connection. The dropped-call rate had been under 1 percent ever since being perfected in December 1998. Colussy wasn't concerned that McCaw had made that mistake—in fact, his exasperation lowered the sale price—but he was concerned that Mark was implying a defect in the Iridium signal based purely on outdated anecdotal evidence. At one point Mark wrote a five-page memo detailing everything wrong with Iridium, concluding that "it's a very tough proposition" because the constellation was "fragile" and had a short life, it would take massive capital expenditures to make it work, and "the entrenched satellite player [Inmarsat] is not going away." Mark reminded Colussy of Sergeant Bilko, from the 1950s series starring Phil Silvers as the charming rogue who was king of the motor pool at Fort Baxter, Kansas. Bilko, like Mark, always confused tactics with strategy.

Mark could occasionally be useful. He looked into the intellectual property issues—the process of transferring a thousand Iridium patents from Motorola to the new company. He was beating the bushes for some new equity investors. He eventually came to the same conclusion everyone did about Stan Kabala—"Why is he hanging around?"—but at a certain point he seemed to start finding reasons to bust the business plan. Motorola had "secret contracts," he told Colussy, and those could cause problems later. There were problems with the North American gateway, he said, both technically and because of its special relationship with Sprint. The new company would need to renegotiate every single Motorola contract, and that would be costly and time-consuming. He had spoken to people at TRW, and they thought the Iridium system was defective. (Mark apparently didn't understand corporate jealousy and rumormongering.) He kept finding insurmountable obstacles to moving forward.

When Mark called Colussy one day to say, "We should sell the customer list"—something you do when you're closing your doors—Colussy had finally reached the limits of his patience. It was almost as though Mark was recommending that they locate assets and auction them off—before they even owned the company! For a merchant bank that had $100 million on the line, or even half that, Castle Harlan was being outright negligent. Colussy placed several phone calls to John Castle, asking his old friend to send more support his way. He needed lawyers; he needed consultants. He needed more than just Jonathan Mark flying around the country touring Motorola facilities and deciding they were hopelessly complicated. In late June Castle was aboard the *Marianne*, somewhere in the South Pacific, and Colussy left messages with his secretary, telling him, "It's getting lonely out here." Finally he decided to write a letter. It had worked in the past. His assistants seemed to have some way to get printed materials to him—by carrier pigeon for all he knew—and he felt like Castle Harlan was futzing around too much and jeopardizing the deal.

"Dear John," the letter began. "We have reached a critical phase in the Iridium project and I have a major concern about Jonathan Mark and more specifically about the soundness of our due diligence process that I need to share with you. I don't relish this type of report to you but I feel it is necessary if we are to keep this potentially high return project on track."

Colussy then went on to outline the "disturbing situation" he was faced with—outstanding issues with Motorola at a time when the bankruptcy judge was running out of patience. He had painstakingly earned the credibility of everyone, including the Iridium board of directors, the FCC, the Pentagon, the service providers, the judge, and even Motorola. But that was all being jeopardized by Jonathan Mark. Mark said disparaging things about the deal in front of outsiders. He was disorganized. He insisted on working out of his home in Boston and had yet to spend a full five-day week on Iridium. His letters were sloppy. His follow-up was bad. And he exhibited "an obviously negative attitude that is noticeable by everyone on both sides of the table."

The biggest problem, Colussy told Castle, was that Mark knew nothing about revenue projections but insisted they were his sole responsibility. He would get anecdotes from service providers and then arrogantly assume he knew how things worked and, more important, didn't work. The reason Colussy had to hire Carter Phillips was that the principal deal attorney being used by Castle Harlan was a bankruptcy expert who lived in Park City, Utah! He was competent enough and a nice guy, but he sent junior partners to court instead of showing up himself. Nobody from Castle Harlan was available to help write the final term sheet that went to Motorola. "In my view," Colussy concluded, "this deal cannot be brought to a satisfactory conclusion with Jon Mark as the senior Castle Harlan representative."

The letter worked—at least so far as getting Castle's attention. Castle called the next day. The real problem, he said, was that Castle's partners didn't believe Iridium fit the profile of companies they traditionally invested in—namely, companies that were already operating and had cash flow. "We don't do venture capital," said Castle. "We do private equity." Colussy was shocked. Castle was treating Iridium as a start-up?

This was the first time Colussy had heard there was any problem at all with Castle's partners. Colussy pointed out that the only reason Iridium was not still operating was that Motorola, for unknown reasons, had cut off the phones, but as soon as the deal was done, those phones would be turned back on and there would be cash flow. Castle said he didn't need to be convinced—he still loved the Iridium deal—but he just needed some signed contracts to show his partners, proving that Iridium could make money during its first year of operation. Colussy promised to put some figures together to try to prove up first-year profit, but inwardly he knew that was completely unrealistic. The Iridium business plan at the time showed a $10 million loss for the first year, and that might have been optimistic. You don't turn around a company this huge overnight.

Nevertheless, Colussy called Staiano, told him what they were up against, and they hired a financial analyst named Buck Hollister to move into a cubicle in Reston and start crunching numbers. The best estimate of the cost of running the Iridium system was $7.5 million a month—7.5 percent of what the old company had spent. That number alone should have given everyone cause for optimism. Then lists were made of potential customers and how much they might send Iridium's way in the form of revenue. Was ParView good for $1 million a month? How about World-Tel, the Rainbow Coalition company that wanted to provide phones for Africa? The team projected another million for them. Colussy spent a lot of time talking to Eric Benson at GlobalTrak, and Benson claimed to be good for $2 million a month. The Frenchman with his secure e-mail system—another million. The United Nations turned up again in the form of the Office of the UN High Commissioner for Refugees—they wanted twenty-six phones right away and possibly more later. At least that was a start. Colussy knew he might have $3.5 million a month from the Pentagon. It was coming together. Even if all these customers didn't pan out, there were enough of them to make up the beginnings of a business case.

And then there were the Brits. One day in early July Mark Adams called to tell Colussy that the irrepressible Chalkie White was coming to town. White was a retired Lieutenant Colonel now working for the British Foreign Service who was well known to the American intelligence community as something of a larger-than-life Gilbert-and-Sullivan character. He was being sent to Washington to evaluate the Iridium system—and, Adams added, Colussy would want to take this meeting. Legendary as a former boxing champion for the Coldstream and Scots Guards, star of the British game show *The Krypton Factor* (where he ran the grueling army assault course), White was most famous among his Royal Army Ordnance Corps comrades as the cadet who appeared in drag before the officer corps at Aldershot, singing "Moon River" as he pranced down the row of medals and ribbons, concluding his performance with a kiss directly on the bald pate of a Brigadier General. White and his more somber colleague, Paul Wells, flew in for an Iridium presentation in Reston, pronounced themselves delighted with the system, and hinted at vast United Kingdom contracts in the future. What could that be? Another million per month? More?

Sometime in late July Colussy called Jonathan Mark to say, "Okay, we now have $7.5 million in revenues per month." Mark asked how confident he was of that, and Colussy could only fudge so far. He was not at all confident of that. Castle Harlan's expectations for first-year break-even were totally unrealistic.

But nothing could have prepared Colussy for the call he got on Thursday, July 27. It was a name he didn't recognize—Chuck Storers, who identified himself as being with Robert Marston Associates, a public relations firm. Storers had been assigned to write the press release.

"What press release?" asked Colussy.

The press release announcing that Castle Harlan was dropping out of the Iridium deal.

Colussy didn't hear the news from Jonathan Mark. He didn't hear it from John Castle. He heard it from the PR guy.

He could have waited another day and simply read it in the paper. According to the Reuters dispatch published in the *New York Times*, "The merchant bank Castle Harlan said yesterday that it had dropped its $50 million bid to buy Iridium L.L.C. because it doubted the bankrupt satellite telephone company would be able to produce a steady revenue stream. 'Although Iridium provides a magnificent international point-to-point telephone service, our due diligence and marketing studies were unable to confirm that Iridium would generate even low levels of revenue with a high degree of certainty,' Castle Harlan said in a statement."

Throughout the summer Colussy had been surrounded by angry people. Now it was his turn. Not only was Castle Harlan withdrawing its bid, but they felt they needed to go out of their way to tell the entire international investment community that Iridium was a bad deal and a bad company, and that Castle Harlan had done some massive study that resulted in certain knowledge that it couldn't make money. Castle Harlan had no massive study. It had done no study at all. It had sent a part-time analyst to collect opinions.

But that didn't matter now. Jonathan Mark had done what, in the back of his mind, Colussy had always known he would do. He had turned in a report that said, "Do not invest." And John Castle had looked the other way. For four months Colussy had been happy to let everyone refer to this as "the Castle Harlan deal." There were people on the fringes who didn't even know who Colussy was. It was one thing to back out, but why the press release? They didn't have to say anything.

Colussy's money was gone. His credibility was gone. The purchase was now portrayed publicly as a deal so bad that John Castle himself, the founder and namesake of Castle Harlan, wouldn't fund it for a man he'd known for thirty-five years. Colussy was so mad at Castle that he couldn't even pick up the phone to yell at him.

And then Ted Schaffner called: *The due diligence period is up. We gave you ninety days. It's time to either buy Iridium or crash the satellites. De-orbit will be August 11, in two weeks.*

Colussy thought: *No money. No friends. No viable business plan.*

"Sure, Ted," he told Schaffner. "We're ready to buy it."

Chapter 12

THE FIXER

AUGUST 8, 2000
BLACK ENTERTAINMENT TELEVISION,
W STREET NE, WASHINGTON, D.C.

The setting was beyond unlikely; it was bizarre. Ten days after losing all his financial backing, a day when Motorola issued its tenth "final de-orbit deadline," Dan Colussy pointed his Porsche toward the mean streets of northeast Washington, D.C. It was a part of the capital rarely seen by tourists, comprising forgotten neighborhoods like Ivy City, Brentwood, Langdon, and Brookland that were centers of genteel social life in the nineteenth century, but had since turned into scarred, pockmarked war zones full of iron-gated strip malls, trash-strewn streets, and abandoned houses. These were the forbidding parts of the inner city where gunfights were related by the *Washington Post* in two-sentence summaries and where the crack cocaine trade had boosted the District of Columbia into America's murder capital. But it was here, in the ghetto, that some very strange bedfellows were gathering to determine the future of Iridium. Colussy and two other men were wending their way to the office of Bob Johnson, founder of the Black Entertainment Television network. Johnson's executive suite sat high atop the BET studios, located in a lonely section of Brentwood noted mostly for its railroad yards and for being in a slough where residents' basements regularly filled up with water from flash flooding. With a postal sorting facility as one neighbor and a sketchy McDonald's as another, BET headquarters was cut off from the reality of the streets by high security walls. Beyond the gate, it was like entering Oz: rising skyward was the gleaming mirrored facade of a ninety-thousand-square-foot pin-striped tower set in an immaculately landscaped campus. Colussy parked his car in one of the VIP guest spaces and was soon ascending to the opulent office of the charismatic dealmaker who had launched BET in 1980 and built it into a media empire traded on the New York Stock Exchange. Bob Johnson wanted to help. Bob Johnson, a friend of every President since Carter, could get to the people involved. Bob Johnson had information from Jesse Jackson and, more important, from the President. Bob Johnson was about to put Colussy's quest back on track.

How Dan Colussy became intimately tied to the black power structure of Washington, D.C., was a lesson in long-shot networking. Colussy had first met with some African American businessmen in April after Dorothy Robyn passed along their names as part of "the Jesse Jackson group." Colussy was talking to anyone and everyone who had the remotest interest in helping him, so he arranged a meeting almost immediately. Colussy was skeptical of their Third World business plan, but he liked all of them personally, especially Herb Wilkins, a big, somber grizzly bear of a man who was known in D.C. as the first investor in the Radio One syndication network and the owner of cable TV franchises in major cities across the country. Wilkins and his partner, Terry Jones, were both Harvard Business grads who had met in the late seventies and scraped together money from the Ford Foundation, the Office of Economic Opportunity, the Presbyterian Church, and Aetna Insurance to form a company called Syncom Capital devoted to minority-owned media investments.[39] What the two friends raised was a mere pittance—only $1.75 million—but since then they had invested in more than 150 companies with a combined market value in the billions, and most of those companies were devoted to media ventures in inner-city neighborhoods not unlike Boston's Roxbury, where Wilkins had grown up in a large, close-knit family in a housing project next door to St. John's Episcopal Church, where he spent most of his free time as a boy.

Wilkins was a man on a mission—a mission inspired by affirmative action programs that had taken him out of the inner city and put him through Boston University and Harvard Business. The turning point in his life came when he took a course devoted to what was considered a new concept at the time: the venture capital fund. The idea of putting together an investment fund that would dole out start-up money to promising but high-risk businesses seemed to Wilkins exactly what the black community needed. It was, in fact, the only way he could envision any kind of future for black entrepreneurs, since the very fact of their being black meant they were classified as high risk. A year after his graduation in 1968, he convinced Bache & Co. to back a $10 million venture capital fund devoted to minority businesses, but Bache reneged when the stock market "went sour." By 1973 Wilkins had become a senior investment advisor at a firm financed by J.P. Morgan and Morgan Stanley, but their idea of a minority investment was a white firm with black corporate managers, which angered Wilkins as it seemed to him patronizing. "They were turning down the deals I brought them in a very sarcastic fashion," said Wilkins. Tensions built until Wilkins abruptly quit in February 1975—"the best thing that could have happened to me," he would say later. He would no longer ask white men to finance black businesses. He would raise the money himself. And so Syncom was born.

Syncom specialized in taking over ventures that the mainstream business establishment had abandoned or avoided, the most notable being the cable television franchise in Newark, New Jersey, where white businessmen feared to tread. Wilkins and Jones built out a cable system there that became much more profitable than the systems in wealthy suburbs that were considered more desirable by the media tycoons of the day. "We knew our audience," said Jones. "The traditional model for making money on cable TV was to go into a high-income suburb and get 30 to 40 percent penetration at $19.95 a month. But in those suburbs, there are only 80 to 100 homes per mile. In Newark, there are 350 homes per mile. We could make twice as much money with 10 percent penetration. Yes, the construction costs are higher, and yes, the collections have to be done a different way—you need a lot of cash windows on payday—but for low-income families cable TV is often the *only* entertainment they can afford, so demand is very high. Blacks and Latinos overconsume technology. Fortunately we knew this. Unfortunately, the companies we tried to partner with simply studied our business model and stole it."

Despite their bruising battles with media conglomerates, making it impossible for them to grow as rapidly as they would have liked, Wilkins and Jones were often profiled as being among the most successful African American entrepreneurs in the country. That's why they were approached in 1993 as potential investors in Iridium's African gateway. Jones was intrigued by Iridium even then. A small-town Kansas native with a master's degree in biomedical engineering from George Washington University, he had grown bored during his first jobs out of college with Westinghouse and Litton—"I realized I wasn't an engineer," he said—then spent four years in Kenya in the seventies, creating Kiambere Savings & Loan and several other East African businesses that eventually had to be abandoned for reasons he blamed on weakening currencies and endemic corruption. While living in Nairobi, he met his future wife, an organizer for a Chicago-based nongovernmental organization called the Institute of Cultural Affairs, and when they got back to the States, Jones went to work for the Booker T. Washington Foundation for Excellence, trying to obtain minority-owned cable franchises in major markets. That's how he met Herb Wilkins, who had just formed Syncom.

The two men pooled their resources and were soon investing in FM radio stations, cable, and any other media property that seemed to have value overlooked by the mainstream. Wilkins hated taking money from the government—it led to invasive regulations, limitations on how much you could pay people, and constant oversight by bureaucrats—but was coaxed by Jones into facing reality and realizing minority contracts were a resource they couldn't afford to ignore. Syncom ended up relying on several federal programs, essentially increasing the capitalization of the company fivefold because of its ability to borrow, but all that money came with strings. "We're making money," Wilkins would tell friends, "but at every turn we're hamstrung on what we can do." He spent enormous sums on lawyers who did little more than fill out government reports. He had to get permission to reinvest his profits, and that could take years. He was often drowning in "arbitrary and unnecessary" paperwork required by local governments. He resented being "constantly asked to do what is socially significant instead of what is economically important for success." But he never wavered from one core belief: black people should sell black products into black markets. He was convinced that was the way to solve the problem of perpetual black poverty. It was not a popular view, even among many of his friends who spoke about "integration into the larger American economy."

"That's Republican laissez-faire bullshit," he would shoot back at them.

Meanwhile, Jones remained tied to the African continent by serving on the board of the Southern African Enterprise Development Fund, an agency run by Andrew Young that was empowered to dispense $100 million in USAID funds to "newly democratized" countries. That's how Motorola found him when it went looking for early partners. Though sentimental about Africa, Jones and Wilkins turned down the deal. The Iridium phone seemed to Wilkins a high-end toy for rich people, and the start-up costs for the gateway seemed excessive. It also seemed like an example of rich white men coming after poor black money, and Wilkins insisted above all that the black business community hang together. "The black middle class can't go home and enjoy the fruits of the labor they have supposedly earned by their own initiative," Wilkins told his colleagues, *while leaving that kid in the ghetto unable to compete.*" So there were a lot of reasons Iridium didn't fit Syncom's objectives in 1993, but the two partners continued to follow the ups and downs of the company, and when it started to issue junk bonds, Wilkins even bought a few with his own money. It was only after Iridium started making headlines in 1999 as a spectacular business flameout that they looked at it again.

"We were sitting around waiting for a partners meeting to start one night," recalled Jones, "and Iridium came up and Herb said, 'We ought to buy it.' And we all broke out laughing." Iridium was a $6 billion company, whereas Syncom was virtually unknown at those levels of American business—and yet the more they looked at it, the more it seemed not that different from what they had done with cable TV. The big boys at Iridium had targeted the wrong customer—rich business executives—and ignored the millions who had no alternative service. "We're into the underserved," Jones said. "That's what Syncom is all about. It can be old people. It can be marine vessels. We always had limited capital, so we have to see the gold in the hills that others missed." Bringing telephones to Amazonian outposts and tiny Indonesian islands didn't seem that different from bringing cable TV to neglected inner-city neighborhoods. They thought, at the very least, that they could be a minority investor with a bigger partner, and that they could open doors in Africa and Latin America that others might have trouble penetrating. They cautiously started looking at the numbers, and when they mentioned it to Bob Johnson, the idea started making its way through the black power establishment, eventually reaching Jesse Jackson, who buttonholed President Clinton at a White House party. Within days the idea of "a Bob Johnson bailout of Iridium" had percolated through the White House, and allies at the FCC even agreed to "exercise leverage" over Motorola so the new coalition would have time to put together a business plan. Less than a month later Ty Brown, Herb Wilkins, and Terry Jones were settling in for dinner at the venerable Metropolitan Club as the guests of a guy named Dan Colussy.

Now, ten weeks after that dinner, Castle Harlan had announced to the world that Iridium wasn't worth investing in. Colussy's money was gone. His credibility was gone. His business plan was portrayed publicly as toxic and misguided. But Colussy remembered Herb Wilkins. He remembered Wilkins as the silent, hulking presence at every due diligence meeting he'd attended. He remembered Wilkins as forthright, honest, good as his word, following through on everything he said he would do. He remembered Wilkins swallowing his pride when he attended Castle Harlan meetings in New York, only to be stung by dismissive remarks indicating that "the black guys" were only grudgingly tolerated on Fifth Avenue. Colussy hadn't noticed the snubs himself, but he had to admit that, yes, that was just the kind of thing that might happen at a company run by John Castle. Even if it wasn't overt racism, it wouldn't be surprising for a Washington, D.C., minority-owned firm to be treated as a lightweight junior partner. But now Castle Harlan had been even more overtly rude to Colussy, and Herb Wilkins started to feel the kind of bond you feel when you're kicked out into the alley by a bouncer and notice that someone else has been kicked out with you. It was a bond that would turn out to be crucial for the future of the satellites.

The day after the bombshell Castle Harlan announcement, Colussy's phone had gone quiet. People had seen the newspapers. Iridium was dead. Craig McCaw had pulled out in March and now John Castle had pulled out in July. Obviously the whole satellite phone thing was a white elephant. The sole exception was a woman named Ginger Washburn, a Motorola sales and marketing veteran who now worked with a service provider in Austin, Texas. She was apparently oblivious to the fact that the financing for the deal had fallen through, and she called Colussy to tell him that several Alaskan Indian tribes were in love with the Iridium phone, perfect for use on their remote reservations above the 65th parallel, and that they might want to invest. A day earlier Colussy would have penciled in her call on a list of potential revenue sources for Jonathan Mark's due diligence report, then forgotten about it. Now he eagerly took down the information and told her he would call back. He doubted any Indian tribes in Alaska had $100 million, but miracles do happen. And he needed one.

If Colussy had known what was going on at the White House, his depression might have turned to despair. When Colussy's name appeared in government documents, it might be spelled "Colucci," "Kaluchi," or "Colussi," the partner of "Stan Kapalla" in a project backed by "Castle Harwin." This was because the only names that mattered among the handful of people trying to save Iridium were John Malone, Craig McCaw, and John Castle. Once they were gone, Dorothy Robyn got on the phone with Iridium lawyer Tom Tuttle to find out how much time there was for plan B—Colussy acting on his own—but Tuttle was pessimistic. "It would be nice for the Iridium estate to have ten to fourteen days to patch together a deal," he said, "but Motorola is not going to wait because they feel like they keep getting sucked in. Iridium can't object to a de-orbit because of the legal issue of who bears the costs. It's a $40 to $50 million issue and the creditors don't want to be responsible—they want that on Motorola. Fortunately this judge does not want to de-orbit. But Motorola has taken this all the way up to Chris Galvin and they want it to be over, and the banks may be feeling the same way."

Robyn then called Ty Brown, the lawyer for "the Rainbow Coalition guys," and asked him if there was still hope for a deal—but he, too, was downbeat. "Castle Harlan wouldn't work with us," Brown told her. "Herb Wilkins even told Castle Harlan, 'I'll guarantee the revenue,' but they weren't interested."

What about the international aid organizations he was pitching? "Their response was 'We can't commit until we know the system will be there,' and it's a nine-month budget process."

Robyn was barely starting to process all this gloom when she heard that the Department of Justice was preparing press releases to deal with the possibility of public panic. She warned the lawyers to hold off, reminding them that Motorola still had to coordinate with NORAD, and it also had to produce the proof of insurance it promised back in March.

Almost alone among government employees, Robyn refused to admit the battle was over. She came into the office on the weekend to send out a "bad news" e-mail to everyone involved in the Iridium situation, hoping to control the panic—and Gene Sperling noticed.

Clinton's top economic advisor, and Robyn's boss, scribbled across the e-mail: "We need to do last minute assist—is there anytg fed gov can do?" *Come to his office*, he said. *Let's get going on an Iridium salvation plan. Let's start fresh Monday morning.*

On that same Sunday morning, the time Dan Colussy usually reserved for making notes to himself, he, too, underwent a sea change in his thinking. After going over all the possible ways to deal with the situation, he took out a fresh legal pad and made a list of potential investors:

- *Saudis 50*
- *Syncom 50*
- *Boeing 25*
- *Alaska 40*
- *John Malone?*
- *Bob Johnson?*

It was a fantasy fund-raising list. Those wishful-thinking numbers beside each entry were millions, of course, but the only principal at those companies he'd actually spoken with was Wilkins. Malone and Johnson had already bowed out—Colussy was just holding out hope that Wilkins could execute some kind of last-minute turnaround in their thinking. Boeing, on the other hand, seemed like a natural investor since its rockets had been used to launch much of the constellation and it stood to make ongoing profits from a contract to operate the system. Colussy did know the Chief Operating Officer of Boeing, Harry Stonecipher, from his days as a defense contractor—but so far his old friend was not returning his calls. "Alaska" was the recently discovered Alaskan Indians, and Colussy had them down for $40 million, so obviously he wasn't in complete possession of his faculties yet. But the key entry came on the last line of the legal pad.

"By the end of the week," Colussy wrote, "I need to have a new company."

But there was still one place where the dramatic Castle Harlan exit couldn't be avoided. On Monday, July 31, Judge Cornelius Blackshear of the Southern District of New York U.S. Bankruptcy Court settled into his sizable chair behind his sizable bench in his ornate courtroom facing Bowling Green at the foot of Manhattan Island. He was expecting to hear the final offer of Castle Harlan. All summer long there had been a steady stream of lawyers to the old customhouse, many of them nominal buyers, most of them creditors, but now Blackshear was startled to find just five people in his courtroom. There were two lawyers for Motorola, one lawyer for Iridium, one lawyer for Chase Manhattan Bank, and one lawyer from Baltimore representing Dan Colussy. Something had obviously changed. So Blackshear's first question was "Where is counsel for Castle Harlan?" The Motorola lawyer quickly stepped up, telling Blackshear that Castle Harlan was no longer interested and therefore the case had returned to the same place it was five months previous. There were "no qualified bidders" and the de-orbit of the Iridium constellation should be allowed to commence.

Fortunately Blackshear was seasoned, skeptical, and rarely impressed by the first thing he heard out of a lawyer's mouth. Knowing he would get a lot of posturing in open court, he invited the lawyers into his chambers and, off the record, asked them the equivalent of "What the hell is going on here?" Isaac Neuberger told him the media had it wrong, the Castle Harlan withdrawal had been a minor event, and that Colussy's consortium of investors was prepared to buy the satellite constellation with or without the help of an investment bank. The judge didn't hear anything that gave him confidence in what Neuberger was saying, and he was not in the mood to let the case drag on much longer, but he considered the satellites such valuable assets that he decided to give it one more try. He told Neuberger to get back to him with a concrete proposal and prove he had the money, and get back to him quickly. When Motorola's lawyers objected, he told them he couldn't do anything to prevent them from destroying the constellation, since Iridium was no longer objecting to it, but he appealed to their common sense. Those sixty-six satellites were worthless unless they were orbiting the planet—Motorola shouldn't do anything stupid if there was even a remote chance of a sale. Motorola agreed to wait, but only forty-eight hours. The hearing concluded on a decidedly dismal note. Hurry, the judge told Neuberger.

The brief hearing was widely seen as the death knell for Iridium, especially since the press treated it that way. "The events yesterday effectively end the tumultuous, under-achieving existence of Iridium," reported the *Financial Times*. For the first time, Motorola had a green light to de-orbit anytime they felt like it, without having to go back to court. Dr. Hans Binnendijk, the most senior analyst at the National Security Council, informed National Security Advisor Sandy Berger, and Berger prepared to inform the President. Colonel Victor Villhard in the Office of Science and Technology Policy and Bretton Alexander, his space policy analyst, put their "inform the public" plans into overdrive and circulated a draft news release.

"For the complete Iridium system," it read, "there is a 1 in 250 chance that some person on Earth would be struck by debris. Thus the risk of an accident is not insignificant, and if a de-orbiting satellite hit a population center, the level of damage could be substantial."

This was too much for Dorothy Robyn, who became alarmed that they would panic the public before the de-orbit was even a certainty. "Just be careful not to get ahead of the process," she told Villhard. "There's still an active effort on the part of Dan Colussy and others to get the funds to keep the system in operation."

Two days later Robyn and Mike Greenberger at Justice presided over a meeting requested by Motorola. Oddly enough, the request for the meeting came not from the executives in Schaumburg or the government affairs people in D.C., but Dannie Stamp at the Chandler Lab. "Everyone was gone on the technical side except me," said Stamp—and so he had been charged with getting the system ready for the de-orbit. "But I had all these technical approvals I had to get from various agencies and I couldn't get these blokes in Clinton's office to coordinate it for me, I couldn't get any one person to say, 'Okay, fine, you're clear.' So I went to DOJ to get all of Clinton's minions together." Stamp's idea was to gather in one room everyone who had authority over Iridium—at the Pentagon, the FAA, the FCC, the State Department, the Commerce Department, the Department of Transportation, and the Justice Department—and get an unequivocal statement from the government that a de-orbit was legal. Robyn saw this as an opportunity to delay the process, but Greenberger was downbeat: "We don't have a very good story to tell." Robyn and Greenberger went over all the insurance issues but didn't think they could get any more mileage out of that. They considered buying time by telling Motorola that its press statements were misleading and it needed to coordinate with them on that. Ultimately, they simply asked Motorola to delay the de-orbit until August 16—another two weeks—in order to give the government time to inform everyone. The way Robyn worded it to Bruce Ramo, the Motorola lawyer, was that "you've been very cooperative and we don't think that, ultimately, we will stand in the way if there's no deal, but we need a little more time to talk to our principals. It's in your interest to have the government stand down, and your de-orbit is too soon."

Greenberger went further with Ramo, saying, "Having the U.S. government be able to stand down on this would be very helpful. Conversely, we don't want to have to seek an injunction. I'm not saying we would do that." Perhaps where persuasion didn't work, an implied threat would.

Neither Robyn nor Greenberger was surprised when Motorola, impressed by neither argument, refused to commit to any more delays.

Back home in Maryland, Colussy took the call from Neuberger, reporting on the judge's comments, and knew he was out of time. He called Herb Wilkins and said, "Herb, I need to know what you can do right now." And Wilkins came through. He had always thought he would be the junior partner of Castle Harlan, or Bob Johnson, or John Malone, but he was angry about how this whole thing was coming down and he stepped up. He said he could raise $25 million in seventeen days and could pledge another $25 million for later. But that was all he had. Colussy's research now indicated it would take $84 million to run the system for a year, plus $50 million to $100 million to purchase assets owned by Motorola and Iridium. If Colussy ran everything on a shoestring, the total bill for year one still stood to be $200 million, and he didn't know where that would come from.

Fortunately Wilkins, who had sat like a silent Buddha at most of the due diligence meetings, now became proactive. Wilkins picked up the phone to call Ty Brown.

"Ty, you're not gonna believe this," said Wilkins, "but for the first time in my life, I'm gonna back a white man."

Brown was indeed stunned. Brown had known Wilkins for twenty-five years, since the late seventies, when Brown was a commissioner at the FCC and Wilkins was a hustler trying to land cable franchises, and the two men had become even closer friends while both were serving on the board of BET. Brown knew Wilkins as a man who had deep-seated resentments toward the white business establishment. Herb Wilkins avoided doing business with *any* white people, much less financing white-run ventures. Just the mention of John Malone's name could send Wilkins into outraged rants, as he believed that Malone had pretended to be a loyal partner, only to steal his business plans and contacts.[40] But Brown also knew that Wilkins had built a small media empire backing start-up companies that no one else cared about. He had put up the money for the most successful black radio network in America at a time when the founder was still operating out of her basement. In the early nineties, when you still needed a credit check to own a cell phone, Wilkins and Brown had figured out a way to provide phones to people in the inner city who didn't qualify. "If you wanted to get something done," said Brown, "show Herb Wilkins a situation where there's a service or a product available to the haves and not the have-nots. He would always say, 'Throw me in that briar patch.'" Since that partners meeting when they joked about buying Iridium, Wilkins had turned fierce in his position. Ty Brown, in fact, had been one of the first people he called once he got the fire in his belly, and Brown still remembered the earlier conversation.

"Motorola is getting ready to deep-six Iridium," he had said.

"Why not?" said Brown. "It's a piece of junk."

"No, it's *not* a piece of junk. I want to buy it."

"What will you do with it?"

"It's a telephone system for the Third World. We gotta fight for it and we want you to be the lawyer."

Brown knew that Wilkins' instincts were rarely wrong.

"I don't wanna be the lawyer," said Brown. "I wanna be a partner."

And so WorldTel was born.

Colussy remained unimpressed by the whole sub-Saharan angle. He had vast experience dealing with African governments—Ugandan dictator Idi Amin once put out a contract on Colussy's life after he canceled 747 service out of Kampala—and he knew that any government deals would be hard to acquire, harder to maintain, subject to nepotism, impossible to enforce, and likely to be attached to kickback schemes that were illegal for an American company. Then there was the fact that Africa hadn't shown much interest in Iridium the first time around. Fully 41 percent of the Iridium users in Africa were in the mining industry, and most of those probably weren't even African. Still, he had to pay lip service to Iridium as "the ultimate solution for the Third World" since that's what people like Jesse Jackson and Bill Clinton wanted to hear.

Meanwhile, Mark Adams was checking with his Pentagon spies every day, and he called to say that Motorola was now trying to give Iridium away. Nick Negroponte, founder of MIT's renowned Media Lab and a Motorola board member, had called Dan Goldin, the administrator of NASA, to ask if there was any way Motorola could simply donate the system to the government for civilian use. NASA turned the gift down, having no budget for operations and no real use for it. Running out of easy options, Motorola decided to give Colussy one more chance. "Come to The Tower, let's talk," said Keith Bane. "Come to The Tower," said Ted Schaffner. The Motorola executives wanted this to be over, and they wanted it to be over now.

As Colussy boarded a flight for Chicago on Tuesday, August 1, he carried a prospect sheet that had exactly one confirmed name on it: Herb Wilkins. Everyone else was vague to the point of being infuriating. Ed Cerny and Tom Sperry, two investment banker types at UBS Warburg who worked for Prince Khalid, called to say they heard rumors that Craig McCaw might still want in on the deal. They were implying that the Prince might invest, too, but only if "someone major" like McCaw were involved. Colussy was skeptical of McCaw since he'd come and gone so many times, but he still thought there might be a shot with another "someone major": Boeing. The senior Boeing executive involved with the Iridium launches was a man named Don Hull, and he had been very helpful when Colussy was doing his due diligence in Arizona, so Colussy called Hull to find out why Boeing COO Harry Stonecipher was not returning phone calls. It took about five minutes for Hull to get the answer: the company wanted no part of an equity investment in Iridium. Boeing had already gotten involved in the stillborn Ellipso project and then pumped $100 million into Teledesic for a 10 percent stake, and it now looked like all that money was up in smoke. The company would no longer be buying any equity in satellite ventures. Meanwhile, Ginger Washburn kept calling to say she was sure the Alaskan Indians would invest, but Colussy didn't really have time to fly to the Arctic Circle to talk to them unless he could be certain they were real.

In combing back through the original Iridium investors, two prospects seemed most likely. Carlton Jennings, an American who had been CEO of Iridium South Pacific in Australia, was a fast-talking sort who kept referring to the licenses he controlled (including the "island king-doms" of the South Pacific), saying he wanted in—maybe not at the $50 million Syncom level, but at some level. Jennings was a former employee of Motorola, hired by Jerry Adams in 1994 to find investors in Asia because of his experience in Japan, Australia, and New Zealand. Colussy told him to get his checkbook ready and optimistically penciled him in for $25 million. Another promising lead was Kazuo Inamori, the Japanese tycoon who had founded Kyocera Corporation in 1959 as a ceramics manufacturer, then turned it into one of the most powerful electronics companies in the world when ceramics turned out to be ideal for semiconductors and cell phones. Kyocera had been an inves-tor in the original Japanese consortium and had built handsets for the service—handsets universally considered superior to the ones built by Motorola—but the Japanese had been especially stunned when Iridium went bankrupt, since they had invested $581 million, making them the largest shareholders after Motorola. Now, thanks to overtures by Tom Tuttle, the courtly Japanese magnate said he was open to a meeting dur-ing his upcoming visit to the United States. Colussy decided to treat all these potential investors as solidly "in"—at least while he was talking to Motorola—but he knew he needed to get some real commitments within a matter of days or else the whole thing would crumble into dust.

Colussy got another big surprise when he touched down in Chicago—a new date for crashing the constellation. Motorola had issued a press release this time, announcing that the satellites would come down on August 9, eight days away, and it had already hit the *Chicago Tribune* in a story headlined "Burn-up of Iridium Satellites to Begin." It was one thing to ignore the pleas of Judge Blackshear, or bluff "de-orbit deadlines," but going public with it? Why did he even bother to fly in? (The only satisfying part of the *Tribune* story was that it referred to Castle Harlan as an "asset stripper" and "buyer of last resort.") Colussy decided to ignore the news—it would just sidetrack his ability to get anything done—and drove straight to The Tower, where he and Schaffner instantly plunged into a meeting about insurance issues. Meanwhile, in another part of the building, Motorola spokesman Scott Wyman was telling the Dow Jones News Service that the satellites would be destroyed "as soon as possible." At the end of the day, Colussy was nervous. When he left The Tower, the August 9 deadline was still in place. The only thing that could dislodge it was an agreement with Boeing—an agreement that Boeing had already ruled out. Motorola made it clear: he had exactly one week.

Meanwhile, in Washington, the Iridium Interagency Working Group had given up on Motorola and was preparing for de-orbit. Staffers dusted off the government press announcement that had been circulating for four months, saying that a) they were not government satellites, b) de-orbiting was an imprecise science, c) there was a slight possibility that some debris would reach Earth with a "minuscule" risk of casualties, and d) the government was monitoring the situation. When it came to the actual risk-of-casualty statement, they chose new, even more confusing language: "NASA calculates there is less than 1 chance in 18,000 that any person on Earth will be hit by debris from each Iridium reentry and that there is less than a 0.5% chance that any person on Earth will be hit by debris as a result of all 74 Iridium reentries." The eventual press document ran to dozens of pages of "if asked" responses to theoretical questions from the media. It didn't help that since the whole Iridium mess began, a 70-pound piece of debris from a Delta II rocket had landed on Pieter Viljoen's vineyard in Worcester, South Africa, and a 110-pound metal ball from the same rocket had landed on a farm in Durbanville.

Once again Colussy bunkered down in his wood-paneled office overlooking the Severn, immersing himself in phone calls at all hours of the day and night, frequently getting up at three or four in the morning to write notes to himself. Helene—sensing a work mode to end all work modes—sharply abbreviated their social calendar and did her best to make it easy for him to take meals in his office and run what had become the Iridium war room. Even at this late date, he couldn't always tell the real from the unreal. A man named Keith McNerney called from the West Coast representing Venture

Partners, the company run by Gene Curcio from the Los Angeles suburb of Rolling Hills Estates. McNerney now claimed to have $350 million in committed funds. Lockheed Martin and General Dynamics were both involved with the deal, McNerney said, and General Dynamics had agreed to run the constellation for $5 million a month. Frank Morton, a descendant of the founder of Morton Salt, was investing as well. Maybe, McNerney said, Colussy wanted to join them instead of fight them.

Even though he was desperate for financing, Colussy had a sense that Curcio had nothing to offer. He asked a few questions and told McNerney he would call him back, but McNerney apparently interpreted this as a snub. The next day Curcio turned hostile, sending a letter to the Iridium board of directors calling Colussy "a Castle Harlan splinter group" that is "neither operationally or financially capable of acquiring the satellite assets or indemnifying the Iridium estate." Curcio went on to say, "I have spoken to Boeing Corporate in Seattle, and I can assure you that Boeing has no intention of operating a satellite system, nor will they foot the bill, and they certainly will not indemnify Iridium." The only thing that gave Curcio any credibility at all was the letter he had produced from General Dynamics. Apparently General Dynamics executives had gotten wind of Colussy's deal with Boeing and were simply taking a competitive position so that, if Colussy failed, they might be considered in line for the operations contract.

Curcio was right about one thing, though. As of August 4, Colussy was not yet "financially capable." That's why the team was spending 24/7 on finding money. Ed Staiano called to say that Bain & Company of Boston might be interested in investing. Colussy was skeptical—that was the firm that Jonathan Mark had come from, and it was indeed an "asset stripper." Mark Adams called with bad news from the Pentagon. Art Money, the Assistant Secretary of Defense for Command, Control, Communications and Intelligence, was the key official with the portfolio for Iridium, and he was only lukewarm about whether the military needed it at all. Then more depressing news from Motorola: the board of directors had met the previous day and drawn a line in the sand. Either "total indemnification from Boeing" or else the satellites would be destroyed. Staiano called to make the bad news worse: Motorola had sent out an official letter to all employees, saying there were no buyers for Iridium and that the constellation would be de-orbited. For some reason the company not only wanted to get rid of the satellites, they wanted to be as public as possible about it.

And then Colussy's caller ID lit up with the last name in the world he expected: John Castle.

Castle was calling from the Harlem River Drive in Manhattan, on his way to work, and he had just gotten off the phone with Prince Fahd, eldest son of Prince Khalid. Prince Fahd had called Castle from Paris after weeks of silence. There had been intermittent static on the line as the two men spoke briefly, but Castle was able to make out that Fahd was giving him a green light for an investment in Iridium. The Saudis would like a chance to recover their original investment. (Although Fahd didn't state a number to Castle, their total losses on two gateways would have been about $250 million.) The call caught Castle off guard since it was a belated response to a presentation earlier that summer at Castle Harlan headquarters—but he went into none of this during his call to Colussy. He simply said he was calling to help. "You know I've always been the biggest supporter of the Iridium phone," he told Colussy. "I tried to do it myself, but it was outside the scope of my investment criteria. I couldn't justify it to my partners."

Colussy knew that Castle could do whatever he wanted with Castle Harlan money, partners or no partners, because he had an iron control over the voting shares, but he didn't say that. Colussy also knew that Castle could have withdrawn from the Iridium effort without announcing it to the world, but he didn't say that either. What he did say was "How can I help you, John?"

"I want to give you Prince Fahd's phone number in Paris. It's the only way to get to Prince Khalid, and I think he might invest."

Did Castle feel guilty? Was this a peace offering? Colussy didn't have time to ponder the imponderables. He just knew that, when it came to the Saudis, Castle always knew what he was talking about. Prince Khalid and another Middle Eastern investor named Sulaiman Olayan had jointly owned 22 percent of Donaldson, Lufkin & Jenrette when Castle was running it, and since 1975 the Prince's son and three other family members had all trained under Castle as DLJ interns. Prince Khalid was such a private man that Castle had met him only twice during the two decades he worked with the family. Once was for dinner at the historic Saratoga Springs racetrack in Upstate New York, where the Prince and Castle both had horses running during the August resort season. And the second time was when DLJ was sold to Equitable Life Insurance Company in 1985 and Castle personally delivered Prince Khalid's check for $180 million to the lobby of the Waldorf Astoria Towers. Castle knew the Prince to be a gentle, thoughtful man who was a rare bird in Saudi Arabia—a man with a royal pedigree who chose not to be involved in either oil or government, even though his wife was a full sister of King Fahd and his father had been an older brother and trusted advisor to Ibn Sa'ud, founder and first monarch of modern Saudi Arabia. Since the last time Castle saw him, Prince Khalid had become even more isolated and reclusive as his business passed into the hands of professional managers, but Castle had remained close to all Prince Khalid's children and their extended families and especially to his former intern, Prince Fahd.

Colussy took down Fahd's phone number and found himself softening a bit toward his old friend. Castle was just not a risk-taker, he thought. He was so ultracautious that he became his own worst enemy when it came time to make an actual investment. Colussy thanked him, got off the phone as quickly as possible, and immediately dialed Prince Fahd.

The prince answered his cell phone, acknowledged that he was expecting the call, then said, "I'm on the train, may I call you back?" It was the kind of phrase that strikes fear into a cash-finder's heart, especially when dealing with the famously elusive Saudis. But, miracle of miracles, Prince Fahd did indeed call back a few minutes later.

"As you know," he began, "my father had a very bad experience with Motorola, but we still believe a partnership is possible if the terms are right. I have strong faith in the Iridium system, and I know that John Castle has a strong confidence in your abilities. At this point we need to be made more comfortable."

So Colussy proceeded to make him more comfortable, quickly reviewing his plans to buy the company out of bankruptcy, run it with the help of Boeing, and reconfigure the marketing goals so as to make it profitable. At the conclusion of the impromptu presentation, Prince Fahd said, "All right, that makes me very happy. That's a very exciting proposition. The marketing is very important, and of course our advisors at Warburg will play a role for us. Could you by any chance meet us in London on Wednesday so that we can discuss this further?"

Wednesday was only four days away, and it was the day Motorola had announced to the world as the de-orbit day, but Colussy said, "Of course, I'll be there."

Instantly he dropped everything and put into motion plans to join Prince Khalid, Prince Fahd, and their armies of advisors at Cadogan Place in Knightsbridge, just off the gardens behind Harrods department store, where a town house functioned as Prince Khalid Central.

But that was the last time Colussy would ever speak to Prince Fahd. By the following day Prince Khalid's majordomo, a punctilious, somewhat demanding lawyer named James Swartz, called to say, "No need to come to London. I'll be coming to the U.S. We'll talk there."

Colussy was learning the first lesson of dealing with the Prince—namely that you never deal with the Prince. There were CPAs, lawyers, bankers, household employees, and all of them jealously guarded access to the Prince and the members of his family. The Prince himself was said to be a charming guy, capable of perfect English, always impeccably dressed in Savile Row suits—but few people had ever met him.

"Of all the wealthy men I've met in my life," said John Richardson, "he was by far the most pleasant of all." But the Prince also preferred his investments to be anonymous. The important thing now was that his son had already gone on record, encouraging Colussy to seek an investment, and so the family's word had been given. Everyone working for them accepted it as a fait accompli, but each now wanted to control it. James Swartz, the functionary intervening to make himself the chief intermediary between Colussy and the family, was in fact the ultimate insider. After graduating from Yale College, he acquired three law degrees (from Oxford, George Washington University, and the University of London), worked briefly for the venerable London investment bank Morgan Grenfell, then spent most of his life handling the Prince's foreign investments. That job included managing the Prince's residence in London as well as managing the financial aspects of 250 racehorses and 300 employees at six stables in Kentucky, England, and Ireland. Swartz actually lived in the Cadogan Place town house and had somehow acquired an English accent, although he was an American who had grown up in Farmer City, Illinois. From now on Colussy would deal with Swartz and anyone Swartz assigned to the Iridium matter—and he would assign many people.

Colussy knew that a Prince Khalid investment could still fall through, though, primarily because, in his opinion, the Prince's advisors were determined to nitpick it to death. The sole exception was a pale, thin Australian named Tom Alabakis, whose job was to manage the original Iridium investment from the Middle Eastern gateway sales offices in Dubai. Alabakis was interested in seeing his job continue, so he flew to Washington to try to influence the deal, but his personal manner was abrasive, especially toward Colussy. Alabakis made it clear that the Prince didn't invest in small-time operations—he invested with the blue-chip corporations of the world. He had a full-time Chief Financial Officer, David Hall, working out of his London town house. He used Price Waterhouse Coopers, the most prestigious accounting firm in the world. He had two lawyers—Steve Pfeiffer and Larry Franceski—on call at Fulbright & Jaworski, one of the largest law firms in the world. All of these people had to be convinced that reinvesting in Iridium was the right thing to do. Everyone employed by the Prince, from the guys at UBS Warburg to Swartz himself, kept insisting, "We have to partner with a major player." By major player they meant Castle Harlan, or Craig McCaw, or Boeing, or anyone *other* than the company that was referred to simply as "Newco" because it had yet to be named. Anyone other than Dan Colussy acting alone.

Colussy was just starting to wade through the intricacies of the Saudi business world when he got an excited call from Herb Wilkins.

"We can get to Cohen," said Wilkins.

By Cohen he meant Secretary of Defense Bill Cohen, the only guy in the world, aside from the President, who could invoke the national interest as a way of saving Iridium.

"What do you mean you can get to him?" said Colussy.

"I can get a meeting with Cohen."

Colussy could hardly believe it. From that first meeting at the Pentagon, when General Jack Woodward confessed that he was not powerful enough to save Iridium on his own, Colussy had known that getting a military contract would be an uphill fight. "There's support for Iridium within the Pentagon," Mark Adams had told him, "but not *that much* support." And the Pentagon didn't move fast even when projects had *universal* support. Now Herb Wilkins had somehow landed an audience with the man who ran the whole building.

William Sebastian "Bill" Cohen was an odd duck in the Clinton cabinet, a former Republican congressman from Maine who had been drafted for the Pentagon job when Clinton was trying to move toward the center after winning reelection in 1996. Cohen was the sort of patrician New England moderate that had started to become a dying breed after the one-term administration of the first President Bush. At the time he became SecDef, Cohen was known mostly for being the only Latin scholar in Congress (Bowdoin College, 1962), the lawyer who drafted the 1984 Competition in Contracting Act to eliminate Pentagon waste, and, as far as anyone could remember, the only SecDef in history who was also a published poet. This latter skill didn't seem very suited to the personality of a warrior, especially when you consider his staggered meter decrying the Strategic Defense Initiative:

> *Before they unleash*
> *hurricane winds,*
> *Before they breathe*
> *through nostrils red*
> *beyond all Fahrenheit,*
> *Turn them to endless*
> *ash, yes, save us from*
> *their savagery.*

But that was a poem about war satellites—maybe this pacifist warrior had a soft place in his heart for the peaceful satellites of Iridium. Besides, Cohen had no more time for poetry. He had become better known as the guy accused of running interference for the Monica Lewinsky sex scandal, first by firing missiles at terrorist camps in Afghanistan one day before Lewinsky's testimony, then for bombing Iraq one day before the President's impeachment hearing. The only reason any of this mattered at all was that Cohen had been reamed by the press several times and was unlikely to touch anything controversial. But Wilkins had a secret weapon.

"First of all," he told Colussy, "Cohen is a Democrat in Republican clothing. He'll want to help us. More important, his wife works for Bob Johnson."

And indeed, as Colussy was soon to learn, Cohen was married to a flamboyant television diva named Janet Langhart, now Janet Langhart Cohen after the Secretary became her third husband in February 1996.

The former Ebony Fashion Fair model was finishing up a long broadcasting career that began as the weather girl at WBBM in Chicago in 1966, continued with regional morning-host jobs on *Indy Today* and *Good Day in Boston*, and included stints as a spokeswoman for Avon Cosmetics and a correspondent for *Entertainment Tonight* before Bob Johnson brought her to BET. She'd made her presence known at the Pentagon, first by masterminding a $52,000 redo of her husband's office, then by anointing herself "First Lady of the USO" and printing her new nickname on personal stationery—until the head of the USO complained to Cohen and she was told to stop. She also created, produced, and hosted *Special Assignment*, a weekly television program that aired on the Armed Forces Network during her husband's tenure as SecDef. Herb Wilkins didn't come right out and say it, but the implication was that Bill Cohen would listen to Janet Cohen, and Janet Cohen would definitely listen to Bob Johnson. Even though Johnson had decided not to invest, Wilkins had prevailed on their long-standing friendship. The Syncom guys had known Johnson ever since he was an obscure lobbyist for the National Cable Television Association in the seventies, coming to Ty Brown to ask for assistance in getting things through the FCC. The bottom line: *Come to Bob Johnson's office—he can fix everything.*

And so it was that, on August 8, twenty-four hours before the publicly announced de-orbit, Colussy found himself sitting in the waiting room at the Pentagon with Ty Brown, Herb Wilkins, and the genial Bob Johnson. They had all come from BET headquarters in Johnson's limo and would return there later to rehash the events of the day. They cooled their heels for a good part of the morning, as people tend to do when they're waiting for an audience with the cabinet member charged with the defense of the free world, but once the young officer escorted them on the long walk to Cohen's inner sanctum, it went off like any other meeting. There was the usual chitchat, and then Bob Johnson took control.

"Bill," he said, "we need your help." And he nodded toward Colussy. "Dan, would you fill in the Secretary?"

Cohen then listened patiently as Colussy began outlining the situation, emphasizing the indemnification problem, the hostility of Motorola, the support for the Iridium phones on both the civilian side in the form of the CIA and the DEA and the military side in the form of the Air Force and Marines, and making the point that, even as they spoke, Motorola was threatening to de-orbit the constellation within twenty-four hours.

"Well, I've certainly heard about this situation," said Cohen at last. "And what you're saying makes a lot of sense. I'll look into it."

There was a pause. Colussy added a few pertinent facts. Cohen seemed to be mulling it over.

"So you need me to say that saving these satellites is in the national interest," he said.

Was it a question? Colussy decided to answer it: Yes, that would certainly help.

"And . . . it *is* in the national interest, correct?"

Colussy assured him that yes, it certainly was.

"I'll get back to you one way or the other," said Cohen.

As they made their way back to BET headquarters to get their cars, the men agreed that the meeting had gone well, but they probably didn't know how well.

An hour later, the phone rang in the Pentagon office of Dave Oliver, whose official title was Principal Deputy Under Secretary of Defense for Acquisition, Technology and Logistics. His unofficial title was "the Fixer." It was Secretary Cohen on the line, and his message was clear: *The President wants Iridium saved.* The reason is that he thinks the system would be useful for the economic development of Africa, providing phone links between villages, but that's not a good enough reason for Pentagon intervention. *Your job,* Cohen told Oliver, *is to find a reason the military can't live without Iridium. And do it now. Oh, and one more thing: you can't use the President's name. He doesn't think he should be directly linked to this effort.*

It was not uncommon for Dave Oliver to get a call from the Secretary, but it *was* rare for him to get one with this degree of urgency. As the number five official on the civilian side of the Pentagon, Oliver was known as a bulldog—the guy who could cross any line and bring together any warring factions. When General Woodward told Colussy he didn't have enough juice to save Iridium on his own, he was hoping the case file would pass to someone exactly like Oliver. Oliver had spent thirty-two years in the Navy, retiring as a two-star Admiral after running the nuclear attack submarine program during the Cold War and then the Navy space program in the eighties. He then spent several years working for Westinghouse and Northrop Grumman before coming back to the Pentagon on the civilian side, partly because he believed in Bill Clinton. He liked to compare Clinton with Admiral Elmo Zumwalt, with whom Oliver had served in Vietnam—the kind of leader who wasn't afraid to be unpopular in the present when he had a clear vision of what needed to be done for the future.

Oliver, like Clinton, was willing to run against the conventional wisdom—and Dave Oliver loved the Pentagon. He was a true believer, an appellation sometimes used derisively, but in Oliver's case usually spoken with respect: he honestly thought what he was doing was important to the free world. And there was one more thing about Dave Oliver that made him perfect for the Iridium situation: his specialty was the dirty job. He had smuggled arms into Israel in the eighties, keeping everything under the radar. After the Soviet Union crumbled, it was Oliver who got the nuclear weapons out of Kazakhstan. He headed a secret mission that located ninety-two Scarp ballistic missiles and two thousand pounds of enriched plutonium, then found a way to bring everything back to the United States. ("We fueled Detroit with it," he said later, without irony.) Oliver was used to keeping secrets and dealing with brinksmanship situations—skills he learned as a nuclear submarine commander in the Sea of Japan, where he frequently played chicken with the Soviets, and then honed while handling every kind of secret backdoor assignment since then. He thrived on this stuff.

Secretary Cohen knew Oliver could work fast and get results, because one thing he'd heard loud and clear during the Colussy briefing was the time element: this had to get done before Motorola destroyed the satellites. It was a threat that Oliver was slow to accept. He thought the de-orbit deadlines were simply negotiating tactics—"but then I learned real quick to take them seriously." He decided he needed to somehow neutralize Chris Galvin, so he placed a call to The Tower, leaving the message that he had been assigned to the Iridium problem and was hoping Galvin would help him "get up to speed on it." But Galvin was in no great hurry to get back to him. "Not the most dynamic guy in the world," Oliver said to colleagues, "and every time I try to call him, he's at a party."

It took a day or two, but Oliver did get up to speed on Iridium, and what he discovered served to motivate him even more. Like most people who studied the engineering of the Iridium constellation, he was in awe. The Motorola handsets, on the other hand, were "dogmeat." If the system was going to be saved, it would need new handsets, but Motorola didn't want to make any new ones. "I could also see that in a couple of generations the Iridium system was going to be really, really good," he said. "So I thought a big company would want to buy it. If you were a big company, and you took over Iridium, you could blame all these current problems on a previous chairman, and in fact Chris Galvin was a great guy to blame it on. But that wasn't happening."

In other words, Oliver was stuck with Dan Colussy, Herb Wilkins, and whoever else showed up on the "Newco" side. He was not in control of that, though, so he proceeded on the assumption that *someone* would buy it. He just had to make it easy for that someone to make it survive, hopefully with a government contract. Oliver was an old hand at Pentagon politics, so his first effort was devoted to building support within the Building. That meant physically walking from office to office, talking to every service branch and especially the senior communicators. "Wouldn't you like to have this system for nothing?" was his message to all of them. And all of them gave him the same answer: *No, not interested.* These were the stovepipers Jack Woodward had warned about.

Increasingly disgusted, Oliver found himself going higher and higher in the hierarchy—after all, he had carte blanche from the SecDef to get this done—and one day he happened to see the Vice Chairman of the Joint Chiefs of Staff leaving his office. Oliver hustled over and collared him in the hallway. He then launched into a passionate pitch for the Iridium phone, telling him how they could take the existing device, "amp that baby up and have connectivity in places we've never had it before."

The general was unmoved by Oliver's pleas.

"The phone is a piece of crap," he said. "We don't need it."

"You dumb fuck," said Oliver, and turned on his heel.

Both men had been speaking loudly, so Oliver walked across the hall to the office of Hugh Shelton, the Chairman of the Joint Chiefs and, more to the point, a decorated Green Beret in Vietnam—exactly the kind of Special Forces soldier who would be apt to use an Iridium phone.

"Did you hear that conversation?" asked Oliver.

Shelton said that he had.

Oliver asked him if there was any reason the Pentagon shouldn't take advantage of the only truly global handheld satellite phone system ever put into space. And the chairman said he didn't see any reason why the government wouldn't do that.

"I'm going to take that as a yes," said Oliver to the highest-ranking military officer in the United States, "and I'm going to run with it."

"Fine," said Shelton.

But Oliver very quickly picked up on the major fear of both Motorola and the government—that a de-orbit of the constellation, even if delayed for years, would create a nightmare scenario of burning satellite components crashing to Earth and killing people. Oliver had university-level training in statistics and ran the numbers himself, concluding that the

NASA study estimating a 1 in 249 chance of injury was ridiculous. "But it was out there and I couldn't destroy it," he said. "I couldn't get rid of that stupid number, and the fear was real, the fear was tangible." It wasn't that the government cared that much about falling space junk—168,000 kilograms of space debris already entered the atmosphere every year—but everybody cared about the kind of debris that gets attention.

Bureaucrats have a long memory for anything that makes headlines. In 1978 and 1983 there had been crashes of Russian spy satellites with radioactive components aboard, one of them strewing debris all over northern Canada. The Salyut 7 space station had gone into premature orbital decay in 1991 and hit a town in Argentina, panicking the local population. Most famous of all was Skylab, the space station that broke apart in 1979 and plummeted to Earth in Western Australia, creating an international media event and comedy fodder for radio disc jockeys the world over. (The Australian ambassador to the United States presented Jimmy Carter with a ticket for littering.) But these events were rare. Most of the fourteen thousand reentries ended with a quiet plop into the ocean.[41]

Oliver eventually gave up on the statistical analysis altogether. He had worked for years with nuclear submarines and knew that, no matter how safe you could prove them to be with numbers, the word "nuclear" made people irrational. He had to face the fact that the words "crashing satellites" had the same emotional effect. So he started using the following argument instead: Every time an airliner takes off, that plane is insured for seventy bucks against things falling off it. Why? Because things fall off airplanes all the time. Yet here are all these insurance companies competing for this business, and we've all seen airliners flying over cities. There are many parts that can and do fall off airplanes, but there is only *one part* from each of these sixty-six satellites that can reach Earth, and when it does reach Earth it will probably be microscopic. So the value of the insurance should be much less than seventy bucks per satellite. Add to this the fact that 70 percent of Earth is water.

A bunch of it is desert. A bunch is like Northern California—it's land but nobody lives there.

This folksy argument eventually carried the day with most of the people Oliver had to convince, and he even found a promising rivulet of support: the officers in charge of Command, Control, Communications and Intelligence, or "Three See Eye." People like Jack Woodward. People like Colonel James Mattis, senior military assistant to Rudy deLeon, the Deputy Secretary. DeLeon was the ultimate powerbroker inside the Pentagon and the person Oliver most wanted to convince. Unfortunately, that was going to be a tough sell. DeLeon was a civilian and a classic Pentagon politician, and after "taking the temperature of the Building" on Iridium, he was not inclined to help. That's why Oliver went to work on deLeon's assistant, Colonel Mattis, who had been a front-line war-fighter, the commander of the Seventh Marines, and the leader of a rifle company, a weapons company, and an assault battalion called "Task Force Ripper" in Operation Desert Storm. Oliver's best friends in the Building were people who had been in the trenches—and they wanted that handheld satphone! But in the meantime he had to find a law authorizing the Pentagon to do what he wanted—accept liability for the safety of the satellites.

It's not clear exactly who unearthed Public Law 85-804. Dorothy Robyn first heard about it from Rick Stephens, head of the space communications division at Boeing, who said it was used all the time for rocket launches. ("That's when the lightbulb went on in my head," said Robyn.) Colussy was aware of it, remembering the time during the Vietnam War when the government agreed to indemnify Pan Am. He had passed along that suggestion to Isaac Neuberger, his personal attorney, who was working virtually full-time now on figuring out how to buy Iridium, and Neuberger had passed it along to lawyers in the White House. As luck would have it, one of those lawyers was Linda Oliver, wife of Dave Oliver, who worked in the Office of Federal Procurement Policy. So Dave told his wife to stand by, he was going to need her for a special assignment.

The call Dave Oliver was waiting for—from Chris Galvin—finally came on a Friday night while Oliver was attending a wedding. He had left numerous messages for Galvin and felt like he was being ignored, so he excused himself from the ceremonies and picked up the call. Then he got right to the point.

"I need to meet you in your office tomorrow," he told Galvin.

"Tomorrow is Saturday."

"I need to meet you in your office tomorrow. It's very important to the government."

The next day, Dave and Linda Oliver took an early-morning flight to Chicago and a cab to The Tower. They were quickly ushered into Galvin's office, and again Dave Oliver was abrupt.

"Look, if I can get you immunity from lawsuits, will you give me this thing?"

Galvin hesitated, then suggested that perhaps Oliver's wife would like to wait in the outer office while they talked. Oliver didn't like his tone. It sounded like he was suggesting "the little woman can read some magazines while the men are talking."

"That wouldn't be a good idea," said Oliver. "She works at the White House. She's the number two person in federal procurement policy. I brought her here because only she can sign off on this."

Galvin then proceeded to tell Oliver how expensive it was to keep flying the Iridium constellation and how unfair that was to Motorola. Oliver kept reemphasizing why he was there: *Will you play ball with us if we get you a solid buyer?* But Galvin said he'd already been through that—there were no buyers. Oliver said he could probably come up with a government contract that would bring some new buyers out of the woodwork. Galvin pushed back again, saying he had already wasted too much time. And then there was the insurance issue—Galvin didn't want Motorola to get sued, and destroying the satellites now meant his company would be fully covered. Oliver said he would get him the same insurance, somehow, someway. "Will you play ball with me?" he repeated.

An hour later, Oliver had an agreement in principle with Galvin, but Galvin wanted it in writing. He wanted a decree, or an authorization, or a letter, or whatever document Oliver could produce for Galvin to take to the Motorola board of directors. Oliver said fine, he'd get it.

But he couldn't get it. At least not immediately. When Oliver got back to Arlington and placed calls to his counterparts at the Justice and State Departments, he was not happy with what he heard. Lawyers working for Attorney General Janet Reno and Secretary of State Madeleine Albright were skeptical. It sounded like a corporate bailout. They were not sure it was legal. Not only that, the Iridium Interagency Working Group had been laboring for five months to force Motorola to do exactly the opposite—buy additional insurance to indemnify the government! The committee preparing for the Mir de-orbit had been so disgusted by government exposure to a private initiative that it had been studying the issue with an eye toward requiring *all* corporations in the future to buy additional insurance especially to indemnify the United States. Now the government was supposed to give *them* insurance? Oliver still had a lot of work ahead of him.

In the course of his full-court press, Oliver discovered his old friend Dorothy Robyn was already on the job at the White House, and he knew right away that she would understand—that this was a war where they were going to need to cross lines and kick ass. Oliver told her there were logjams in the Building, and he wasn't sure he could get past the cautious Deputy Secretary Rudy deLeon. "Let me assist on that," she said. "I'll get Gene to call him." But Robyn was worried on other fronts, she said. She was following the Colussy negotiations through reports from Tom Tuttle, and he had told her that Motorola had turned flaky. "It will seem to be going well," Tuttle told her, "and then in classic Motorola style, they ask for more. Two executives keep coming up with more and overruling the negotiating attorney." Those two executives, she had found out, were Motorola attorneys Bruce Ramo and Kathleen Massey. Both seemed spooked by the idea of the Colussy group taking over the system and then being unable to manage it. Mike Kennedy, one of the Motorola lobbyists, had been especially blunt. "It's not about money," he told her. "It's a legacy problem for us. We're being hit hard by the Wall Street analysts. The company would very much like to keep the Iridium system going, but we're fighting this mind-set. It's about finding a replacement company that can keep the system up there but distancing ourselves from it so Wall Street won't continue to hammer us."

Meanwhile, Colussy continued to field thirty to forty calls a day from the Iridium war room in Glen Oban, Maryland. Many of those were strategy calls with Isaac Neuberger. Neuberger had tapped into his network of rabbis and—as luck or divine intervention would have it—found a personal friend who worked as outside counsel for Motorola in Chicago. Neuberger now knew every move Motorola made, which created some confidence that, as each day began, they could at least know whether the de-orbit had been ordered or not. Also helping was Dannie Stamp, the Motorola employee who had supervised the launch of the constellation and who had already made it known to Colussy that, if the satellites were saved, he would like to work for the new company. Slowly Colussy was building a team, and what all the players had in common was shock and outrage that Motorola might bring the constellation down.

When the August 9 de-orbit deadline passed, Colussy assumed it had been yet another example of Keith Bane crying wolf. Ted Schaffner didn't even mention it in his daily phone call to Colussy, and Colussy didn't bring it up either. He told Schaffner he would be ready to close, with all his investors in place, no later than August 31. He was nowhere near being ready to close, but he felt he had to tell Motorola whatever Motorola wanted to hear.

Fortunately the Japanese were acting friendlier in regard to a deal. In this case the go-between was Yoshiharu Yasuda, the former President of Nippon Iridium, who had a good working relationship with both Ed Staiano and Tom Tuttle, and Yasuda-san was working hard to bring aboard Kazuo Inamori. At last Colussy got the breakthrough call he was waiting on: Inamori would be at the offices of one of his companies in Irvine, California, on August 22, and he would be happy to meet with Colussy and hear his ideas. Colussy immediately started making travel arrangements. But then Ed Staiano called from somewhere in Pennsylvania to say, "I hear that pressure is building in The Tower. Call Keith Bane."

When Colussy did call, Bane was curt and to the point: "Should we de-orbit now or risk continued operations?"

Colussy had been through this dance before. "Everything is almost finished, Keith. We're less than two weeks away."

"I want a letter from Boeing telling me that they are responsible for everything."

"I thought we'd been through that."

What was going on? Kathleen Massey, Motorola's senior litigation counsel, had been at meetings in London with Isaac Neuberger, the underwriters, and Boeing lawyers, and all the details had been hammered out to make Motorola happy. Why was Bane going back to this tired refrain of wanting indemnification from Boeing? The insurance was in place. It was more insurance than Motorola had ever bought. It was more insurance than was needed. It was as much insurance as any company would ever write on any satellite constellation.

"From our standpoint," said Bane, "there's no reason to take any risk. We're the manufacturer. Manufacturers of guns get sued. Manufacturers of tobacco get sued. The reason companies take a risk is to make money, but there's no money to be made here. Right now we have an advantage because we have a mass de-orbiting policy that's good for eighteen months. We benefit from a policy written for Iridium that was a mistake. It's broad and it includes all liability. But the message we're getting is that the insurers have learned from Iridium and they don't want to be stuck on this. That's why your policies are not as broad. Last year Motorola wrote off $2.119 billion, and we're continuing to pay $2 million a week, and no one wants to help us pay for this."

Bane had turned the call into a lecture on Business 101, and Colussy didn't like his tone. He was silent, letting Bane go on for a while, and then finally agreed to talk to Boeing one more time.

After a few minutes they conferenced in Rick Stephens, Vice President and General Manager of Boeing's space communications division. Stephens—a veteran Boeing executive, an ex-Marine, and the tribal chairman of the tiny Pala band of Luiseño Mission Indians—had been talking to Motorola for several weeks, and he knew this conversation was going nowhere. ("Keith Bane and Ted Schaffner kept saying to me, 'Rick, you have to indemnify us or this is not going to work. Get it off our plate! Get us out of this!'")

Stephens repeated that Boeing had accepted liability for everything it could accept liability for.

"Your company is thinly capitalized," said Bane to Colussy. "Your business plan is not demonstrated. I have to assume, Rick, that Boeing wants the satellite business to be successful. What you're doing can be a road map for the next generation of satellite systems. I know you want success for the launch industry. The way you can be certain to do that is to get me this indemnification."

"I can't do that, Keith. Not what you're asking for. You want us to pay for everyone's mistakes including yours. No one can do that. We could be sued if we did that."

"Tell me what you *can* do."

"Okay, I'll ask Harry."

When they hung up, Colussy was troubled but not *that* troubled. The most annoying thing about the whole conversation was that it was going over issues that had first come up six weeks earlier and had already been disposed of.

And then, two days later, on August 22, Schaffner called with the bombshell:

The Motorola board had just ordered Chris Galvin to destroy the constellation.

Motorola was giving formal notice to the government.

It could happen within forty-eight hours.

There was nothing anyone could do to stop it.

Iridium was finished.

Schaffner expressed no emotion as he ticked off the points.

Just like that, it was over. Prince Fahd didn't matter. Prince Khalid didn't matter. Secretary Cohen didn't matter. Public Law 85-804 didn't matter. Nothing from the White House or the Pentagon or anywhere else had made the slightest bit of difference. Nothing mattered except the single-minded determination of Motorola to get what it wanted and everyone else be damned. All the work, all the accumulated goodwill, all the promises of help had come to naught. Iridium was history.

Chapter 13

CHICKEN LITTLE

AUGUST 24, 2000
SIX-CONTINENT CONFERENCE CALL,
IRIDIUM COUNTRY CODE 8816

"There's no better time than now to de-orbit," Ted Schaffner kept saying.

As Colussy listened to Schaffner's deadpan recital of decisions that were fatal to everything he'd worked for the past five months, he went into a slow burn but didn't let Schaffner hear his anger.

"There's too much risk for us," Schaffner repeated. "There's no better time than now."

Colussy hung up the phone, slightly stunned.

Dannie Stamp called to confirm the worst. He had been ordered to Schaumburg and installed at a desk in the reception bay outside Ted Schaffner's office so that he could supervise the destruction of the satellites, and technicians had already started writing the suicide software. "Chris wants to be sure he can grab my nose in five minutes," said Stamp.

Colussy called Schaffner back and got Keith Bane on the call as well. The Motorola board had made the standard very simple, Bane said. If it was safer to de-orbit now, using the existing insurance policy, then Motorola would do that rather than risk lawsuits later. Chris Galvin was leaving on a business trip to China and, as he left, issued an order to "decommission the constellation." An official news release had already gone out, and Reuters was reporting that Motorola was "finalizing a schedule" to destroy the satellites. Bane sounded like he was finished talking and wanted to get Colussy off the phone.

"We very much appreciate the time and effort the Colussy team has put into the process," he said, "but no business plan has come into being that makes any sense."

Colussy quickly placed calls to Ed Staiano, Tom Tuttle, Herb Wilkins, Ty Brown, and Mark Adams, wondering how this could happen when they'd just made such massive inroads with the government. Not only that, they'd made inroads with Kathleen Massey, the feared Motorola in-house lawyer who had refereed all the arguments about indemnification.

Tuttle said, "I'll schedule an emergency meeting of the Iridium board."

Ty Brown said that the FCC would be willing to go into emergency session so that Chairman Kennard could discuss whether there was anything the government could do.

But these maneuvers sounded like Hail Marys. For the first time Colussy thought he might be licked. He started making notes about getting his expense money back from Motorola "for leading me down the garden path."

But then he reconsidered. The Motorolans were blowing him off, so what did he have to lose? He was tired of their tough-guy blather. If Motorola wanted to play hardball, he'd play hardball, too. He called Tuttle back and said, "I like your idea. Get the Iridium board together and let me speak to them."

A day later, Colussy was patched into a conference call with the Iridium Board of Directors and their assistants, some fifty people ranged across the planet, representing virtually every time zone and six of the seven continents. Colussy had decided to come out swinging.

Motorola has announced that it's going to destroy the assets that you have paid for, he told the gateway owners. *They say that we won't be allowed to save those assets because we're undercapitalized. But there are only two* *reasons why we're undercapitalized. One is Motorola's constant demands for liability protection, demands that become more and more outrageous the longer they talk about them. And the second reason is that Motorola has obstructed us at every turn by being on-again, off-again: stating the criteria that would be needed for an asset purchase, then changing the criteria; talking, then not talking; constantly raising the bar of what it expects from us. In such an environment, no wonder it's difficult to get investors to commit. I've gone through three months of increasing demands from Motorola, and each time they increased the demand, I've met the demand, and each time we've reached a plateau, Motorola has created a new mountain. Boeing has indemnified Motorola for everything that is within Boeing's control. I have gotten additional insurance for everything the constellation will need in the future. The cost to the new company is $1.1 million per year for insurance alone, insurance that doesn't even include the operating company. It's insurance for the benefit of Motorola long after Motorola has ceased to have anything to do with the satellites. Motorola has acted in bad faith. I will continue to fight as long as a fight is possible, and any gateway owner who is considering an investment should know that his investment will be safe.*

When Colussy finished his remarks, he realized that Keith Bane was on the call—and Bane was furious.

"Motorola was clear from the start!" he barked back. "We cannot accept an expansion of liability!"

Colussy usually took extensive notes, but this time he just listened to Bane's increasingly emotional rant, imagining his face turning beet red, and absentmindedly wrote the words "CHICKEN LITTLE" on his notepad. They enjoyed it, he thought. These Motorola guys enjoyed controlling when the sky would fall.

But Bane's decision to speak was like throwing blood into shark-infested waters. Every gateway owner blamed Motorola for the failure of Iridium, and so they all started spitting questions at Bane. Tom Alabakis, the board member representing the African gateway, demanded to know when the de-orbit would occur.

"It's imminent!" said Bane.

You won't give us a time and a date?

"We will not," said Bane.

"Will you give us twenty-four hours' notice before the constellation is destroyed?" asked Tom Tuttle, the Iridium lawyer.

"No, we will not," said Bane.

Atilano de Oms Sobrinho, the amiable Brazilian universally liked by the gateway owners, tried to restore calm to the proceedings, pleading with Bane to delay his decision for a few more days. Bane grumbled that he was sorry, but the decision had already been made.

If any performance was designed to portray Colussy as the white knight and Motorola as the evil empire, this would be it. Later that day an open letter to the "Valued Motorola Customer" appeared on Motorola's website. The pertinent sentence read, "Please be aware that any and all remaining Iridium service could end at ANY TIME—WITHOUT ADVANCE NOTICE." But the letter went on about how hard Motorola had worked to make sure the Iridium constellation was saved, and how "going forward, Motorola will continue to look for new opportunities that will provide a path to the future."

Provide a path to the future? What did that even mean? Motorola was piling sandbags across the road to the future. Motorola was making any kind of future impossible.

After the meeting, Colussy had an empty feeling. He may have won the battle and lost the war. Rick Stephens faxed a letter from Boeing about how much respect he had for the professionalism of Colussy and Staiano. "While I personally feel bad that this has resulted in a non-resolvable issue, I can't accept the intent of Motorola's terms," Stephens wrote.

Colussy called Stephens. Of course he understood—*no one* could accept the intent of Motorola's terms—and then he continued to make calls for a while, talking to his usual team, thanking all of them, feeling like a football coach whose players just lost the national championship in the last three seconds of the game.

But one person refused to accept his resigned mood: Dorothy Robyn. Robyn had spent the three weeks since the Castle Harlan announcement networking, scheming, and working the bureaucracy. As soon as she got wind of Dave Oliver's involvement, she had dropped his name all over the Department of Defense—sometimes annoyingly so, she would find out later, since calls from the White House were not always welcome at the Pentagon—but she had continued to fight for the system even when she was warned to back off the "touchy area" of seeming to support a commercial entity.

"Are you sure we're doing the right thing?" she asked Oliver one day in the midst of a strategy session.

"Trust your instincts," said Oliver.

And her instincts told her that Chris Galvin was on the wrong side of history.

When Colussy called Robyn after the disastrous meeting with the Iridium board, she said, "Don't despair. Don't worry. We've got things working. It's not over."

Colussy didn't really believe her, but she did have things working. During the week leading up to the Labor Day weekend, as Dannie Stamp sat in The Tower waiting on a call from Galvin ordering him to throw the death switch, Dorothy Robyn stayed on the phone with her counterparts at the FCC and Justice, trying to get through to someone who could stop Motorola. She also stayed close to Motorola's Washington office, because she wanted to know the exact status of the de-orbit process. (Motorola was wiring the funds to the insurance underwriters, indicating that its de-orbit plans were real.) When Barry Lambergman, Motorola's Director of Government Relations, informed her that "we will be taking the first irrevocable step in the implementation process" within forty-eight hours, she even went so far as to ask about the specifics of the suicide software. After it was uploaded, how long before the first satellite started to tumble? Could the software be canceled? If it were activated and then canceled, how long would they have? "We might lose one satellite if we did that," she was told. Okay, she said, she just wanted to know how long she should keep fighting, and whether they could lose a few satellites but still save the rest.

Strategically, Robyn knew that the national security card was now the only way to stop Motorola, but tactically she was still using safety and Africa. Somebody at Motorola talked to the press on August 23, resulting in a *Washington Post* reporter turning up at NASA, and that ended up causing confusion that she used to her advantage. The press tended to call NASA to ask Iridium questions, even though NASA had almost no involvement, but it gave the government the chance to talk about risk to the public. Kathy Brown, chief of staff at the FCC, was still helping on that front by combing through regulations for anything that would require government approval prior to a de-orbit. Brown even urged her boss, FCC Chairman Bill Kennard, to place a call to Chris Galvin directly. Kennard was well acquainted with the Iridium problem by that time, having gotten an earful from Jesse Jackson earlier that month when the two men shared a ride on Air Force One. Kennard did call Galvin to express his concern—"and I was somewhat surprised when he pushed back—it wasn't a pleasant conversation." Kennard pointed out that the FCC had licensed Iridium for a specific purpose and that the de-orbit was not a part of that license, but Galvin replied, "You may have licensed it, but it's private property."

Somewhat alarmed by Galvin's intransigence, Kennard authorized Kathy Brown to call an immediate meeting at the FCC for the next day, requesting the attendance of Motorola—specifically Keith Bane and Rick Severns—as well as Colussy, Randy Brouckman, Rick Stephens from Boeing, Tom Tuttle, and Kennard's old friends, Ty Brown and Herb Wilkins. But Robyn was upset when Motorola seemed to be "making a conscious decision to send people to the meeting who are not decision-makers." Robyn found Dave Oliver at home and told him Motorola was stonewalling, and during the few remaining days before the Labor Day weekend, the three coconspirators—Robyn, Oliver, and

Kathy Brown—came up with a plan. They didn't quite come out and say it, but they tacitly agreed: Motorola was now an aggressor and a foe that had to be eliminated. As one Iridium executive put it, "Motorola is taking a calloused approach. They've concluded that this is a no-win. They look bad either way. If Iridium succeeds after they're gone, they look bad. So when they take this up to the board, it's not a huge priority." When one of the Motorola executives said, "We've been at this since March *out of the goodness of our hearts*," it made everyone's blood boil.

The FCC meeting slowed Motorola down, but only slightly. Colussy prepared a chart laying out the six types of insurance he had agreed to purchase, amounting to $1.1 million a year in premiums, while Brouckman expressed surprise and dismay that all the positive meetings of the last three weeks were being jettisoned at the last minute. Rick Stephens had flown in from his Boeing office on the red-eye and had to leave the meeting at 11:45 A.M. to fly back to the West Coast, but he made good use of his time, going down the list of Motorola's most ridiculous demands, saving the best for last: Motorola's insistence that Boeing take legal responsibility for Motorola's own simple negligence. Ted Schaffner then described situations in which Iridium would need to be de-orbited but various government agencies would refuse to give permission, and Iridium attorney Tom Tuttle said, "Those scenarios are wild and ridiculous, Ted." Nevertheless, Stephens volunteered that Boeing would agree to remain in control of the satellites until either an "acceptable replacement operator" was found or until de-orbit. Boeing would even agree never to sue Motorola—the only legal action Boeing could take would be against the government. None of this was good enough for Schaffner.

"Why would anyone come to Motorola with these ridiculous requests," Stephens asked him, "if we, Boeing, are operating the system?"

The FCC regulators tried to paint Motorola into a corner, but Mike Kennedy, Motorola's Director of Global Telecommunications Policy, kept insisting on Motorola's right to de-orbit. At one point he even brought up the Smith & Wesson argument: "Who knew they would get sued for manufacturing guns?" Motorola believed that there was some way to protect itself from any lawsuit past, present, or future, including theories of legal liability that had not yet been thought of—to the point that Tuttle finally said, "Mike, there are some things that can't be put between the four corners of a contract." But that was really Motorola's point. "There will never be a better time to de-orbit than now," said Kennedy, referring to Motorola's insurance coverage for the next eighteen months. "Every other scenario puts us in a worse position."

"There will never be a better time"—this had become the refrain of Bane, of Schaffner, of Galvin, of presentations to the Motorola board. It was the way every meeting started and ended. And ultimately it was what turned everyone against Motorola. Because the unspoken conclusion of the sentence "There will never be a better time" was "for Motorola." There's a certain way businessmen speak to one another to show how tough they are—they often take a delight in their hard-nosed lack of sentimentality—so saying "There will never be a better time" was fine for the boardroom and for internal Motorola meetings. But they were saying it to government regulators, and career military men, and people like Dorothy Robyn and Kathy Brown who had dedicated their lives to service. They were saying it to people who really meant it when they talked about the humanitarian goal of providing phone service to the poorest parts of the world. They were saying it to people who were not going to gain a single penny from the success or failure of Iridium, but nonetheless felt it was important for soldiers and spies and scientists who otherwise had no advocates. For Motorola executives to constantly set the standard for action as "what's good for Motorola" was a course guaranteed to turn them into the enemy. It was a mistake Bob Galvin never would have made. Every speech he ever gave included love of country, even when that country was China. It was a mistake Paul Galvin wouldn't have been capable of. He provided police radios because he thought the police were the finest patriotic servants in America. He built Walkie-Talkies so that America's communications would be better than Nazi communications. So for Motorola executives to wade into Washington, D.C.—against the advice of their own lobbyists—and say, "You don't understand, it's all about money for us," was a position that turned them into the ugly corporate guys. That was why the results of the FCC meeting were so positive for Colussy. It made Kennard a true believer in Iridium, and it made Chris Galvin a pariah.

The other new wrinkle from the meeting was that Isaac Neuberger brought up Public Law 85-804, which allowed the government to directly indemnify private parties in the case of "extra-hazardous" situations that were in the interests of national security. Dorothy Robyn now believed in 85-804 as *the* solution. "It was tricky politically," she said, "but if you could get the Department of Defense on board with it, it was doable." To get that ball rolling, she called the two top defense specialists at the National Security Council and asked them if they thought it was feasible. Unfortunately, they both told her that they thought the situation fell short of the legal definition of "extra-hazardous." "I'm not convinced that the national security argument is strong enough," said Ed Bolton, and he added that, based on his conversations with the Pentagon, "there is no support for this approach." Still, just the mention of it required Motorola to spend some time preparing internal memoranda as to whether that would be acceptable. They had bought forty-eighty hours.

Motorola was now losing credibility every time a Motorolan opened his mouth. The number one telecommunications regulator in America was upset with Motorola, and Motorola's whole identity was defined by telecommunications. "My greatest concern was that this would cause a public crisis and pandemonium," said Kennard. "People would be really concerned about news reports that these sixty-six satellites were going to come crashing down. I didn't want a public crisis. I thought the FCC would be vilified by the public for allowing this to happen."

But despite everything Kennard could do, Kathy Brown eventually called Robyn to say, bluntly, "Motorola has decided not to play."

How can we make them play? asked Robyn. This was now a brawl.

And Brown came up with two ways. First, Motorola wanted a waiver from the FCC requirement to inform customers sixty days in advance about termination of service. Motorola had given notice, but it was not necessarily *formal* notice. Second, Motorola's original license called for a "controlled de-orbiting plan," and it would be possible to challenge the word "controlled." This was news that thrilled Robyn. She urged Brown to be aggressive on both fronts—refusing the waiver and challenging Motorola on the basis of the controlled de-orbit. The only drawback, Brown pointed out, was that Motorola would undoubtedly seek a court injunction.

No matter, said Robyn. "It will buy us a few days."

The "Stop Motorola" effort had gathered steam. Neuberger called Colussy with good news from his spies inside Motorola: "There's a power struggle going on there." Unfortunately, the power struggle was mainly about timing, not principle. Robyn spoke to Tuttle the day after the FCC meeting, only to be told, "Motorola is not talking now, they decline to give us specifics, and we have no confidence that they won't take action quickly."

Chris Galvin was not giving in. If anything, he was becoming more defiant.

The forty-eight-hour deadline passed, but it was impossible to find out why. A hasty meeting was called by the Department of Justice that included representatives from the Pentagon, the National Security Council, the FCC, NASA, the FAA, and the State Department, as well as White House counsel and Robyn, representing the National Economic Council. By that time Robyn had talked to Motorola's Washington office several more times and heard even more "ridiculous scenarios." Motorola's latest fear was that Al Gore would be elected President and that he would refuse permission to de-orbit the constellation no matter how much money the operators were losing, simply because he loved the technology. This led to requests for insurance against "irrational acts of government agencies." On another call a Motorola lawyer said, "What if the government requires us to send the Space Shuttle up to remove space debris?" In yet another conversation, Bane said, "Our concern is that, as more commercial satellites are put up, environmentalists will come up with new regulations." At the end of the day, there were three key decisions that came out of the DOJ meeting. The first was to have chief economic advisor Gene Sperling call Galvin and try to turn him around. The second was to take all references to the NASA risk study out of the public news release. (It would be available only via Freedom of Information Act requests.) The third decision was to haul everybody over to the Eisenhower Executive Office Building—Iridium, the Colussy group, Motorola, Boeing, Pentagon people, DOJ people—and "do it the way we do airline mediation." As one mediator put it, "You bring in the CEOs and just don't let them leave the building until it's settled."

What the mediator didn't count on was the obstinacy of Chris Galvin—there was no way he was going to show up for any meeting.

So Robyn went to the top—to Gene Sperling. "We should go to the mat to keep these satellites in the sky," she told him. "We have two choices. We can bring in Galvin and Stonecipher and Colussy and shake them by the shoulders and tell them to finish the deal, or we can deal solely with Galvin, since that seems to be where the problem lies." Robyn believed that the negotiations were being "driven by Motorola lawyers with an overly narrow perspective," but if someone senior in the White House called—someone like Sperling—Galvin could overrule his team. *If you're going to make the call,* she told him, *this is the time.* "Maybe Chris Galvin has made a thoughtful decision," she said, "but it's not clear that this is the case. The government needs to make darn sure that Motorola has thought this through. Has the decision been made in the crucible of all facts, including the administration's interest? We're not sure this has happened. Dan Colussy's problem at this point is that he has to make up the capital he lost when Castle Harlan pulled out, and he feels that, if he could make a solid deal with Motorola, he could raise that capital. That's why we don't understand why Motorola would suddenly change course."

Sperling did agree to call Galvin, but that was more easily said than done. First Galvin said he was unavailable, then said he would be available for the call in three days, then said he thought the call should be dealt with by Keith Bane, not himself. But the White House gave a firm no to that notion. Sperling was the number one economist in the nation: senior people talk to senior people. This call had to be with the CEO. And Galvin finally agreed to talk to Sperling.

The talking points had been worked out long before.

"Chris, we appreciate the cooperation you have shown in working with the federal government on Iridium over the past five months," Sperling began. "I know that Motorola has made a decision to proceed with the decommissioning, and you may well be frustrated that the federal government doesn't seem to want to accept that decision. But let me explain why we have such a strong interest in the future of the system. First and foremost is public safety. Second is the future of the satellite telecommunications industry. Third is national security. The Department of Defense has invested $200 million to build a secure gateway in Hawaii and to develop a secure handset. Can they get by without Iridium? Yes, but we would much prefer to continue to use it. Fourth, the President has very much hoped that Iridium could provide a way to bring telephony to sub-Saharan Africa and other developing regions, including Native American reservations."

But these were clearly not words that had any meaning for Galvin. He may have even felt relieved when Sperling made national security third on his list, since that was the reason that could stop him. The Iridium satellites belonged to a private consortium, said Galvin, adding that he was not inclined to pursue goals that had nothing to do with the goals of his company. "I would like for you to stand down," he told Sperling. The government concerns, said Galvin, were incompatible with the needs of his company. Motorola had wasted too much money on delays. He needed to destroy the satellites and he needed to do it now.

Sperling felt he had to buy time.

"What would it take then," said Sperling, "for you to settle with the guys who want to buy the system?"

"Total unconditional release of liability forever," answered Galvin.

"I need a few days. Can you work with us? We want to leave no stone unturned."

"I can't agree to that."

"Well, I wish you would reconsider."

"I can't. Too much time has passed."

Robyn was listening in on the call and thought that Galvin sounded desperate. She had been wrong. It wasn't lower-level lawyers with narrow perspectives. It was Galvin himself who was the problem.

Two days later, Galvin placed follow-up calls to both Kennard and Sperling to say that Motorola could not accept Boeing's terms and that was that—discussion over. No more negotiating. Asked to wait for the government to research Public Law 85-804, Galvin refused. Kathy Brown felt that was a slap in the face to the government. Motorola had given assurances at the FCC meeting, and those assurances had now been rendered meaningless. The curious thing about Galvin's calls was that somehow he believed he could convince Kennard and Sperling of the righteousness of his position. Instead they just thought he was inflexible and, based on what they knew about Colussy's insurance, irrational.

Meanwhile, Dannie Stamp got the call he had been dreading: "De-orbit all the stuff—Chris says it's over." His satellites were doomed. The only thing holding things up was an insurance policy rider. Motorola's main insurance company was in London, but the policy was jointly held with companies in France, Germany, and Belgium. All of those companies had to send a fax to London saying they had been paid, then a fax had to go from London to Stamp. Once that happened, he was to call the SNOC and start the de-orbit.

Unaware that Galvin had ordered the de-orbit, Robyn pressed on, placing calls to her Motorola contacts, asking them why Public Law 85-804 was being ruled out. The answer was "because it would require an annual renewal"—something that had never before been mentioned and was not true. Robyn insisted on talking to Motorola lawyer Bruce Ramo, and he finally agreed to look at Public Law 85-804 himself, only to tell her that, in his opinion, "it doesn't apply to Motorola because the work is not ultrahazardous and we're not a contractor and the scope of the indemnity we need appears to be outside this statute—we want Boeing to put something on the table."

After Galvin placed the "final" calls to Kennard and Sperling, Keith Bane thought it was over. He called Colussy again to say, "The deal is off. Boeing won't agree."

"You knew they wouldn't agree," said Colussy. "We told you they wouldn't agree. That's not the reason. If that was the reason, you could have called it off long before now. You led us on. You owe us for our expenses."

"I'll think about it," said Bane curtly, and hung up.

Mike Greenberger at the Department of Justice sent out an e-mail to thirty-three people in various agencies and departments reporting the grim reality: "The full court press to try and achieve a sale has now almost certainly fallen through. It is now virtually a certainty that the satellites will be de-orbited. . . . Motorola will do its best to give us six hours' notice before software adjustments are made to commence the de-orbiting."

The State Department followed suit with a diplomatic cable to every embassy in the world, giving official "responses to host governments and media inquiries" once the satellites started crashing.

The NORAD Command at Cheyenne Mountain prepared its formal notification to Russia that nonnuclear objects would be entering the atmosphere.

And everyone moved on to other things.

Everyone, that is, except Robyn and Oliver.

When a sympathetic Motorola executive called Robyn on August 30 with a harrowing message—*"If DOJ is going to enjoin us, better to do it quickly"*—she organized a last-gasp White House meeting with the only five people left on the planet who didn't believe Iridium was dead: herself, Oliver, Colussy, Kathy Brown, and Isaac Neuberger.

Robyn ticked off the grim realities. September 1 was the key date, she told the group. Two days from now a wire transfer would be made to the European insurance companies that had the policy for the de-orbit, and after that there would be no turning back because the $5 million was nonrefundable. Colussy confirmed the $5 million figure, since he'd just talked to Keith Bane, who said, "Somebody give us $5 million or we de-orbit."

Robyn was still incredulous that the whole conflict could hinge on insurance. Maybe, she suggested, there was some eleventh hour way to get insurance that no one had thought of yet. And so Colussy agreed to make one more stab at getting additional insurance from the private sector. He didn't really believe this would work, but he called Jean-Michel Eid in the middle of the meeting, and Robyn took the extraordinary step of getting on the phone herself with experts from Aon, the leading aviation insurance company. What the Aon underwriters told her was that Colussy had already purchased "state-of-the-art coverage" that was "as broad or broader than the coverage that Motorola had." In fact, it was more than they usually wrote on space launches, which were considered much more dangerous.

"Besides," said Eid, "Iridium has been up there awhile and hasn't caused any problems. Satellites are safe."

Robyn's conclusion: this was not about the insurance.

After the meeting, Robyn shifted gears. The Africa card wasn't working. The safety card wasn't working. The insurance card had been tried and found irrelevant. It was time to play the only card she ever thought would work anyway: national security.

She told Oliver, "Get the DISA letter. Let's find out what the Pentagon will pay." She had heard a figure of $3.5 million a month—was that number still in play? But Oliver couldn't be sure.

"Unfortunately," Oliver told her, "no one in uniform gives a shit."

Oliver had been hoping for a Joint Chiefs directive saying, "Save these phones at any cost," but the best he could come up with was a fax from Doug Larsen, Deputy General Counsel for Acquisitions and Technology, containing a tentative promise of $3 million a month.

Oliver had the letter faxed over to Robyn, and Robyn gave it to Ed Bolton at the National Security Council. "Ed, we're in crisis mode," she told him. "I need your help. Here's a letter proving that the government needs Iridium."

Bolton told Robyn, "We're going to spend the weekend developing memos for Podesta"—meaning John Podesta, the White House Chief of Staff.

Then make this clear, she told Bolton: *It's not just that Iridium is needed for national security. The argument should be positioned to say that Motorola is* damaging *national security by the positions it is taking.*

Later that day she called Oliver.

"Motorola thinks the administration isn't serious," said Robyn to Oliver. "But Motorola is very sensitive to customers. And the Department of Defense is a huge customer. You need to address this head-on within the DOD, and the message to Galvin needs to be this: 'I don't know what we need to do to convince you, but this is not *just* your decision.'"

Oliver agreed. Oliver was on it. Meanwhile, Robyn called Leslie Batchelor, Janet Reno's lawyer at Justice, and said, "I'm looking for ways to encourage Motorola." Could Justice do anything else?

"Is there any remaining issue?" Batchelor said. "Does Motorola have everything it needs from the government?"

Maybe not.

On the Sunday afternoon of Labor Day weekend, Robyn called Colussy with the news. The ball was teed up. Everything was in place.

"Be at the Pentagon on Tuesday at 0900 to meet with Rudy deLeon, the number two guy after Cohen," she said.

Colussy flashed back to his conversation with Beverly Byron months ago, before he ever set foot in the Pentagon. "Rudy deLeon is the man you need to get to," Byron had told him.

Also at the Tuesday meeting, said Robyn, would be Dave Oliver, "Three See Eye" head Art Money, and the Pentagon general counsel. Rick Stephens and the Boeing lawyer, Bob Catania, would be there as well. Bob Catania was an expert on Public Law 85-804 and said that "it's used all the time" on Delta IV rocket contracts and that the Space Shuttle couldn't be launched without it. Tom Tuttle would be present representing Iridium. And she would be finding Keith Bane to request his presence.

Of course Colussy said he would be there bright and early.

Next Robyn hunted down Bane at his summer home on Lake Minocqua in Wisconsin. Bane said he couldn't make it but gave her Ted Schaffner's number. She eventually located Schaffner at a White Sox game in Chicago and told him that it would behoove Motorola to show up for the meeting. Schaffner told her that it might be too late—the satellites might be gone by Monday morning. She told him that Monday was Labor Day, and surely Motorola could wait one more day before making that decision. Schaffner reluctantly agreed to come to the Pentagon on Tuesday.

Tuesday turned out to be a long day, as most days at the Pentagon were. Meetings at 0900 never started at 0900. Colussy, Schaffner, and the others sat in the outer lobby, waiting for Rudy deLeon to be available. Fortunately, when lunchtime came, they were able to avoid the annoying food courts and get tables in the Navy Mess Hall, finest of the four Pentagon dining rooms. Afterward they returned to the lobby and waited all afternoon. It was almost dusk when the men were finally greeted by Oliver and ushered into SCIF 3E-928.

DeLeon, as the top decision-maker under Cohen, was a hard man to get to. The Deputy Secretary always was. The Deputy Secretary was the man who pulled the levers of the Pentagon on a daily basis. In many ways a meeting with the Deputy Secretary had an air of seriousness that meetings with the SecDef lacked, mainly because you knew that, at the end of the meeting, the decision was likely to be made one way or the other. The SecDef was the person you talked to at the beginning of a process. The DepSec was the last person you talked to.

DeLeon had a scholarly gravitas about him, enhanced by oversized wire rims that were slightly at odds with startlingly red hair that crested and curled, like the feathers of a cardinal. He was known as a brilliant student of American government and foreign affairs, a Loyola Marymount graduate who went on to the prestigious John F. Kennedy School of Government at Harvard. By the time he became Under Secretary of the Air Force in 1994, he had served on congressional committees in both the House and Senate and then spent six years in Pentagon jobs, where he acquired a reputation as a politically gifted strategist able to negotiate through thorny public embarrassments like the "Tailhook" scandal but, more important, problems with Congress. He eventually ascended to one of the most crucial Pentagon jobs, Under Secretary of Defense for Personnel and Readiness, and probably would have finished out the Clinton presidency there, but in January 2000 John Hamre abruptly resigned as DepSec to run the leading foreign policy think tank in Washington, the Center for Strategic and International Studies. DeLeon got the appointment for the lame-duck year because he was about as seasoned as they come. His soft voice belied his steely resolve to do what he thought was right for the country. And for most of the past few months, the word was that he was opposed to any heroic efforts to save Iridium. He was wary of the political climate, afraid of the precedent, not sure there was enough support within the Building. But Robyn and Oliver had been working on him in roundabout ways—Robyn through Gene Sperling, Oliver through deLeon's senior military assistant, James Mattis. "All Dave's meetings and memos had one overriding purpose—to bring Rudy onboard," said Robyn. That's why Colussy, unaware of the byzantine backroom maneuvering, was surprised to be meeting with deLeon at this late stage.

DeLeon welcomed the delegation into the SCIF and immediately took the floor. He knew everything about Iridium, past and present. He talked about "problems with the Iridium business model that have brought us here today." He noted that the Department of Defense was not the only agency that had an interest in the Iridium phone, but that the Pentagon had become a "clearinghouse" for the concerns of everyone. He mentioned NASA, the State Department, the Treasury Department, the Justice Department, the White House, the FAA, the Coast Guard, the DEA, the Secret Service—as though to say, *I'm not just speaking for the military when I speak today.*

"The Iridium system," he said at last, "does have value—for ships at sea, certainly, and in Alaska and in other places. It's not exclusively for military communications. It has law enforcement value, and it has value in hunting down global rascals. I think we should save it."

So far Colussy was liking what he heard. DeLeon then turned to Ted Schaffner.

"Mr. Schaffner," he said, "I've noticed that the day has all but passed and I'm looking out the window here and I haven't seen any falling satellites and, as far as I can tell, nothing is crashing to Earth, so I have to assume you haven't followed through on your threats."

Schaffner didn't say anything. He was curious to see where this was going.

"I have a message for your CEO," said deLeon. "Tell Mr. Galvin that if I hear one more *fucking* threat to bring these satellites down, then his corporation is going to have a really hard time doing business with the Pentagon in the future."

The room went still.

"I will give him that message," said Schaffner.

"All right," said deLeon, "we should all now work together to get the constellation indemnified so that we can move forward without destroying it."

And just like that, at the eleventh hour, the dogs were called off. Motorola was on notice that it should never again mention the word "de-orbit." The Pentagon started actively working to guarantee that Motorola could never be sued, no matter what happened to the Iridium constellation.

Did Colussy dare think it was over?

The private battle had been won. Dan Colussy's public battle was just beginning.

Chapter 14

FOUR DEAD BIRDS IN THE SITUATION ROOM

NOVEMBER 20, 2000
IRIDIUM SPACE VEHICLE 79,
TUMBLING THROUGH THE MESOSPHERE, APPROACHING THE STRATOSPHERE

And now, like a cosmic joke, Iridium Space Vehicle 79 was falling out of the sky. Other Iridium satellites had failed, become comatose, drifted off course, floated into the nether regions, but none of them had ever actually lurched into a crash orbit. After two years of chaos, after the Iridium system had almost died a dozen times, after thousands of powerful people had written it off or actively worked toward its destruction, the last thing Dan Colussy needed was a satellite defying the odds of a trillion to one and plummeting into a populated area. Sure, it wasn't likely, but nothing for the past eight months had been likely. One particularly eccentric congressman kept comparing Iridium to the Skylab scare of 1979. It was the kind of thing you could brush off if people were acting rationally, but nothing about this situation was sane. Six months earlier the Compton Gamma Ray Observatory—one of the heaviest payloads ever flown at thirty-seven thousand pounds— had been put into a "controlled de-orbit" by NASA and dumped into the Pacific Ocean, but it had gyroscopes to work with. Iridium Space Vehicle 79 had nothing but gravity.

It was a technician at the Johnson Space Center in Houston who first noticed. There was no way for the SNOC to eyeball the Iridium satellites once they left the grid, but NASA had access to spy cameras and monitoring devices that were operated by the Orbital Debris Program Office at Cheyenne Mountain.[42] There was no question about it—Iridium 79 was going to reenter the atmosphere around November 21. And, by the way, Iridium 85 was not looking too good either, although its crash orbit would probably take another month to come in contact with Earth. It's not really possible to predict where a tumbling satellite will end up, but you can do approximations based on trajectory and gravitational force. The most likely place of impact for Iridium 79, according to Aerospace Corporation, was a remote part of the Northwest Territories in Canada. That was some comfort. And the second most likely place was even better: the Indian Ocean, a few hundred miles off the coast of Antarctica. Then again, it wasn't the actual danger of fatalities that Colussy was worried about—this was just a bad time for any satellite, much less an Iridium satellite, to be crash-landing anywhere. It was expected to come down on the very day the National Security Council was convening a meeting in the White House Situation Room to come to a final decision on Iridium. For all Colussy knew, his on-again, off-again luck could end up with Iridium Space Vehicle 79 plunking down on Disney World while the President's advisors were debating.

Dan Colussy was nearing the half-century mark as a military officer, engineer, and businessman, and during that time he'd had a lot of dirty and disagreeable jobs, but this was the single most complicated and bruising of them all. It was worse than his first year in the Coast Guard, when he'd worked the International Ice Patrol out of Argentia, Newfoundland, bundled against frostbite, listening to the roar of the "growlers" as his ship picked its way through the scud ice, pinging radar signals off the deadly white canyons that would loom up just fifty yards away in heavy fog. It was worse than the Korean conflict, when it seemed like his body never got completely dry while patrolling Ocean Station Victor, the most brutal storm region in the world, a lonely grid a thousand miles south of the Kamchatka Peninsula. It was worse than his first civilian job at General Electric, where he would sleep all night on a cot in the control room of a test cell, so he could make sure prototype aircraft engines were thoroughly stress-tested. It was worse than anything he ever encountered in the airline business, where Colussy was always the odd man out, the Ivy League egghead in corporate cultures that never appreciated MBAs. No, Iridium was a whole new level of frustration. Perhaps that's why he couldn't let it go.

And it wasn't as though Colussy was some starry-eyed novice who didn't know how to deal with the shocks and indignities of the real world. As the President of Pan Am, Colussy had dealt with terrorists on three continents and dictators on five. He had gone toe-to-toe with Bill Genovese, head of the airline division of the mobbed-up Teamsters union in New York. He had negotiated with the Ayatollah Khomeini, who chartered a 747 to take him from Paris to Tehran, then negotiated later the same day with National Security Advisor Zbigniew Brzezinski, who convinced him to cancel the flight. Colussy had suffered through what is to this day the worst aviation accident in history, when a KLM Airlines 747 decided to take off from Tenerife in heavy fog and plowed into a Pan Am 747 that was crossing the runway, killing 583 people. The Pan Am copilot bashed out a window and dropped twenty feet to the ground to find a phone so he could call New York headquarters, and the grim aftermath began. Then, as though that wasn't enough carnage, one of the most notorious tragedies in New York City history occurred on Colussy's watch when a helicopter preparing to take off from the roof of the Pan Am Building—ten floors above his office—fell on its side and killed four passengers in the heliport and one person on Vanderbilt Avenue who was hit by a falling rotor blade. So Colussy was intimately familiar with what it took to soldier on in the midst of adversity, and he had faced most of the slings and arrows of public criticism. He had been shunned by his friends and what seemed like the whole city of Baltimore after his Columbia Air start-up had its landing slots pulled in 1981 during the air traffic controllers strike. He was branded as an interloper and insulted to his face when he reported for duty as the CEO of floundering Canadian Pacific Airlines, greeted by Prime Minister Pierre Trudeau with the opening line, "Why do we need an American like you up here?" In the eighties, while dealing with antinuclear protesters and dissatisfied shareholders, he had to close down UNC's nuclear reactor plant in Connecticut, and it took him two years to deal with all the people who lost their jobs, eventually authorizing a year of severance pay for those who found it impossible to be rehired. But still this Iridium mess was worse than anything he had ever encountered, because he was constantly treated as a pest or, even worse, a nonentity. He was the only guy in the room who had no funds, no support, and no corporation to back him.

And the most annoying part of it all was raising the money. It was one thing to stop Motorola, but now that the Pentagon had intervened, he had to put up or shut up—and the fact was he didn't have the money. He didn't have even a fraction of the money he had led people to believe he had—and he had never seen businessmen run so far and so fast at the mere mention of a company's name. Yet he truly believed his business plan made it self-evident that Iridium was a safe and promising investment. He even went so far as to produce a document called the "Finite Life Project," which was a plan B in case it became impossible to raise the money for replacement satellites. Colussy knew that engineers always design redundancies into technology, so the effective life of any device is often five times greater than the official estimate. In a project as complex as Iridium, there were hundreds of these redundancies, and according to Dannie Stamp, the expected life span of the constellation was closer to twenty years, not the five-year average specified in Motorola contracts. Back in 1992 Iridium CEO Jerry Adams had insisted on a

Motorola operations-and-maintenance contract to reassure investors—like having AAA insurance for your car—and since that contract was worth $140 million per quarter, Motorola kept making the system more robust so that they would have to replace as few satellites as possible.[43] The bottom line was that the new Iridium owners could simply run the constellation until it died, meaning until it had so few functioning satellites that calls became impossible. Without ever replacing a satellite, they could get their money back and make a small profit. That's not what anyone wanted, of course, but it meant the investment was safe. Still, Colussy didn't think anyone was listening very well amid the constant bad press and erratic behavior from Motorola.

During the year 2000 Colussy spoke to perhaps a hundred potential investors, everyone from billion-dollar corporations to foreign governments to religious organizations that wanted to use Iridium to propagandize the Third World. But the nadir came in early September, shortly after he received yet another ultimatum from Keith Bane. After the Labor Day Massacre—the meeting where Rudy deLeon said he would call down the wrath of all the Pentagon gods on Motorola if they didn't stop their threats—Bane had finally agreed to put the de-orbit on "pause." But the pause wouldn't last forever, and he was now calling to say that the proposed government contract was only $72 million over two years and, as far as he could tell, the only other money Colussy had came from two black guys with some cable TV franchises. It sounded too puny to bother with.

"I don't believe you have enough investors," Bane said. "Show them to me within forty-eight hours. Prove up $150 million."

Bane also gave Colussy a list of "deal-breakers" and told him that, if he expected Motorola to continue to manufacture Iridium handsets, he would need to produce a $30 million letter of credit, because he didn't intend to get stuck with any unpaid bills when Colussy's little company inevitably went out of business. Colussy responded in his usual way—saying he would produce everything Motorola needed—and then went into a controlled panic. He didn't have $150 million. He didn't even have the $30 million to secure the letter of credit. All he had in hand was $25 million from Herb Wilkins at Syncom. Everyone else was still talking.

Fortunately, a potential investor named Charles Medlin was flying in the following day, and he came with the highest recommendations. Jim Walz, CEO of Iridium's North American gateway in Tempe, Arizona, told Colussy that Medlin had access to deep-pocket investors related to "family winery" money. Walz had Colussy check him out with Laura Lo Bianco, the senior attorney for Iridium North America, and she said Colussy should definitely meet with Medlin—the man was a genius. He was managing director of a company in Porterville, California, called CMC International that designed and sold embedded microprocessors and remote data devices, and, more to the point, he was "a visionary in telecommunications." So Colussy called Medlin and invited him to spend the night at his Maryland home so they would have plenty of time to discuss his proposal.

Medlin turned out to be a personable guy with a fairly complete understanding of Iridium. For several hours he outlined his ideas, even going so far as to propose his own management team, which would have Jim Walz as CEO—attractive to Colussy, since he didn't want to get stuck with being CEO. After a while Colussy said, "All right, this is all well and good, but how much money do you think your people can invest and how quickly?" Medlin put off the question a couple of times, then Colussy became more insistent. "We can't talk about anything else until I know about the money," he said.

At this point Medlin asked forbearance while he made a phone call to his investors. Colussy sat across the room while he tried several times to get a connection to a foreign number. After two or three tries, Medlin started using an operator and asked to be put through to "the minister of finance." Again the call failed.

Finally Colussy said, "What country are you calling?"

"Fiji," said Medlin.

"Is that the source of your funding?"

"Yes, it is."

"You're calling the minister of finance of Fiji?"

"Yes, why?"

"All right, this meeting is over," said Colussy. "You can stay until the morning, and I'll have a car take you to the airport."

Medlin made some weak protests, but Colussy had done business with every country in the world. Fiji was an isolated, struggling archipelago in the Pacific with less than a million in population, and it couldn't even pay its own bills. Medlin was either a smoke-and-mirror man or hopelessly naive. Somehow he had gotten the ear of Jim Walz, but there was nothing there.[44]

And so Colussy ended up on September 8, 2000, the day of the Keith Bane deadline to "show me the money," with only $25 million in the bank, a promise of $25 million more later, and various credible "maybes":

- $10 million to $30 million from Kazuo Inamori, the founder of Kyocera
- $10 million from Atilano de Oms Sobrinho, one of the investors in the original Iridium and CEO of a company in Brazil called Inepar
- $10 million from Stratos Global, the largest satphone service provider in North America
- $25 million from Prince Khalid . . .

. . . and after that the list got dicey. Even if all the maybes became yeses, Colussy was still $20 million short of the $100 million he needed to get through the first year, and that didn't include the money he would have to pay the banks to get the company out of bankruptcy, or the money he would have to pay Motorola to purchase spare satellites, handsets, and facilities. No matter how you sliced it, the number he needed always came out at $200 million, with $150 million as the bare-bones budget and $100 million as the lowest possible start-up capitalization, meaning he could open his doors but would have to seek additional funding right away.

Thanks to the efforts of Ed Staiano, Motorola's purchase price for all the buildings, launch contracts, intellectual property, and equipment had come down from $100 million to $25 million plus a 5 percent piece of Colussy's new company. This was a true bargain, since the "book value" of everything Motorola was transferring to Iridium was around $231 million. Even scrimping on start-up costs, Colussy had to admit that he really did need $150 million more—as much as he disliked Bane's constant badgering, the man was right—so he decided to wait until the end of the day on the eighth and tell Bane a few polite lies. The letter of credit and the investors list would be coming shortly. He patiently ticked off the various maybes, portraying them as certainties, but said he needed a little more time for the paperwork. The investment group, after all, was spread all over the globe. And so the tap dance continued.

After the intervention of the Pentagon, the sales efforts should have gotten easier because the bureaucrats were now indicating they might be willing to give Colussy a guaranteed contract. Mark Adams had been meeting all summer with a Colonel Dick Gaudino at the obscure Defense Information Systems Agency (DISA), a combat-support division of the Pentagon that was widely known as the place you had to go when you needed to pull money out of a hat. DISA's charter involved "common-user information infrastructure," which was sufficiently vague as to be almost meaningless, other than to say this was where the IT functions of the military were funded. No one inside the Building used the term "slush fund," but if you went deep inside DISA, through a virtual library of inscrutable acronyms, to the part of its operations known as the Global Information Grid, you would find provisions for the coveted IDIQ contract, which stood for "Indefinite Delivery/Indefinite Quantity" and was essentially a blank check for supplying goods and services to the government without much oversight. Mark Adams was angling for an IDIQ, and although he fell short, he didn't fall *that* short.

Colonel Gaudino did an in-house investigation into what Adams was looking for—some type of guaranteed pricing—and was told by General Jack Woodward and others that the military's Iridium gateway in Wahiawa, Hawaii, needed to be preserved at all costs. Gaudino was shocked to discover that the government was currently paying through the nose for every Iridium call. The fee for each handset was a reasonable $69.56 a month, but most of the actual calls were from the Balkan theater to the United States—and those were averaging $5.34 per minute. Elsewhere it was even worse. Government users in Africa had made calls to Diego Garcia that cost $8.18 per minute. So although Gaudino eventually ruled out a straight IDIQ, he came back with some pretty simple requirements. The Pentagon wanted twenty thousand phones for a flat fee, with options to buy more later, and it wanted unlimited minutes as part of the monthly price. It also wanted its phones to be fitted with a secure module that would allow their signals to be encrypted. If Iridium could deliver on these requirements, the Pentagon would guarantee a monthly payment. At first that payment was expected to be $3.5 million per month, but it was eventually curtailed to an even $3 million. Changes like that made Colussy wince—that was $6 million a year he lost at the stroke of a pen—but he was still able to tell investors that he had $36 million in annual revenue guaranteed.

Still, nothing had been signed. Everything remained at the what-if stage. And even the what-if stage was causing problems. There were too many enemies of Iridium inside the Building for word not to leak into the Pentagon gossip mill. Admiral Bill Owens had been trolling the halls of the Pentagon for months, pushing for the government to use Craig McCaw's Eagle River Investments to run Iridium, but at the premium price of $120 million per annum guaranteed for four years. General Tony McPeak, a former Air Force Chief of Staff, had been hired by Carl George, the Minneapolis businessman who had been one of the early bidders in bankruptcy court. (George's plan was to invest $18 million of his own money, but only if he was allowed to keep the $155 million in cash that Iridium had on hand.) And then there was Gene Curcio, the California wireless guy who had somehow gotten a letter of partnership with General Dynamics and was also making his presence known within the Pentagon. With rent-a-generals and rent-an-admirals all but bumping into one another in the corridors, news of a potential government contract was bound to travel fast.

It took less than a week for Dave Oliver to call Colussy with the bad news. "Bill Owens is all over the Building," said Oliver, "and Bernie Schwartz is all over the Hill. They're screaming bloody murder about the DISA contract."

That would be the DISA contract that didn't yet exist. And of course it would be Globalstar that was most upset. The *other* satellite phone company. The one that was eking out a hand-to-mouth existence after getting partial service going nine months earlier. Colussy and Mark Adams were trying to get DISA to make the Iridium contract for five years, since the resulting $180 million would make guaranteed revenue enough to borrow against, but now there was political pressure on Colonel Gaudino not to do that. The Pentagon didn't really believe the Globalstar system would ever match Iridium, but they had to make it theoretically possible for the contract to be transferred to Globalstar at some future date. So Gaudino refused to approve any term longer than twenty-four months—the time Globalstar claimed it would take to develop its own encrypted "secure module." And all of this was theoretical anyway, because Gaudino couldn't negotiate anything until the bankruptcy court had awarded Iridium to Colussy.

Globalstar was a formidable opponent. The pet project of Bernard L. Schwartz, chairman and CEO of Loral Space and Communications, it was a co-venture with the San Diego wireless company Qualcomm that, from the beginning, was conceived as a lower-cost alternative to Iridium.

Even after concluding his end-of-career deal in the mid-nineties—selling forty-three companies to Lockheed Martin for $9.1 billion—Schwartz remained chairman of Globalstar, and the only thing that had wrecked his business plan, he believed, was Iridium. Schwartz was very vocal about it, telling the *New York Times* that Iridium was "a three-legged race horse" that had wiped out the equity markets for everyone else's more sensible satellite ventures.

Even though the Globalstar constellation used traditional bent-pipe transponders and was useless across large parts of the globe, and even though Globalstar's reliance on earth stations meant it wouldn't be as secure, Schwartz had to be taken seriously because he was almost ridiculously well connected. He was, in fact, the largest single contributor to the Democratic Party in the nineties and had celebrated his seventy-first birthday at the White House, with Bill and Hillary Clinton both in attendance. Adding to the stakes was the fact that his New York–based Loral was a conglomeration of companies that competed head-to-head with Motorola every day—on radar systems, electronic warfare systems, guidance systems—so the idea of Iridium suddenly landing a government contract at the very moment Globalstar was building out its earth stations would not only be a bitter personal pill for Schwartz but would have the appearance of dirty dealing by Motorola.

Schwartz had made the first Globalstar call to Qualcomm founder Irwin Jacobs on November 1, 1998—the same day that Iridium launched service—but it was purely ceremonial. The company didn't launch real service until February 2000, and after nine months in business had just 21,300 customers and had defaulted on a $250 million line of credit secured by Globalstar's partners, including Lockheed Martin and Qualcomm, who were none too pleased when they had to pay it off. Schwartz also had $1.4 billion in junk bond debt that was trading at 8 to 13 cents on the dollar due to what he called "the Iridium syndrome." Schwartz's original business plan was based on the theory that Motorola had it all wrong—the world didn't need global coverage, and it didn't need a science project. Globalstar executives were openly scornful of Iridium's claims to blanket the planet. "Yes, they dominate the penguin market," said one. All that was *really* needed was a $750 phone at 70 cents a minute, which could be provided by forty-eight Globalstar satellites at 876 miles above sea level. This was the orbit that required flying through the Van Allen belt, which meant Globalstar could be done for a mere $1 billion, not $6 billion. (The actual price tag, years later, was $4.6 billion.) But after word leaked from the Pentagon that Colussy was about to get a DISA contract, Schwartz announced that his plan had changed. He would no longer seek individual customers. In the future Globalstar would go after the military, the government, and "large corporate" customers. In other words, he pursued the original business plan of Iridium for most of a year, then switched to the future business plan of Iridium, all the while decrying Iridium as a failed system.

But it was unlikely that Globalstar could wrest the contract away from Colussy, mainly because Mark Adams had already trashed it in his reports to the Pentagon. Globalstar was one of a half-dozen satellite phone systems studied by Adams at MITRE Corporation in the early nineties, and he had ruled it out for military use, mainly because it was bent-pipe and it lacked the global coverage that was essential for the "disadvantaged user." Apparently none of these shortcomings mattered to Schwartz. The lack of a formal government bidding process was patently unfair. This was a bailout of the spendthrifts at Motorola. Everything about it was wrong.

When word reached Robyn that Schwartz was crying foul, she got an instant headache. Too many Friends of Bill. She now had FOBs trying to buy the system and FOBs trying to kill the system. At first she thought Schwartz's arguments would be easy to parry, since the government owned a functioning Iridium gateway but no secure Globalstar gateway. That was no problem, answered Globalstar—you could just install a Globalstar earth station on a U.S. flagship anchored off the coast of wherever you needed the service. "Very inefficient and it would cost a lot of money," said Robyn as soon as she heard it—and even if the government did that, Globalstar had no secure, encrypted handset. But Robyn thought the main fact working against Globalstar was that it was a bent-pipe system, meaning its calls could be "geo-located" by the enemy. Since Iridium's architecture was all in space, its military handsets could never be traced. Nevertheless, part of her job was to protect Bill Clinton against his friends. She made a decision to "inoculate the President."

She needed a policy paper, and that required hearing out Global-star's case. So she agreed to a briefing by Larry Atlas, a Loral executive who had formerly worked at the FCC. Atlas's argument was that there was room for only one satellite phone company in the marketplace, and the government's actions to save Iridium were patently unfair because the nation would be destroying Globalstar at the same time it was preserving a company that should have already gone out of business. Robyn then took that argument to Bob Pepper, the respected economist who headed the FCC's Office of Plans and Policy. As expected, Pepper rejected the "natural monopoly" theory, saying that the market would sort out how many companies could compete. The Colussy investors had a novel business plan and they may well fail, he said, but they should have an opportunity to try. If Iridium had been a land-based system, reasoned Pepper, it would have simply been auctioned off and then continued to operate in one form or another. In this case a contrac-tor had the power to destroy the cell towers, and so it was appropriate for the government to try to solve that problem so the auction could occur. Robyn summed up Loral's position as "Let Iridium die and then give us all the spectrum"—and sent all Pepper's conclusions to Dave Oliver, who instantly launched a reconnaissance mission inside the Pentagon to locate and neutralize any Schwartz allies with stars on their shoulders.

Meanwhile, Colussy was running out of time in the fund-raising department, so he bunkered down again in his home library, making dozens of calls per day, changing his pitch to fit each potential investor.

The Iridium system was extremely attractive to so-called tracking companies—firms that already used GPS satellites to monitor trucking fleets, fishing vessels, and the like, but could use the Iridium transceiver to get more extensive real-time information. Mark Adams was handling ongoing talks with GlobalTrak and ParView, trying to sign them up as investors but, as an alternative, trying to lock them into contracts as customers. Colussy was talking to the Colorado real estate mogul Pat Broe, who had resurfaced, looking for a communications solution for the eight hundred locomotives he used for his private railroads. Honeywell International, which owned the AirSat system for private jets, also seemed like a natural strategic partner. Colussy knew many of these companies would be future customers, so he played on their desire to dominate their markets and suggested they become owners as well. A man named Henrik Broberg called from Eutelsat, the French communications satellite company, to say he was considering Iridium for a South African project backed by Saab, but then he disappeared like so many before him.

And then there were the gateways. Colussy still had hope for a few of the less alienated gateways. Some of them wanted to save face. Others still believed Iridium would eventually find its market. A few were run by people trying to save their own jobs. And all of them despised Motorola. He had to be careful when he told them that part of their investment would end up in Schaumburg, Illinois.

Colussy had all but given up on Wall Street. He remained in contact with Texas Pacific Group, the huge private equity firm that sometimes took minority positions in distressed companies like J.Crew, Del Monte Foods, and Oxford Health Plans, but otherwise he knew he didn't have time to go through the lengthy process of convincing the boards of directors at blue-chip firms. Allan Holt, a managing director at the multi-billion-dollar Carlyle Group, called to say he wanted to see Colussy's business plan but then never followed up. Better to focus on firms that had a single decision-maker, firms like IdVectoR, run by the strange Parisian

Paul Maruani. IdVectoR was situated in an office off the Champs-Élysées and had some connection to the Quai d'Orsay, where Maruani had once worked. Maruani was paranoid about e-mail communications and contacted Colussy only by voice or fax, but he spoke of a "Project IRIS" that would use Iridium to create the only secure e-commerce system in the world. To do that, he was talking about a "strategic alliance" worth $220 million over ten years, so Colussy continued to take his calls and continued to try to wrap his mind around the arcane uses for Iridium that Maruani had in mind, described in one document as "300 ideas for products and services out of which an anticipated 40 will be launched" and in another as a "worldwide e-commerce platform" that would be used "in remote unwired villages around the world."

Then there was Kazuo Inamori, the founder of Kyocera, who had agreed to meet with Colussy the last week of August in Irvine, California. Inamori was one of the richest men in the world and one of the most legendary businessmen in Japan. Unfortunately, his meeting with Colussy had to be canceled on the day of Motorola's doomsday announcement. During the hectic days following, the meeting had been rescheduled twice more for Kyocera International's U.S. headquarters in San Diego, but those meetings were postponed when Colussy was summoned to the meeting at the FCC. After a fourth attempt to set up an Inamori meeting, his majordomo said there was a problem. "Dr. Inamori doesn't like frequent changes," said Yoshiharu Yasuda, the former head of Nippon Iridium. Apparently Inamori was offended by the canceled meetings, and that offense might be fatal to any investment. Colussy wrote an elaborate letter of apology, but the next contact with Yasuda-san brought the news that "Dr. Inamori is now in deep China where he cannot be reached." Colussy wasn't sure whether "deep China" was a physical place or a state of mind—Inamori had recently entered the Buddhist priesthood—but Dannie Stamp would later find out it was both. Sometime that fall, in the Japanese prefecture of Nagano, on a hillside not far from the arenas and stadiums built for the recent Winter Olympics, three Japanese executives lit a torch and set fire to the gateway of Nippon Iridium, burning it to the ground in a ritual of shame.

More promising were the talks with Toronto-based Stratos Global. Stratos was a $176 million corporation specializing in "remote communications"—data transmission systems for industrial users in isolated areas all over the globe, but especially offshore drilling platforms in the Gulf of Mexico and the North Sea—so it was no surprise that it had been the largest service provider for the Iridium phone. CEO Derek Woods was optimistic about a $10 million investment by the Stratos board. The only other investor that seemed halfway solid was one of the partners in the original South American gateway called Inepar, a conglomerate of eighteen companies based in the remote Brazilian state of Paraná. Inepar had started out in the 1950s providing equipment for hydroelectric dams but had more recently branched out into cable TV, pagers, and cell phones, mainly owing to the enthusiasms of company president Atilano de Oms Sobrinho, who grew up as a ham radio devotee in the Paraná capital city of Curitiba. Like most of the gateway owners, Oms was an outsider on the national level—his company was provincial—but he had already proven immensely valuable to Colussy since he was popular among Iridium board members. Colussy used him to communicate with the other gateways and get them to define what precisely they wanted. More important, he was now entertaining a possible investment himself in the new Iridium—at least $10 million, perhaps more.

Colussy was now telling white lies on a daily basis to the Pentagon, to Motorola, and to potential investors. He had to convince the Pentagon he was well financed. He had to convince Motorola that, regardless of what they might want to do, the government wasn't going to let them de-orbit. And he had to convince every investor that the guaranteed DISA contract was all he needed to jump-start the company and get it into the black by the end of the second year. The truth was if anyone started to sense weakness—if he seemed to be underfinanced, if the regulators didn't trust his ability to run the system, if the Pentagon decided Iridium was not that important, if the White House decided it couldn't risk a Bernie Schwartz public relations offensive—there was a real danger of everyone deciding "The whole thing is just too complicated." Colussy made daily calls full of unbridled optimism. He had figured it all out, he told everyone. The new Iridium would be well financed, all the licenses needed to operate would be easily transferred, he had lined up customers all over the world, and all the gateway owners would be able to save their businesses with all the free minutes and discount handset deals he was willing to offer them. If any of these statements were technically untrue, they were at least possible *once he had the money*, and besides, if he didn't state these goals as though they had already been achieved, then nothing would get done. He didn't stop to think about it, but what he was doing was willing a company into existence.

The pressure from Motorola was relentless. The fifteen members of the Motorola board of directors—including a former FCC commissioner, Anne P. Jones—were all squirming in their upholstered chairs. Chris Galvin was all over Keith Bane to get rid of the Iridium problem. This made Bane and Ted Schaffner even more determined to back Colussy into a corner, so that he either bought the constellation or admitted he had no money—which they knew to be the case. In mid-September Schaffner called to say, "We want this done by next week."

So increasingly Colussy had to work with the art of the possible. He narrowed the prospect list to Prince Khalid, Inepar, Syncom, Stratos Global, and maybe the guys at the Italian gateway, who kept calling to say, "Come back to Fucino, we can get you $15 million." The Italians had devised a plan to get a new investment from Finmeccanica, the Roman aerospace company. All they wanted in return was to keep Fucino open and to let the Italians have the territories of fourteen countries in Europe and North Africa.

Oddly enough, everyone else wanted Africa, too, even though it was the least lucrative part of the planet. The Syncom guys wanted it because it fit their Third World vision. The Italians wanted it because it was traditionally part of Italy's market area. And the Saudis wanted it because they'd owned the African gateway the first time around. In fact, the Saudis resented Syncom wanting to take Africa away from them, which probably helped boost their investment to $50 million, since that was the only way they could make themselves equal to Herb Wilkins and Terry Jones at Syncom. Fortunately Africa was a big continent—fifty-four nations with Iridium licensing agreements—and there were convenient ways to split it up, marketing-wise. WorldTel was primarily interested in sub-Saharan Africa. The Saudis were interested in the Maghreb nations. And the Italians were flexible about everything.

In late September Colussy flew to New York City to attend a dinner at Ristorante San Pietro, an upscale establishment on 54th Street where so many billion-dollar deals had been hashed out that owner Gerardo Bruno had a special seating section called "Chairman's Row." The restaurant had been chosen by James Swartz, Prince Khalid's de facto chief of staff, because he believed that everything about the Prince's image must be elegant and classy. The Prince's retinue always flew first class, always stayed in five-star hotels, and wouldn't think of dining anywhere but a place like San Pietro, which served fairly simple southern Italian food at haute cuisine prices in an environment that glittered with gilt trim and ceramic wall murals. Colussy had been hoping to speak to Prince Khalid, or to his eldest son, Prince Fahd. But Swartz called himself the "Chief Executive Officer of Prince Khalid's Overseas Family Office," which meant he brought a team of hired hands, including the Prince's chief accountant, a Scot named David Hall, and Jeddah gateway manager Tom Alabakis, as well as two analysts from UBS Warburg, Tom Sperry and Ed Cerny. (Cerny had already requested so much information that Colussy was starting to long for the days of Jonathan Mark. Typical question: "Financing from WorldTel—where will it come from?" Answer: from WorldTel.) Coordinating the team's effort was a square-jawed lawyer named Steve Pfeiffer, representing Fulbright & Jaworski, the law firm that had represented Prince Khalid since the early 1980s. Pfeiffer was a Rhodes Scholar and retired Navy Commander with law degrees from Yale and Oxford, not to mention a master's degree in African studies from the University of London. Like everyone else working for Swartz, his time was not cheap.

As white-jacketed waiters loaded up the tables with Neapolitan specialties like sea-salt-encrusted branzino, Swartz selected several bottles from the eight-hundred-vintage-wine list, then pronounced himself unimpressed with Colussy's business plan. Why weren't there any big corporations involved? And, more important, why was Ed Staiano still around? The Prince couldn't work with Staiano. Staiano was an arrogant puppet of Motorola who flew around in a Gulfstream jet and wasted corporate money on fancy parties, and he didn't have proper respect for the Prince. The Prince might come in on this investment, but there was still a lot of work to be done.

Swartz was laying down a gauntlet. Prince Khalid bin Abdullah bin Abdulrahman Al Saud—the entire name was necessary in order to distinguish him from other members of the royal family with similar names—was said to be a charming and courteous man, but his organization seemed to live in a state of paranoia, constantly alarmed that his status would be unacknowledged or not respected. After the Swartz rant was complete, Tom Sperry took the floor to repeat his hostility toward the whole deal; at two prior meetings he'd been nasty to Colussy almost to the point of viciousness. Sperry's usual role at UBS Warburg was to "beat up the creditors," so he was accustomed to focusing on minutiae and going in as the company's pit bull. Sperry was just following orders, but that didn't make him any easier to take.

After Sperry had gotten through his tirade against Colussy's business plan, the conversation turned to Syncom, and Colussy thought he detected barely disguised contempt for a company Swartz and Pfeiffer considered too small to bother with. The Prince was accustomed to partnering with blue-chip firms—and, in fact, Pfeiffer had already talked to the investment bank Lazard Frères, which advised him, "This thing has no chance unless Craig McCaw does it." As a result, Pfeiffer had breakfasted with McCaw's Pentagon representative, Admiral Bill Owens, on the very day John Castle made his bombshell announcement that he was backing out of the deal. Now Pfeiffer was telling Colussy, the Prince wanted to see "more muscle."

The Prince would not be impressed by Syncom. The Prince did not want to collaborate on Africa. The Prince wanted performance milestones that would allow him to pull his investment out if the company didn't make progress. The Prince also wanted approval of the CEO. The Prince had a high regard for Craig McCaw's company—was there any chance that he would be involved? The Prince would be pleased if Boeing were to become an investor. And did they mention that the Prince didn't like Ed Staiano?

The entire time Swartz and his lieutenants were talking, Colussy kept wondering one thing: *If the Prince feels so strongly about so many things, where is the Prince?*

But at that moment, Colussy couldn't have cared less about the eccentricities of the Prince's crew. All he wanted from the Prince was $25 million, with an agreement to provide $25 million more if needed later. These were the Syncom numbers, so he played on their sense of superiority to Syncom to get them up to that level. He also indicated that they could have Africa. Let the various parties fight over Africa later. Better yet, tell all of them they could have Africa, because they apparently didn't understand that there were no more territories, no more regional operations, no more gateways—there was only one call-processing center and it was in Arizona. Sometimes he wanted to scream: *No one will have Africa! We will all have Africa!* Then again, ignorance of how the new Iridium would function sometimes worked to Colussy's advantage.

The Prince's team saw it differently, of course. They were tempted to say, "What's your hurry?" Unaware of the perilous five months during which Colussy had fought two dozen battles just to keep the satellites aloft, they considered him imperious, irascible, and far too impatient for their tastes. They felt like they'd been too passive in 1993 when they accepted the blandishments of John Mitchell and invested in two gateways, so now they were doubly determined not to be rash. Yes, there were a lot of them—eight people so far assigned to due diligence on Iridium—and yes, they were pushing back against virtually everything Colussy said, but this was the Prince's A team. The very fact that they were all assembled in one place meant that the Prince took this situation very seriously. Steve Pfeiffer had been vacationing in Montana when he was summoned to the San Pietro get-together. What looked to Colussy like obstructionism was, to them, simply assembling the team that could get the job done. They had no idea that Colussy had been knifed in the back by several other would-be investors, so they took his insistence on quick action to be insulting. It was the classic clash between an operator and a pure investor, and it would continue throughout the fall of 2000.

The next day Colussy shifted gears to attend a big marketing meeting at Stratos Global, the Canadian corporation that had its operations headquarters in Bethesda, Maryland. The Achilles' heel of Iridium had always been marketing—how do you sell the service and whom do you sell it to?—but the Stratos people were optimistic. They believed, for starters, that Iridium could easily woo back the sixty-three thousand people who had initially purchased a phone, especially since customers continued to flock to Iridium long after Staiano was gone, Sentinelli was gone, and the multimillion-dollar advertising campaign was gone. This indicated that the product was considered brilliant even if the marketing had been all wrong, and, in fact, a poll of the Iridium North America users indicated that 85 percent of the former subscribers would come back. Even more encouraging, Stratos was optimistic about the ability of Iridium to deliver limited data service. When the system was created, everyone assumed it would be voice only—there was no discussion of data at all—and when it was abandoned, one of the reasons was that it was narrowband. But Randy Brouckman and others pointed out that it was capable of handling up to 2.4 kilobytes per second of data transmission, which was all that was needed for short-burst messaging. Stratos recommended dropping the handset price from $3,500 to $1,200 and made it clear that they wanted to be the marketing solution for Iridium going forward. The Stratos investment was looking better every day.

By the end of September, Motorola was saying, "Come to The Tower and let's do this." Which was Keith Bane's way of saying, *Put up or shut up.* When Colussy showed up in Schaumburg for what seemed like the hundredth time, he said he was one week away from closing— not true—then spent most of the day in Ted Schaffner's office going over things like launch contracts and the transfer of property. Motorola still had some future launches of spare satellites that were mostly paid for, and Colussy wanted those deposits released to the new company. (Motorola eventually did agree to release $37 million worth of deposits already paid to Boeing for Delta II launches.) Colussy also decided that he couldn't pay $25 million for the old Iridium property that Motorola controlled in Chandler and Tempe. "You're going to scrap all that anyway," said Colussy. "I should pay the scrap price, which is around $15 million." They talked about how to transfer the licenses in 160 countries from Motorola to the new Iridium, how to handle the transition, and particularly how to handle Dannie Stamp. Stamp had decided he loved the satellites more than he loved Motorola, and he needed permission to talk to Colussy about a job with the new company while he was still working out of the Chandler Lab.

As they talked about each item, Schaffner would work out deal points with Colussy, then take them down the hall to Keith Bane for approval. Bane, kibitzing from afar, would accept some and reject others, but was mostly helpful. Colussy figured this was as close to a good mood as Bane was ever likely to slip into, so near the end of the day, he said, "I need one more thing, Ted."

Schaffner had a sense of foreboding.

"I have $25 million from Syncom, $25 million from Prince Khalid, $10 million from Inepar, and probably $10 million from Stratos. I'm $30 million short."

"And?" said Schaffner.

"I want Motorola to invest $30 million."

A moment to catch his breath, then, "Do you realize how much Keith is already yelling about how much he dislikes this deal?"

"Yes, but I don't have as much cash as I expected to have. I want you to get me $30 million so we can get this done."

"He'll have a heart attack. I can't ask him that. And even if I ask him that, he would have to ask Chris Galvin, and there's a good chance Chris would have to ask the board."

"I don't mind speaking to Keith myself."

And so the two of them went into Keith Bane's office and settled in before the skeptical gatekeeper. A heart attack, in fact, didn't seem out of the question, since Bane was overweight, red-faced, and had mentioned having health problems a couple of times.

Using as calm a voice as he could summon, Colussy said, "Keith, I need you to go in on this for $30 million."

Bane's skin flushed. "Go in as what?"

"As an investor."

"You've got to be kidding."

"I need it to get this thing done."

"You don't know what you're asking. You don't know how this company feels about Iridium."

"I know what I'm asking. I thought I would have more than this, but this is where I am."

When Colussy left The Tower that day, he had an agreement in place with his new investor: Motorola Inc.

What Motorola didn't know—what no one knew—was that, in terms of money in the bank, Motorola was now the *principal* investor. Syncom and Prince Khalid would be contributing only $25 million each in the first year—that is, if the Prince's representatives ever got beyond all the melodramatic due-diligence meetings orchestrated by UBS Warburg. The rest of the Syncom and Saudi money was in the form of later cash calls, if needed.

Colussy was on a roll, so a week later he decided that, if his Italian friends were telling the truth, it was worth a trip to Rome to get another $15 million. Mark Adams asked to tag along on the Alitalia flight, and Colussy was happy to have him, since his knowledge of Iridium was unparalleled. Fulvio Ananasso and Antonio Marzoli were as good as their word, not only getting a hearing with Finmeccanica but setting it up with its most eminent board member, a national celebrity named Pier Francesco Guarguaglini. Guarguaglini had run the company's defense business until 1999 but was currently CEO of Fincantieri, a Trieste shipbuilding company, and he was so revered that they never referred to him by name in messages, using instead "Very Frequent Flying Person" or "(VI)Person." Finmeccanica was a national institution, Italy's version of Boeing and Motorola combined, dominating the region in airplane manufacturing, electronics, and communications equipment. This was the kind of company that could invest in Iridium out of petty cash.

When they arrived in Rome, Colussy and Adams checked into the Hotel Giolli on the Via Nazionale and had a pleasant lunch, spending part of the time talking about what would happen after the new Iridium was operational. Adams, to Colussy's surprise, wanted a job. He, too, had fallen in love with the satellites. In fact, he wanted to run the system, perhaps in some role like Chief Operations Officer. Colussy certainly had confidence in him, but he knew that Dannie Stamp wanted the same job. He would work it out, he told Adams, but it pleased Colussy that the people who understood Iridium best—the engineers, the scientists, the service providers, now even the *intelligence operatives*—were asking to join his team.

Later that afternoon Colussy and Adams arrived early for their appointment with Guarguaglini and decided to grab an espresso at the

San Marco, a sidewalk café next door to an imposing office building on the Via Sardegna. After a few minutes of chitchat, Adams jumped up from the table.

"My briefcase is gone," he said.

Inside that briefcase was his laptop computer. Adams went into a full-scale panic. They searched all around the table, looked up and down the street, but whoever took it was long gone. It had happened so fast that they didn't see or hear anything.

If they had been ordinary tourists, they might have just thought they were victims of an especially adept Roman thief, but they weren't ordinary tourists, and Mark Adams didn't have ordinary things in that briefcase. Adams was, in fact, the supervisor of fifty other engineers with top secret security clearances at MITRE Corporation.

"We have to go to the police," he said.

At the station house, Adams identified himself as an employee of MITRE—not likely to inspire much awe in the average *agente*—and then he insisted on reporting the theft. The report took a couple of hours and was greeted by the reporting officer as you might expect a sidewalk theft to be greeted in Rome—with indifference.

Adams couldn't get over it, though, and was rattled during the two days of meetings with Guarguaglini and Telespazio. On the plane ride home, Colussy didn't know what to worry about more—the decision of Finmeccanica or the loss of whatever Adams had in that briefcase. Both men hoped the computer was in the possession of someone who had no idea what he had stolen.

As it turned out, the trip to Italy was a waste. Finmeccanica would dither around for another month and then, like every other big aerospace company, pass on the opportunity to invest. Adding insult to injury, they would invest a few weeks later in Globalstar, raising suspicions that the whole trip to Rome was just a ruse to inveigle trade secrets out of Colussy.

It looked like Stratos Global was the best option now, and Colussy agreed to go to its international headquarters in a downtown Toronto skyscraper to present the investment to the Stratos board of directors. He ended up staying there three days, so grueling was the decision-making process. Then, at the end of that time, they didn't say yes and they didn't say no. What made it frustrating was that this one had seemed like a slam dunk. These were some of the few people who understood the Iridium system completely, and their investment would allow them to dominate their market. Moreover, they were already in the middle of a shopping spree, having spent $553.6 million in the past year acquiring four companies, and another $10 million would have been enough to buy a huge position for what was chump change. Stratos had a 1.76 percent ownership position in Inmarsat, so perhaps the board members thought they would be competing with themselves. At any rate, there was this hesitance—the same hesitance Colussy had encountered with every large corporation. "Yes," they would say, "we see your numbers and we see your plan and we believe in it. But tell us again why Motorola couldn't do this?" The implication was that if Motorola couldn't make it work, no one could. Investment or no investment, Colussy knew he wanted Stratos to handle the first-year Iridium marketing. He trusted the company's expertise and he knew they could sell the phone. He also knew that he couldn't afford a marketing department of his own.

Meanwhile, Colussy continued to put out grass fires. Apparently Ted Schaffner had a "bad meeting" with Colonel Gaudino and was upset that the government wasn't willing to indemnify new satellites, spare satellites, or lawsuits against subsidiaries. "We want more insurance," he called to say. "And we want it from Boeing." Colussy gave an almost audible sigh.

Increasingly this was a dance in which the competing interests of Motorola, Boeing, the banks, the bankruptcy court, the FCC, the Pentagon, the White House, and the new investors all had to be perfectly balanced, so that at the moment they came into alignment, the door could be quickly slammed shut on anyone's attempts to change the deal. It's the kind of problem best solved by the man who keeps the best notes, and so Colussy started slowly, painstakingly—meeting by meeting, call by call—pushing everyone toward convergence.

Even though the DISA contract looked more and more likely, Colussy kept having to defend himself. Dave Oliver and Rudy deLeon called one day to say they needed some reassurances. "By my calculations the Department of Defense will be investing $114 million over a two-year period," said deLeon. (He was including $18 million for handsets, $12 million for insurance, and $12 million for gateway maintenance, in addition to the $72 million that the Pentagon would be paying to Colussy's company.) "What I want to know is—are you sure you can do this?"

In other words, after all this time, the number two man at the Pentagon was saying, *Who are you and where did you come from?* These were the kinds of deals that were normally done with the TRWs and Boeings and Northrop Grummans of the world, not a guy from Palm Beach who penciled "Newco" into the contract he printed out that morning on his home computer. Colussy was used to explaining who he was by now, and he didn't resent it. He thought it was understandable for someone like deLeon, who had never met Colussy until Iridium happened, to ask questions about his competence and financial health. It was less understandable, on the other hand, for John Castle, who called around the same time to ask for $10,000 for the Harvard Business School reunion, to inquire, only as an afterthought, "How's your Iridium deal going? Are you going to be able to get that done?" It was going quite well, answered Colussy, so well that he probably wouldn't have time to attend that year's reunion.

Now that the grouchy Keith Bane was an investor in Colussy's company, Bane wanted to meet the other partners. *Be careful what you wish for,* thought Colussy, and set up the meeting. There were nine people on the conference call—and four of them worked for the Prince. The other three shared an equal dislike for Bane, especially when he outlined his plan for letters of credit. Motorola and Boeing had so little faith in the ability of this pack of misfits to run Iridium that they now wanted $60 million in advance letters of credit, meaning Colussy would have to park $60 million of his scarce resources in a bank somewhere in case either company had claims for unpaid handset bills or delinquent operations-and-maintenance bills. The only thing more controversial than that provision was Colussy's suggestion that Ed Staiano be appointed CEO. Bane heartily endorsed Colussy's suggestion—and that was like being sponsored by Satan. Absolutely not, replied the Saudi delegation. Staiano should retire to one of his four homes, preferably the one farthest away. The meeting was full of friction. Steve Pfeiffer, the Prince's lawyer, was so annoyed that the next day he called Isaac Neuberger to tell him the Prince was out of the deal. Neuberger apparently misunderstood—Pfeiffer was just making it clear that Staiano would not be acceptable—but he reported this alarming development to Colussy, and Colussy called Pfeiffer to find out what was going on. A few minutes later Colussy called Neuberger back. "They're still in," he said, chuckling. "Apparently they just don't like some of the other lawyers involved. That would be you, Isaac."

Colussy decided he needed to spray some flame retardant on the embers that were starting to glow among the various would-be partners, so he spent a day placing calls to each investor calling for compromise. *We can all share Africa,* he said. *You don't have to talk to Keith Bane if you don't want to—I'll handle that. And I'll tell Ed Staiano that we have to move on without him.* Colussy actually thought this last decision was unwise—Staiano had proven himself to be incredibly dedicated, influential with Motorola, and willing to work without salary—but he couldn't fight the three years of pent-up frustrations and personal slights. When he called to tell him, Staiano was devastated. "We'll work out some payment for you," Colussy said. "The Saudis have my hands tied. I can't do this without their investment." Staiano said he didn't blame Colussy and wished him luck. The next week Colussy hired an executive headhunting firm so that someone could be found before he was forced into the CEO job himself.

By early October Dave Oliver had identified "nine areas of disagreement" among Motorola, Boeing, and the government, and he flew to Chicago to try to hammer them out with Bane. But it proved impossible. The complexity of the upcoming interconnected deals had become so byzantine that phone calls were constantly flying back and forth from Boeing to Motorola to Colussy to Iridium to the SNOC to the gateway in Tempe to the various investor groups and the various legal teams, not to mention the Washington regulatory agencies and the Pentagon. With the DISA contract still not issued and the indemnification not hammered out, with Motorola being increasingly regarded as a schizophrenic cousin who had to be watched at all times, the same conversation would have to be repeated thirty times to get everyone informed and up to speed. Finally Keith Bane suggested they get everyone into the same room on the same day. All parties should come together on the forty-seventh floor of 35 West Wacker Drive in downtown Chicago, world headquarters of Winston & Strawn, which Motorola sometimes used as outside counsel. Winston & Strawn was the kind of law firm that had $3,000 leather chairs in the conference rooms, so Colussy knew the Saudis would be pleased. Dave Oliver would be there representing the Pentagon. Linda Oliver would be there representing the Office of Federal Procurement Policy. Boeing and Motorola would have their legal teams in place, ready to draft any necessary documents. Isaac Neuberger would be present, annoying Steve Pfeiffer and Larry Franceski. Herb Wilkins would represent Syncom. And, of course, the Prince would send a workforce suitable for negotiating the end of the Hundred Years' War. Anyone who had to sign off on anything had to be present in the room on the day of the meeting, because Bane asked everyone to agree: "We will meet until finished."

But before that could happen, international events conspired to make the meeting both more urgent and more complex. On an otherwise uneventful day in October, USS *Cole*, a guided-missile destroyer out of Norfolk, Virginia, was going through a routine refueling stop at the port of Aden, in Yemen, after coming through the Suez Canal. It had pulled up next to a "dolphin" platform—the equivalent of an offshore gas station—and the ship's mooring operation had been assisted by several small craft that worked for the port. The expansive Aden harbor was, in fact, full of small tugboat-type vessels due to the large number of ships that used the port every day. Forty-eight minutes after refueling began, an inflatable small craft manned by Ibrahim al-Thawr and Abdullah al-Misawa was seen pulling alongside the *Cole* by the duty officer. Al-Thawr and al-Misawa both stood at military attention and saluted as they approached. What was impossible to see from the deck was that the two men were standing on top of 250 kilograms of composition C-4 plastic explosives, and once they pulled alongside, they used a detonator to blow a sixteen-hundred-square-foot gash in the side of the *Cole*, directly opposite the mess hall, while most of the crew was at lunch. The two al-Qaeda bombers died instantly, and the Navy casualty toll would end up being seventeen dead and thirty-nine injured.

Back at the Pentagon, it was 3:30 A.M. when confused messages about the explosion started coming through the National Military Command Center, which was operated by the J-3 Directorate of the Joint Chiefs of Staff—Jack Woodward's bailiwick. Woodward was awakened by his staff and started trying to find out exactly what went wrong so that he could send out Emergency Action Messages (EAMs) to all the other vessels operating around the world. But the communications center aboard the ship had been disabled by the explosion, so his only contact with the *Cole* was through a single phone, borrowed from the U.S. embassy—an Iridium phone. A week later he went before the Joint Chiefs and said, "Do you need any further example of why we need Iridium phones?" Every other communications system needed a box and a handset. When the box got blown up, the handset didn't work. With the Iridium system, all you had was a handset. The box never got blown up, because *all the boxes were in the sky.*

The attack on USS *Cole* had a sobering effect on the Pentagon even greater than that of the "Black Hawk Down" massacre in the streets of Mogadishu in 1993. Mogadishu had been a total surprise, the first encounter with "asymmetrical warfare." But now, after seven years of preparation for just this sort of guerrilla tactic, they had been beaten by two punks in a rubber raft. With FBI agents swarming all over Yemen, many of them sleeping in their clothes with loaded weapons at their sides, the realization sank in that no Middle Eastern port was safe. These guys were everywhere. And that may or may not have had something to do with the Pentagon legal office suddenly becoming fixated on the possibility of a Middle Eastern Arab about to become a part owner of Iridium.

Dave Oliver got wind of a flurry of questions coming down the pike from the Pentagon legal office, all of them directed at the background and sympathies of one Khalid bin Abdullah bin Abdulrahman Al Saud, owner of the Mawarid Holding Company, better known as "the Prince." Oliver asked what Colussy knew about him, and Colussy replied that he had never met him. He referred Oliver to the Prince's attorney, Steve Pfeiffer, and Oliver indicated there was going to be a thorough vetting of Mawarid's sixty-five companies and possibly an objection to foreign ownership. That would not be good, said Colussy—he had Brazilians, Australians, Canadians, and Italians who were also being considered for equity positions, and he had outstanding licensing issues with gateway owners in China, Russia, Korea, and India. He was purchasing a global corporation—some foreign ownership was inevitable. Oliver said he thought the problem could be worked out if the Prince would be willing to file some formal disclosures, but when that idea was broached with James Swartz, the Saudis dragged their feet. The Prince would prefer not to have his name registered anywhere, said Swartz, much less his address or any other identifying information. He would prefer that his ownership position be as discreet at possible. If there was any behavior more designed to arouse suspicion, Colussy couldn't imagine what it would be. By the time all the parties gathered at Winston & Strawn, it had become a full-blown issue.

But now there was an even bigger problem. Colussy stopped at the Pentagon on his way to Chicago, and Dave Oliver told him that Rudy deLeon and the SecDef had gotten wind that Motorola was about to become an equity partner, owning 2 percent of the new Iridium with an option to buy 5 percent more later.

"That's not gonna fly," Oliver told him. "No ownership for Motorola. They've burned their bridges. They're not allowed to own anything."

Oliver had just gotten off the phone with Dorothy Robyn, who had been on calls with Motorola discussing this very issue. "Any equity interest is unacceptable," she told Ted Schaffner. "That would allow Motorola to take advantage of the upside gain even as indemnification eliminated the downside risk."

Schaffner offered nothing in response, so she called Motorola lawyer Bruce Ramo—and Ramo was inclined to be reasonable. "He stressed those features of the deal were not locked in," Robyn said, "and that Motorola would do what was necessary."

What was necessary was for Motorola to give up any hope of having any income from Iridium at any point in the future. Apparently the intransigence of Chris Galvin had been branded into the collective psyches of everyone at the Pentagon and the White House. They wanted to be rid of Motorola just as badly as Motorola wanted to be rid of the satellites. Mark Adams called later to confirm the news: "Motorola can't own a single spare part on a single satellite."

When Colussy got to Chicago, Keith Bane didn't seem that upset about the Pentagon's position. Perhaps he was relieved by the prospect of never having to deal with Iridium again. Colussy didn't care about that so much as he cared about protecting his $30 million.

"We'll just convert the investment to a loan," he told Bane.

"I think you should give us a commitment fee."

"A what?"

"You need us to commit to this deal. You should pay us $4 million to commit to the loan. A commitment fee."

Colussy mulled it over for the rest of the day, then came back to Bane and Schaffner.

"Loan me the money and I'll give you a success fee."

"A what?"

"If the company survives and thrives, you'll get a fee after ten years."

A success fee didn't sound as desirable to them as a commitment fee. But Isaac Neuberger had also been working overnight to find out what loan terms might be acceptable to both sides.

"I'll give you 11.5 percent if you meet certain conditions," Colussy told them.

"What conditions?"

"We have a right to another $80 million loan later."

"You've got to be kidding."

Colussy never got his additional $80 million loan, but he did get the $30 million—with "success fee" attached.

The Winston & Strawn meetings were not unlike the Paris Peace Accords of the sixties, in which the various warring parties spent weeks arguing about the shape of the table. (The Vietnamese negotiators ended up using two tables, one circular and one rectangular.) In this case, Colussy made it clear that unless everybody backed off on demands for guarantees and letters of credit and escrow accounts and up-front payments, the whole thing was going to fall apart and the last year of agony would have been in vain. He said, "It will simply not be possible to achieve every single element needed to give all parties the level of certainty and security that every advisor or every law firm believes is necessary." He then went down a list of concessions he expected from Motorola, Boeing, the Pentagon, and the banks, telling them that if they required him to spend $60 million up front on the equivalent of insurance policies, then the whole deal was going to die. He didn't get everything he wanted, but all parties except the Department of Defense did back off their original demands. A $12 million letter of credit went away. A $20 million escrow account disappeared.

But now the parties were far past the point of deal fatigue; they were reaching deal exhaustion. Back in Washington, Boeing kept calling about the DISA contract. When would it be signed? Why was it taking so long? And when Colussy asked Dave Oliver for a progress report, the answer was alarming. Colonel Gaudino was no longer involved. The contract was out of his hands. They were now talking to General Harry D. Raduege Jr., who had become director of DISA in June but was just now getting around to looking at the contract. The good news was that Raduege was a satellite man, having served in the Cheyenne Mountain bunker at NORAD, heading up the command and control systems for the ICBMs. The bad news was that the head of DISA would not normally be personally assuming control of a contract so small. The even worse news was that the Attorney General's office had also called to get a copy of the contract. Janet Reno wanted to see it. Most people who work in Washington, D.C., are policy wonks—they studied graduate-level civics, that's what they know, that's what they love—and the Iridium problem had so many policy implications for the United States and the world, implications analyzed by two dozen domestic agencies and five international ones, that everyone wanted a piece of it. Everyone had an opinion. And every time a new opinion was expressed, the whole process slowed down by a day, or two, or a week.

The days started to pass in a blur.

"We need to get this wrapped up," said Schaffner.

"The presidential election is coming up," said Oliver, "and we need to be finished before then."

Bane chimed in with "We're at the bitter end."

But after all this time, Colussy still didn't have his money. Stratos was being flaky, Inepar was hesitating, and Prince Khalid's hired hands were impossible. Tom Sperry and Ed Cerny, the UBS Warburg bankers, kept coming up with reasons to delay or obstruct any potential deal with the Prince. Colussy couldn't have known it at the time, but Sperry and Cerny were new to the Prince's team and were probably showing off in an effort to prove how tough they were. But Colussy didn't have time or patience for that anymore. He needed cash and he needed it yesterday. He had enough for the Iridium purchase price and for the Motorola assets, but very little more. He wouldn't have anything left over to actually operate the satellites if he paid it all to the banks. On the other hand, if he didn't pay it to Chase, the bankruptcy judge could lose patience and the satellites could be awarded to someone else or, more likely, be destroyed. No matter—he had bought some time. It was time, not the details of how he would pay, that mattered right now. Colussy knew that if he could hold out another week, or another month, he would become inevitable. He was on his way home.

But the worst things can happen when you're on your way home. On November 2, when all of Washington was preoccupied with the upcoming presidential election, Bernie Schwartz struck at Colussy's most vulnerable spot. Representative Jerry Lewis of California and Senator Ted Stevens of Alaska coauthored a formal letter to Defense Secretary Bill Cohen, demanding that the Iridium deal be stopped. It was bad enough, they said, that DISA was about to award a contract without competitive bidding. But now they had discovered that the government was planning to indemnify Motorola. This cannot happen, they were saying. There were "important policy and funding issues" being raised here. They expected the Pentagon to send them every bit of information on whatever deals it was making with "the so-called Iridium commercial communications satellite system." This was un-American and possibly illegal. They would fight this.

More important, they had the firepower to fight it. Jerry Lewis was not just any congressman. He was the chairman of the House Defense Appropriations Subcommittee. He controlled the specific purse strings that Iridium needed to unfasten. Ted Stevens was even more powerful— he chaired the Senate Defense Appropriations Subcommittee *and* the larger Senate Appropriations Committee, as in *all* appropriations. At the Pentagon, no one is feared more than congresspeople with spending authority. Dave Oliver had already reacted to the new offensive, and he called Colussy to say, "They want us on the Hill. We have to go. Bernie has a whole Barnum and Bailey show going on here at the Pentagon."

Lest the message be misunderstood, a woman named Betsy Phillips, on the staff of the House Defense Appropriations Subcommittee, called to say the representatives expected a formal Pentagon explanation of both the DISA contract and the indemnification agreement. "Rudy deLeon needs to get up here to the Hill" was the way she put it, naming the powerful Deputy Secretary. It was the specificity of the request that made it so chilling. Obviously Schwartz knew everything.

Dan Colussy was an amateur student of history, especially the history of his ancestors in the Republic of Venice, and as a sailor, he knew that the Venetian fleet was most often attacked by the Ottomans when it was battle-fatigued and heading for its home port. At the Battle of Lepanto, perhaps the fiercest battle of the Renaissance, the Ottomans attacked just as the combined naval forces of the Roman Catholic maritime states turned into the open waters of the Ionian Sea on the edge of the Adriatic. John of Austria and Agostino Barbarigo, the two Christian commanders, could have sailed straight for Venice and avoided the Ottomans, but that would have emboldened the Turkish admirals to try the same tactic in the future. The other choice was to form all 212 vessels into divisions and face the 300 galleys of the Ali Pasha right there in the Gulf of Patras.

Colussy chose to turn his ships toward Asia Minor.

First he called Ginger Washburn, the woman who was soliciting him on behalf of the Alaskan Indian tribes. How was it possible, he asked her, that a senator from the state where Iridium was most popular could be opposing this? Didn't the Native Americans say they couldn't live without Iridium? Washburn said that yes, that was true, the members of the Tlingit, Haida, and Tsimshian tribes had a corporation called Sealaska that ran a massive timber operation in areas beyond the reach of any phone except Iridium. Iridium was also being used by the Council of Athabascan Tribal Governments to stay in contact with Canyon Village, an abandoned settlement that was being repopulated by latter-day pioneers. But she could go one better: the Alaska State Troopers said that Iridium saved lives every day in the regions of the state where no other phone worked. The phones were also used heavily by the Bureau of Land Management, the National Park Service, the Alaska Fire Service, the Lifeguard Air Ambulance Service, and University of Alaska geologists in the Arctic National Wildlife Refuge. Quickly the word went out: all Alaskan users needed to send messages to the office of Ted Stevens. It was not hard to motivate this crowd. Within two days there was a "Save Iridium" website and an organization dedicated to doing just that, and Senator Stevens was soon feeling the wrath of mining companies, construction workers, bush pilots, wilderness guides, hunters, fishermen, stevedores in the Chukchi Sea, and hundreds of ordinary people whose lives had been changed by the introduction of the Iridium phone, only to be devastated when their service was cut off.

Jerry Lewis was a tougher nut to crack. He was either a friend of Bernie Schwartz or truly believed that the contracts with Iridium would be illegal, because he was serious about calling the Pentagon on the carpet. This was far beyond the capability of Colonel Gaudino to handle on his own, so a few generals, including Jack Woodward, were rounded up by Dave Oliver to do the briefing for an executive session of Lewis's powerful subcommittee. "We're going to have to go through the Eight Cardinals," Oliver explained to Colussy. That was Pentagon jargon for the eight subcommittee chairmen of the House Appropriations Committee. What Oliver wasn't revealing was that he had ended up on a congressional blacklist—the lawmakers didn't trust him because "they felt like he was trying to sneak something past them," as one Iridium executive told Robyn. That meant Oliver couldn't go alone; he needed some blue-suiters with him.

The night before the briefing, the Pentagon contingent decided to ask Mark Adams to come along in case they needed technical data on Iridium. Then, when they arrived at the committee room, one of the

generals said, "Why don't you just make the presentation, Mark?" Adams
had no notes or prepared remarks, so he was a little nervous, but agreed
to give it his best shot.

When the signal was given for the Pentagon to proceed with its
presentation, one of the generals nodded to Adams and he began.

"Good morning, my name is Mark Adams—"

Jerry Lewis raised his hand, indicating that Adams should stop
speaking.

"We know who you are, Mr. Adams," said the chairman. "And we
know what you're here for. I want to save as much time as possible. If
you can't answer this one question, then this meeting is over. Iridium
LLC and Motorola went through $6 billion in one year and couldn't make
this work. What's gonna be different? How can you assure the government you're not flushing money right down the toilet?"

Adams swallowed hard. When he answered, it was slowly and
deliberately. He explained, first of all, that most of the cost of Iridium
was building and launching the system and that it didn't cost that much
to operate once the satellites were flying—so the new owners' costs
would be only a fraction of what the old owners were spending. Then
he explained that the government would be spending only $3 million a
month for services and would obviously not be required to spend that
should the system stop functioning. Finally, he said, the life of the satellites was much longer than five years. At some point, as Adams continued to talk, all those years of hanging out at the Chandler Lab kicked
in and he reviewed the whole history of his evaluation of the Iridium
system, how it differed from every other system in the world, and how
the government was finding it indispensable.

Adams was constantly interrupted for questions, and a lot of the
questions involved references to the government bailout of Chrysler in
1980. The first time it came up, Adams answered the question directly.
"There is no comparison," he said, "because in that case the government
was authorizing a loan. This is not a loan. This is, in fact, a no-risk bargain

for the government." But once the word "Chrysler" had been mentioned, it spread like wildfire through the committee, with one member stating, "I was opposed to bailing out Chrysler and I'm opposed to bailing out Motorola!" Adams wanted to say, *It has nothing to do with Motorola and it has nothing to do with Chrysler,* but he thought better of it, bit his tongue, and let the congressmen rail. Eventually the "Chrysler bailout" line of questioning played itself out, and as the meeting progressed, the questions became less hostile. Some of the committee members even said they believed there was a reasonable chance of success. Mark Adams' answer to the "one question" had gone on for two hours.

"We shook hands when we left," Adams recalled. "And it had not started out that way at all. Iridium had huge brand recognition and awareness, and at that point all that awareness was negative. I had to overcome that perception. I counted it as a victory."

But now the Saudis were apparently gone from the deal. They had gone strangely silent, leading Colussy to wonder whether they'd really gotten spooked by the Pentagon's interest in the personal details of Prince Khalid's life. So on Election Day 2000, Colussy was struggling once again to find the $100 million he needed, even though the recent publicity about Congress demanding explanations had, ironically, brought more potential investors out of the woodwork. Colussy fielded all the calls but told everyone there was no time for due diligence. Anyone wanting to invest was going to have to make a seat-of-the-pants decision and do it now. The more time they wasted, the more time they were giving Globalstar to destroy his plans. One day Mark Adams ran into Bernie Schwartz at the Pentagon. Schwartz was personally roaming the halls, demanding $40 million for Globalstar. If Iridium was going to get money, Schwartz was saying, then Globalstar should get money, too. Iridium was bankrupt and Globalstar was not. The field belonged to Globalstar. Iridium should be nothing more than an "interim solution" until Globalstar was fully operational worldwide, and then the contract should be shared.

"We need to move before this whole thing blows up," said Adams. "The two-star is getting involved for a reason." (Since no one was quite sure how to pronounce General Raduege's name, he was referred to as "the two-star." Only at the Pentagon would this be considered somewhat of a snub.) At any rate, the message was obvious: Iridium was turning into a white-hot career-killer of an issue.

It was now well over sixty days since the Pentagon said it would work with Colussy, but after all this time there was still no invocation of Public Law 85-804 and still no DISA contract. Increasingly it seemed like the Pentagon was some kind of shifting ice shelf of loyalties and interests or, more appropriately, unstable tectonic plates presaging an earthquake. Every time the approvals seemed imminent, the whole problem got kicked up to a higher level. Bureaucrats treated it like a dead skunk and passed it to their bosses, and those bosses passed it ever higher. It was such a small contract, but in the surreal world of Washington, a $36 million purchase could be highly toxic on the same day that a $36 billion purchase was invisible and routine. Schaffner was still working both sides of the deal, telling Rudy deLeon one day in November that "Colussy's little peanut of a company" wouldn't be able to handle the job.

DeLeon sighed. "Before the Clinton administration is over, I've got three problems to solve," he said. "One, figure out how to respond to the attack on the USS *Cole*. Two, bring North Korea under control as a rogue state. And three: What do we do with Iridium?"

"Well then, Rudy, we should take care of Iridium right now," said Schaffner. "The other two things are much more important."

But Iridium had become too big for any one person to solve. Mark Adams was hearing ominous noise from the State Department. Iridium was now scheduled for a showdown in the White House Situation Room. The technical name for it is a "deputies meeting," in which each department and agency involved in a problem is required to send its number two official. Rudy deLeon, the Pentagon representative, would be arguing strongly for invoking Public Law 85-804, while the representatives of Attorney General Reno would be arguing strongly against. Strobe Talbott, representing Secretary of State Albright, was expected to vote no as well. Talbott was the ultimate FOB, having roomed with the President during his Oxford days, and he had been a bulldog helping Iridium get licenses from countries that were dragging their feet in 1998. But now the political landscape had changed and the State Department was going to oppose any efforts to save Iridium on the grounds that it was unfair to Inmarsat. The National Security Council had decided to stay neutral because it was still troubled by the statute being used.

The problem with Public Law 85-804 was that it was intended to be temporary. It had been enacted in 1958 as an emergency measure made possible only because President Truman had, years earlier, put the nation into a state of emergency. It was invoked because of "the increasing menace of the forces of communist aggression"—and communist aggression was today nonexistent. In addition, you were only supposed to use it for risks to government contractors that were "unusually hazardous or nuclear in nature." First of all, Motorola wasn't a government contractor. Second, the government wasn't the only customer of Iridium. Then, even if you could overcome those two objections, the idea of the government giving Motorola indemnity for lawsuits caused by falling satellites any time in the future seemed quite a broad interpretation of "unusually hazardous or nuclear." What was being proposed by the Pentagon would oblige the government to pay claims that wouldn't have been paid by Motorola's current insurance. "Motorola has not yet identified and defined the unusually hazardous risks for which indemnification is requested," said a Justice Department lawyer. "And the threshold here cannot be ignored—it would take either the Secretary of Defense or POTUS saying that it facilitates the national defense." Whenever Public Law 85-804 was used in the past, it was for operations so dangerous that insurance wasn't available at all, whereas in this case insurance was available—it just wasn't enough to satisfy Motorola. If the government gave this insurance, the policy experts were saying, it was a horrible precedent for the future. What if some other corporation wanted the same protection?

"Well, what if they do?" said Colussy to Adams once he was briefed on the battle lines. "You don't have to give it to them!"

The whole purpose of Public Law 85-804 was to be *an exception to the rules* when nothing else would work. By definition, it created no precedent!

Adams agreed, but this was far beyond Adams' ability to exert any influence. This was one step away from the President, always referred to in memos as "POTUS." Adams was still talking to various congress-people who kept asking the somewhat bizarre question of why they shouldn't consider this a bailout of Motorola, just like the unpopular bailout of Chrysler in 1980. He had no answer for that. Congress had authorized $1.5 billion to go directly into Chrysler's coffers. That was the equivalent of $3.44 billion in 2000 terms. So the amount being talked about here was *1 percent* of that, and it was not a loan at all, and it was not going to Motorola, and it was paying for an actual service that the government needed, at bargain rates. It was not just apples and oranges; it was apples and fried tarantulas. Asking why the government should pay it was like asking a man renting a used motorcycle why he was bailing out Harley-Davidson.

By the time the deputies met, Dave Oliver had come up with four reasons why Iridium satisfied the definition of "national security": 1) it was the only worldwide phone with secure capabilities, 2) the Pentagon had a shortage of bandwidth and capacity, 3) it provided instant commu-nications for special-ops forces like the Navy SEALS, and 4) there were no other global commercial satellite systems currently available. As to the foreign ownership issue—which kept being posed by various junior bureaucrats—Oliver pointed out that the original Iridium had been 65 percent foreign-owned and the government had no problem doing busi-ness with it. Of course, the real problem wasn't foreign ownership, it was *Saudi ownership*, and it was a particularly bad time for the issue to come up because Senator Ernest Hollings of South Carolina was pushing for a new law requiring the FCC to deny permits to any telecommunications company that was more than 25 percent foreign-owned.

And the policy wonks were proliferating. An FAA administrator named Patti Grace Smith weighed in with a whole series of legal objections, the most important of which was: What about the fact that Public Law 85-804 was intended only for "unusually hazardous" situations? Dorothy Robyn had spoken to three Pentagon lawyers about this precise point, but she had her own unique take on the issue: "The fact that the commercial insurance industry is not willing to provide additional insurance is, to me, the definition of 'unusually hazardous.' Stated differently, almost by definition, if you can't insure for it, it's 'unusually hazardous' from an economic standpoint. That, rather than a technical definition of 'unusually hazardous,' is what matters here, I think." (Interestingly, this turned Motorola's insistence on "complete and total indemnification" into an asset, and created a situation where bureaucrats who despised Motorola were working to achieve exactly what Motorola wanted.)

Dave Oliver's official position was even simpler: de-orbiting seventy-four satellites at the same time was unusually hazardous—NASA's top orbital debris expert already said so. Most of these issues were worked out in interagency memo wars and preliminary meetings, so that by the time of the actual Situation Room meeting, the Justice Department's resistance had been greatly lessened—but now the Treasury Department was aroused. In the course of answering the question "Where does the money come from if the government has to pay off on this insurance?" the answer came back, "We'll take it out of the Judgment Fund." The Judgment Fund was an account administered by the Secretary of the Treasury so that, when the United States lost a lawsuit, the Secretary could pay the damages. He was not *required* to pay the damages, though. If he thought the damages should be paid by a particular department, he could send it the bill—and in this case the Department of Defense looked like a pretty good place to send the bill. At any rate, the issue was now something that could end up on Treasury Secretary Larry Summers' desk, adding a new layer of complexity to something that was already byzantine.

Until the "Sit Room" meeting, Colussy's plan was to stay on the phone, poking at the "maybe" investors, hoping one or two of them would fall in at the last minute. His standard appeal was for $10 million, with a $10 million cash call later, but, to his surprise, some of them would say, "Can we do more?" Carlton Jennings kept calling from Melbourne, representing a Perth-based health care magnate named Michael Boyd, and he was one of those inquiring about more equity in the future. Meanwhile, Stratos Global kept requesting more documents and going over the proposed DISA contract with a magnifying glass. Bob Kinzie would occasionally call with a suggestion—he was still working on the United Nations—but Colussy didn't really take that seriously. Two or three more days passed with no progress on the list—and then the unpredictable lone wolf Craig McCaw poked his head up out of the haystack of potential investors and offered Colussy more money than all the others combined.

McCaw—after fourteen months of mind games with the bankers, tentative deals with Motorola, starts and stops, feints and demands, and private conversations with the Pentagon through his on-site admiral— apparently decided, at the eleventh hour, that he could live with Dan Colussy. Or did he? Was he friend or foe? Was he coming with olive branches or false promises? Ever since the initial bankruptcy of Iridium, McCaw had been alternately bullish and bearish on satellite phones. Motorola had been encouraging McCaw to buy Iridium, but McCaw's engineers were always afraid of the system, mainly because it was nar- rowband. But now McCaw was making it clear—he wanted in. McCaw had called his personal friend Paul Sarbanes, the Maryland Senator, and asked him to ferret out copies of the contracts among Colussy's company, Motorola, Boeing, and the Department of Defense. Sarbanes said he couldn't do it, so McCaw called friends at the Pentagon. McCaw was apparently irked that someone else was about to conclude a deal. Of course that's not what Eagle River President Dennis Weibling said when he called Colussy the next day. Eagle River wanted to invest in Colussy's deal, Weibling said. They wanted to be partners. And not only that, they thought they could bring Boeing along with them. Boeing was a close friend, tied to Eagle River through the Teledesic project. And, by the way, it wasn't Teledesic anymore, it was ICO-Teledesic Global Ltd. Colussy listened carefully, took notes, and was amazed when, within a matter of hours, everyone wanted to be his best friend.

An hour later, Keith Bane called to say he would highly recommend working with Craig McCaw. "Come to The Tower," he said, "I've got it all set up for Monday morning."

Bane then went on to talk about his long friendship with McCaw, how they were in Nextel together, and how McCaw could bring capital and operating experience to the deal.

Colussy said he would agree to meet with McCaw's people, but he couldn't do it Monday morning because he had a possible meeting in Baltimore with the Saudis.

"Forget the Saudis," said Bane. "You don't need 'em. They're giving you foreign ownership problems anyway. This is the best thing that could have happened. You don't have to be a poor boy anymore. Just talk to McCaw. He can provide everything you need."

Of course, Bane had less than pure motives. Motorola had a long-standing arrangement with McCaw—he owned the systems, Motorola provided the infrastructure. In Mexico City, Motorola had built out the entire wireless system for a metropolitan area of eighteen million people, but the system belonged to Nextel. McCaw could put Motorola back into its comfort zone—selling gadgets and systems but *not owning anything*.

A few minutes after the Bane call, Steve Pfeiffer called Colussy from Fulbright & Jaworski. "Is it too late to get back into the deal?" he asked.[45]

Obviously, when Craig McCaw spoke, people listened.

And Colussy thought it was worth a shot. Stranger things had happened. So five days later he boarded a plane with Herb Wilkins and Isaac Neuberger for a meeting with the three top Eagle River executives. Oddly enough, the meeting wasn't in Kirkland, Washington, where McCaw's company was based, but in Schaumburg, at The Tower. Motorola wanted to keep everything close.

On the plane to Chicago Colussy penciled in his new funding plan:

- *Motorola 30*
- *Boeing 30*
- *Syncom 30*
- *Stratos 30*
- *Inepar 10*
- *Saudis 30*
- *Eagle River 30*

. . . for a total of almost $200 million. He was liking Craig McCaw more and more. At last this thing was starting to come together. At last people were coming to their senses.

Dennis Weibling came on strong at the meeting. "We're here to take all your risk away," he said. He then went into the history of Teledesic—hundreds of millions spent creating the future space-based broadband Ferrari of all communications networks, bolstered by Motorola's technical know-how, and combined with the 36 ICO Global satellites that were originally owned by Inmarsat but were now being converted from narrowband to broadband and would function as a start-up constellation until the 288 Teledesic satellites could be launched. Colussy asked if he could pencil them in for $30 million. Weibling said, "Not enough." Eagle River was willing to put *$130 million* into Colussy's company so that the two companies could be intertwined and, two to three years from now, merged into the ultimate satellite broadband communications system.

After puttering along with $25 million for most of the past three months, Colussy was now being offered the carrot of a company capitalized at $300 million, which was three times what Castle Harlan ever promised. (It's amazing what a guaranteed government contract can do for your prospects.) Not only that, Eagle River would help out with a $250 million secondary stock offering so that the whole system would be heavily capitalized for a fast marketing ramp-up. If Iridium started to "wind down," added Weibling, then Eagle River would buy out the Iridium shareholders with Eagle River stock.

Colussy listened to the whole spiel, backed up by Keith Bane, and then said he needed time to think about it. Why did the Eagle River executives keep talking about Iridium "winding down"? McCaw's people were assuming that Iridium would quickly become obsolete and need to be phased out. But what would replace it? ICO-Teledesic consisted of three drawing-board companies (four if you counted Celestri) that together had launched a single space vehicle, and the 288 satellites that would someday make up the dream constellation had not even been fully designed, much less manufactured. Increasingly the deal looked like a tactic to use Iridium to lock in a relationship with the government so that Eagle River could then buy out Colussy, scrap Iridium, and appropriate the spectrum. Randy Brouckman had heard a version of this plan earlier in the summer when executives referred to Iridium as a "backup system." One thing was certain about the draft agreement they gave to Colussy: they would have the right to buy him out in two to three years at a predetermined price based on various milestones such as number of subscribers, amount of revenue, and the like. Keith Bane kept pushing Eagle River forward as an alternative to the problematic Saudis, but Craig McCaw had more in common with Prince Khalid than anyone would ever admit. Like the Prince, he never came to meetings. The only access to McCaw was through Dennis Weibling.

Back in Maryland, Colussy kept mulling over the buyout provision. Why would Eagle River insist on that? Why wouldn't McCaw just take an equity position like everyone else? If it made sense later for Iridium to merge with Teledesic, then the deal could easily be done. Why make it inevitable?

Meanwhile, Mark Adams called to say that this was the week the Pentagon would be finishing up with the Eight Cardinals. For the key meeting, Rudy deLeon was going to carry the ball, and he was taking 3CI chief Art Money with him. The two-star would be there, too. General Raduege had reviewed the DISA contract and was worried about where to find the $3 million a month—he needed Money to say it was okay.[46] There were also annoying new wrinkles. Adams said someone needed to talk to Charles Rangel, the Harlem congressman, who was making noise about "bailouts." (Herb Wilkins called him, played the Africa card, and calmed him down.)

Weibling kept calling from Eagle River, calls that were usually followed up by pressure from Bane. Colussy thought McCaw was trying to have it both ways. If Iridium prospered, Eagle River could buy the company and the customer list without any risk. If the company failed or puttered along without profit, Eagle River could just jettison the arrangement and pursue its original plan for Teledesic. Isaac Neuberger spent three days going over the deal and said, "These terms are impossible." But Colussy didn't want to give McCaw an outright no.

"Look, Dennis," he told Weibling, "the problem is this part about being able to buy us out after two years. I'll give you that, but only if we have the right to invoke the opposite. In other words, we want a 'put.' After two years we have an option to *force* you to buy us out."

As badly as Colussy wanted to hold on to that $300 million investor sheet, the McCaw deal didn't smell right. It smelled like a way to buy Iridium and then get rid of Iridium. He wasn't going to save the satellites this year so McCaw could destroy them next year.

A day later Weibling called back. "We don't like the 'put' requirement," he said.

"Well, we need the 'put' requirement."

"Think about it."

Was he going away? Was that a withdrawal from the deal? When you were dealing with McCaw, you never knew. Weibling was on his way to New York for a meeting with Merrill Lynch to talk about the secondary offering of $250 million. *Look at all this money you're throwing away,* Weibling seemed to be saying. *How can you pass this up?*

Unfortunately, Senator Ted Stevens and Representative Jerry Lewis were *not* going away. Keith Bane, newly energized by the prospect of McCaw owning Iridium, had Motorola lobbyists in Washington working on Stevens, and even volunteered to make a call to the senator himself. The grassroots efforts in Stevens' home state were starting to work as well. Sam Romey, a rough-and-tumble mountain man and service provider in northern Alaska, was so infuriated by the Stevens-Lewis letter that he had rounded up a contingent of Alaska State Troopers to talk to Stevens. But Bernie Schwartz wasn't defeated so easily. One of the most powerful Democrats in the Senate—Ted Kennedy of Massachusetts—was now lobbying for Globalstar, calling the Pentagon to ask questions, but even that was not as surprising as the fact that one of the most powerful *Republicans* in the Senate had jumped onto the Schwartz bandwagon as well—Trent Lott of Mississippi. "We need to get this done," Mark Adams told Colussy. "Yesterday one of these guys called the Pentagon to find out when the rest of the Iridium satellites would be launched. They were under the impression that only 40 percent of the constellation was airborne." Colussy wondered: Was false gossip like that even effective? "We need to get this done," repeated Adams.

On November 7, General Raduege finally approved the draft of the DISA contract, and Colussy prepared to go back to the bankruptcy court for the formal presentation of bids. The Colussy bid was the only one approved by Iridium and the banks, but the court allowed "stalking-horse" bids. That meant the final sale could be stopped by anyone offering a minimum of $7.5 million backed by a 20 percent deposit, or $1.5 million in cash. Colussy called for a meeting of all prospective Iridium shareholders at the Millenium hotel, across the street from the World Trade Center, so that he could get checks and give deadlines. But Dave Oliver called to say, "Not yet. Don't go to the court yet. I'm on my way over to brief Senator Stevens. Let's get him off of this thing first." And there was yet another problem: the DISA contract had been approved in draft form, but there was a process by which the final form was worked out by a "contract officer." Her name was Sharlene Capobianco, and she had become the personification of bureaucratic hell. For starters, she refused to put the number $72 million—the amount of the government's total commitment—in the contract. "We'll reserve judgment for later" were her exact words.

Reserve judgment for later? Colussy was furious. Mark Adams was furious. Dave Oliver was furious. This was the money that the whole deal revolved around. With the DISA contract up in the air, all the lawyers agreed to ask Judge Blackshear to delay the hearing for one more week.

The next night, wearied by the constant "now you see it, now you don't" jockeying with the government, Colussy sat down at the Metropolitan Club on H Street with all his investors. Of course, there weren't enough of them. Herb Wilkins and Terry Jones were there for Syncom. Swartz and Pfeiffer represented Mawarid. Atilano de Oms Sobrinho was present for Inepar. And the purpose of the meeting was for Derek Woods, president of Stratos Global, to meet all the others in advance of his imminent $10 million investment. *Here's the situation,* Colussy told everyone. *We've got to get all the contracts signed, and all the money to the bankruptcy court, to finish this thing up by Tuesday—five days from now. Dorothy Robyn at the White House is spending all her time on this. I was up past midnight on the phone with Dave Oliver and Keith Bane, and in the midst of the discussion, Bane gave me another de-orbit deadline! The deadline is the twenty-second, the day before Thanksgiving. I don't think the deadline is real, but what he's saying is that he doesn't like the way this thing is trending. Rudy deLeon told me point-blank yesterday that the Pentagon is now officially spooked. They don't like being on the wrong side of the*

Appropriations Committee. Everybody needs to move fast. Everybody needs to get in or get out. Everybody needs to write checks. And, by the way, we can forget the Italian gateway—the Germans want $3 million for the facility, or about ten times more than we're willing to pay.

The next day Colussy found himself in Isaac Neuberger's office in downtown Baltimore, meeting with the mercurial Carlton Jennings in an attempt to squeeze $10 million more into a business plan that had returned to pre-McCaw levels now that Dennis Weibling had disappeared. Jennings had worked for a partnership made up of the Bakrie brothers of Indonesia, Kyocera of Japan, and the Taiwanese gateway during the first incarnation of Iridium, but now he was representing Michael Boyd, best known for founding a medical diagnostic company called Sonic Healthcare that had grown to be one of Australia's largest multinational corporations. Colussy told Jennings it was now or never: if he wanted to attend the investors meeting later that day, he needed to give proof of funds and commit.

Even as they spoke, lawyers and bureaucrats were running amok all over Washington. Steve Pfeiffer sent the deposit agreement from the Prince, and it had a weird provision that Neuberger interpreted to mean that the Saudis could get their money back if performance goals weren't met. Unacceptable, Colussy told him, barely disguising his exasperation. Dave Oliver called to say that Rudy deLeon was on his way over to meet with Senator Stevens and Representative Lewis—this should be the final showdown with Globalstar. (General Raduege had issued a report comparing Iridium with Globalstar and finding Iridium far superior.) Ted Schaffner called to say that Motorola needed $375,000 from Colussy to cover the cost of canceling its de-orbit insurance policy. (Sigh.) The FCC called for the umpteenth time, asking for the list of company owners, obviously concerned about Saudi ownership. By this time Steve Pfeiffer had prepared the necessary disclosures on the part of the Prince—Pfeiffer had been through this drill many times over the years since Saudis were always under scrutiny—but for some reason the issue continued to percolate.

And then the bombshell: Derek Woods called to say the Stratos Global board had decided against the investment. Woods was profusely apologetic. "They're just afraid," he said.

Don't worry, said Colussy, *I'm used to it.*

So that was it. Stratos Global was the last possibility of getting a publicly traded mainstream company to invest. And that was $10 million gone from the most recent plan, with two days left before everyone had to settle up in bankruptcy court. The new Iridium was going to be the polar opposite of the old Iridium. A communications company once run by the most powerful corporations and prominent nations of the world would now be run by two African American cable TV guys who felt beaten up by the white establishment, an extremely private Saudi prince who seemed to care more about his racehorses than his sixty-five companies, a ham radio–loving industrialist from a remote part of Brazil, and two obscure guys in Australia who were coming in at the last minute. Whatever might happen from now on, this was likely to be the wildest ride of Colussy's career—a career he thought was over five years ago.

Colussy's immediate problem, though, was Sharlene Capobianco, the contract officer who was not easily persuaded to give Colussy the language he wanted. At first she refused to put the number 20,000 in the contract, denoting the number of handsets the government would buy, because one of the draft e-mails had a typo in it stating the number was 22,000. She still refused to put $72 million as the total commitment by the government, but *did* agree to put 24 months at $3 million a month, leading to discussions among lawyers as to whether that was the same thing. In one note she agreed to remove a clause from the contract, then in the same note said she would reserve judgment until later. Eventually she said that if Dave Oliver would give her the go-ahead, she would put $72 million and 20,000 handsets in the contract, but only if he initialed his e-mail. "That's a request that may send him through the roof," Neuberger told her.

Finally, on November 15, Colussy and five other bidders for Iridium showed up in the New York bankruptcy court for what Judge Blackshear said was the final chance to put down deposits and offer bids. Gene Curcio was there, offering a $27.5 million purchase deal, as were church consultant Carl George and the Fiji-loving Charles Medlin. William Griffith, chief executive of a previously unknown Nevada company called Global Development Concepts Power Corporation, offered $100 million in gold. (It wasn't explained why the payment would be in gold.) The Paperless Newspaper Company of London had an $18 million offer on the table. Many of the formal offers stated that their bid was for assets that included the SNOC, which was, of course, the property of Motorola, and not Iridium's to sell. Curcio showed up with a new partner, Silver Inc., a company that announced plans to acquire Iridium and sell it to yet another company called Parallax Corporation. Parallax professed to be the owner of an aircraft telemetry system designed by one Bryan Zetlen of Seattle Scientific Corporation, and Zetlen complained that he had been shut out of the bidding process. Silver Inc. executive Terry Martin sometimes seemed to be in partnership with Curcio and other times seemed to be making a bid on his own. At any rate, Martin failed to tell the court that his available funds came from a phony municipal-bond scheme he had recently completed in Everett, Washington. (He would eventually end up in federal prison after being convicted of conspiracy, securities fraud, wire fraud, and money laundering.) The offers before the court ranged from zero to $100 million, but Judge Blackshear ignored all of them and said, "Who is prepared to write a check today?" One by one they told the judge that, if their application were approved, they would pay the deposit on a particular date. When Blackshear asked Neuberger about the Colussy group, Neuberger said, "Mr. Colussy has written a check for $1.5 million."

Blackshear checked with his clerk, verified that Colussy had submitted a certified check, confirmed that no stalking horse had matched that deposit, and then admonished the others. "I told you to bring money. Being prepared to bid is having the deposit. Iridium is awarded to the Colussy group for the sale price of $25 million. The cash on hand is awarded to the banks."

When the bankruptcy ruling reached the financial press, there were two reactions to the $25 million sale price. One opinion was that everyone from Chase Manhattan to Barclays to Motorola to the Pentagon were incompetents who had been fleeced, because how could a $6.5 billion company be worth that little? Clerks at the bankruptcy court joked that if the satellites were that cheap, they should have been offered as the luxury gift in that year's Neiman Marcus Christmas catalog. Surely the hardware alone was worth fifty times that much. The second reaction was "This proves that satellite phones were never meant to exist." There was no market for Iridium. It was a colossal mistake. The company had been proven worthless. "The new investors have unrealistic expectations for the system," said Roger Rusch, inventor of the stillborn Odyssey system. "The satellites are well into their design life and a number have failed or are in the process of wearing out. The new investors have high hopes based on unrealistic expectations." It didn't help that Globalstar lobbyist Bill Adler was still characterizing the deal as a bailout. "Iridium is essentially a failed technology," Adler told the *Fairbanks Daily News-Miner*, "and in the long run it would be hugely expensive to keep it operating, and that certainly is not in the interest of the Department of Defense." Industry analyst Karekin Jelalian was even more brutal: "Why are we still talking about this? Investments to 'save' Iridium have been discussed for months and nothing has come to fruition because even at $25 million, there isn't a good business case to buy the Iridium system."

But the cash up front wasn't $25 million; it was actually only $6.5 million. A few days before the hearing Colussy had polled his investors and come up with just $102 million promised, with at least $10 million of that being kind of shaky. That would normally be enough to open for business, but he owed $25 million to the banks and $15 million to Motorola for the property in Tempe. That put him back at $62 million in working capital, give or take, and he really needed $80 million to guarantee he could operate the satellites for a full year while the company started to build up revenue. Time was up. He had to go to the bankruptcy court, and he was $18 million short.

Colussy's solution: he told the banks that, if they wanted the deal to close, they needed to lend him $18.5 million of the $25 million he would owe them at closing. In return, the banks would receive 5 percent of the new company. Since the main goal of the banks was to get control of the $155 million in the Iridium cash drawer, and since the other bidders for Iridium were unlikely to be approved by the bankruptcy judge, what other choice did they have?

Ironically, the total cash outlay of the Colussy group—$6.5 million—was the precise amount that Bob Galvin had authorized in 1989 to fund the Iridium brain trust. Colussy had to swallow hard before writing the $1.5 million deposit check out of his personal funds, because even at this late date, his investor agreements had not been signed. There was some concern going into the hearing that Gene Curcio would write a stalking-horse check and hold everything up, and some surprise when he didn't bring any money at all. But now the only thing that could delay the process was government approvals. Surely by Thanksgiving everything would be done.

Or so they thought.

Back in Washington, Dave Oliver called Colussy to say, "I have neutralized the Democratic side in Congress"—but now Globalstar had a new ally in the form of the Office of Management and Budget (OMB). For weeks Oliver and deLeon had been coordinating all the approvals Iridium would need through a group of twelve agencies and departments, and there had been dozens of e-mails, memos, and PowerPoint presentations as they got ready for the showdown meeting in the Situation Room. (Dannie Stamp called it "organizing the circle jerk.") But that debate had occurred without involving OMB. Now OMB had announced its intention to oppose the Iridium deal. Bringing OMB up to speed was like starting the whole process all over again, with every policy wonk in the city now alerted to a unique situation that read like a case study at the Kennedy School. All over Washington—at the National Security Council, the FCC, NASA, the Office of Science and Technology Policy, and a dozen other agencies—bureaucrats were doing LexisNexis searches and pulling up everything from the War Powers Act of 1941 to the Antideficiency Act of 1884 to *Hercules Inc. v. United States*, the 1996 lawsuit in which the manufacturer of Agent Orange tried to get the government to pay for the liability suits filed by Vietnam War victims. (Hercules lost.) Dorothy Robyn alone found forty-three legal cases addressing the indemnification issue, "but none of them on point." The closest, she thought, was the 1976 emergency act passed when, in order to avoid a swine flu pandemic, the United States insured those entities involved in manufacturing or distributing vaccines, regardless of whether they were at fault or not. She also discovered that NASA had invoked Public Law 85-804 at least fifteen times since 1983—and never paid out a single dollar.

The Iridium deal had now been studied by NASA, the State Department, the Justice Department, the Pentagon, the FCC, the FAA, the Department of Commerce, the Department of the Interior, and the National Security Agency, but the first question asked by OMB was "Why not procure insurance from a private company?"—indicating OMB officials had been asleep during the previous nine months. Their first official e-mail to the Pentagon contained forty-six additional questions, including "Why isn't this considered a bailout?" Another dog-and-pony show would be necessary. Dorothy Robyn set about writing yet another position paper addressing OMB's concerns.[47]

Eventually the National Security Council changed its mind and decided to take an active role in the Situation Room meeting, which was now likely to be lengthy, complex, and full of pushback in the form of Jack Lew, the powerful Director of OMB. Chief economic advisor Gene Sperling would end up chairing it, along with Mara Rudman, NSC Chief of Staff. As it turned out, the NSC became a fierce advocate of Iridium. Dr. Johannes "Hans" Binnendijk, the leading defense policy specialist at the NSC, turned in a report in advance of the meeting, saying "we ought to support Rudy" on Public Law 85-804 because financial risk to the government was very low and potential benefit to the military very high. He admitted it would be "a unique first-time use of this authority" but recommended the government do it anyway. Then, right before the deputies met, Rudy deLeon increased the stakes: "If there is no Iridium package approved, I want it decided at this meeting who is on the hook for all the consequences. Is it Justice or is it NASA or is it FAA? Or is it somebody else?" He was essentially saying to the committee members, *Proceed with caution if you're even thinking about denying this authority.*

And then, as if the story couldn't get any stranger, word came from Isaac Neuberger that weekend that President Clinton and Chris Galvin were playing golf together in Vietnam.

No, there's no problem with the phone line. Golf. In Vietnam.

Apparently Galvin had been traveling in Asia at the same time that Clinton was on his historic journey as the first American President to visit Vietnam since the end of the war. Clinton's first two days were spent in Hanoi, but on the final day the Clintons flew to Ho Chi Minh City, where Gene Sperling conducted a press briefing at the Caravelle Hotel, summing up the newfound friendship between the two countries. Careful listeners might have thought it odd that, in the midst of the briefing, Sperling made a reference to Iridium, citing it as an example of expensive technologies that took so long to be developed that they became outdated before they could become profitable. "Motorola, one of the best companies in the world," said Sperling, "their CEO would tell you that their Iridium project will be the last time any major company will ever try to plan something in this area five years in advance.

It just—things move too fast. They're now dealing with what to do with the satellites in space. The notion that a government can predict these things is very suspect."

Sperling was actually contradicting himself—his point was that private investors know how to stimulate the economy better than government-directed investment—so the whole digression was a "huh?" moment. What he didn't say was that on the golf course earlier that day, Galvin had spent most of the round telling Clinton that it was better to let the satellites be destroyed, because they had a life span of only five years anyway and some of them had already been flying for three years, so it was silly to go through all these machinations in Washington just to keep a constellation alive that couldn't be replaced as satellites went out of commission. After all, who was going to throw another $6 billion after the $6 billion that had already been spent? Clinton was "spooked" by the remarks, according to Neuberger, and passed them along to Sperling.

Why would Galvin do that? Why would anyone mess with this deal when it was on the verge of being done? What's up with Chris Galvin? How could anyone be this determined to destroy the satellites? Colussy wanted to ask Keith Bane for an explanation, but apparently Motorola was having one of its periodic bipolar episodes—because the next day *Bane announced a de-orbit*!

Lieutenant Colonel Kaye Martin, head of the Space Systems Division at the Joint Chiefs of Staff, faxed the news over to the White House: "Motorola has issued an ultimatum to the Department of Defense. Finalized document by the 22nd of November, or Motorola will prepare for de-orbit of the satellites."

What kind of alternate universe was this where, a week after the bankruptcy court had awarded the satellites to Colussy, Motorola was telling the world they might get destroyed anyway?

Galvin apparently had no idea how unpopular he was within the halls of state. Even as he tried to persuade the President, government officials were calling Neuberger to make it clear one more time. *We don't want this Motorola loan being converted into equity at any time. It needs to be written into the loan agreement. Motorola can have no ownership of the new Iridium. Not now, not in a year, not in a thousand years.* Neuberger passed the message along to Keith Bane, who got on the phone with the appropriate government officials, and the next day they discussed how to structure the "Ban on Motorola Equity" language, further delaying the final draft agreement.

Sperling was still jet-lagged from his 3:00 A.M. touchdown from Vietnam when he arrived at the West Wing to chair the Situation Room meeting on November 20. Keith Bane's de-orbit deadline was two days away, and the battle lines had been clearly drawn. There were twenty-eight people present from nine departments and agencies, but only two mattered. The Office of Management and Budget was determined to change this deal or get rid of it entirely, and the Pentagon was determined to make it happen. That meant Sylvia Mathews, Deputy Director of OMB, was the prosecuting attorney, and Rudy deLeon was the defendant.

OMB Director Jack Lew had thrown down a gauntlet in a memo: "What's being proposed here is highly unorthodox, precedential, and there's a disproportionate gap between what we're giving away with the indemnification and the benefit to the government. The risk on this doesn't have a cap on it, and the government is getting two years on a service contract for ten to one hundred years of liability. I'm concerned about the moral hazard." Then Bob Kyle, OMB's national security and international affairs expert, took over the formal cross-examination and forced the plaintiff, Rudy deLeon to make his case.

It was true that this was a novel interpretation of Public Law 85-804—all the lawyers on both sides agreed—and the Pentagon even conceded that Iridium was "valued but nonessential." So Kyle challenged deLeon to prove that the military really did need Iridium and that the need had nothing to do with White House pressure. As one Justice Department lawyer put it, "National defense has to be a stand-alone reason." DeLeon also had to show that it made economic sense—and he did, arguing that an investment of an additional $100 million would save $9 billion the government would otherwise spend building its own system. He then had to show that Public Law 85-804 was the only solution left after every other possible solution had been considered. Mike Greenberger from the Justice Department helped with that argument, outlining his six months of talks with Motorola. Finally deLeon had to prove that indemnification of a third-party corporation was legal, that the whole idea was good policy for the future, and that the benefits of having Iridium far outweighed the cost in case the whole thing went wrong.

But the crucial part of the meeting came when addressing the language of the law itself. The issue of "unusually hazardous risk" had to be addressed, since that was what the statute required—and it was addressed directly by Dave Oliver. Forget these imaginary people on the ground—when you brought down ninety-two satellites, you ran the risk of hitting the International Space Station, and "the damage that can be caused by contact with a satellite in space or on de-orbit is potentially catastrophic."

As expected, the chief government user of Iridium phones, the State Department, failed to show any enthusiasm. Deputy Secretary Strobe Talbott was a no-show, and his lieutenant said there was "not a strong case" to preserve the system. Asked to at least admit that the system had value in Africa, the State Department representatives said that they were "not previously aware of White House interest in Africa." Everyone was baffled by their apparent indifference—what about Madeleine Albright's love of the phones?—but their position was assumed to be the handiwork of junior-level functionaries trying to enforce a strict interpretation of the U.S. treaty with Inmarsat.

Had deLeon proved any of his premises? The debate kept circling back to a single fact. There were eighteen satellites that were already in uncontrolled decaying orbits—why should the government be insuring those when Motorola would have had to deal with them anyway? What OMB was pointing out was the blackmail aspect of Motorola's position, the idea that it should be insured far more heavily than it deserved for the simple reason that it had the power to destroy the satellites. DeLeon finally gave a little ground when he promised to go back to Motorola one more time.

"I'll put the dead satellites on the table," he said, "and walk out if Motorola doesn't give us a compelling case for those."

Somewhat surprisingly, this single promise by deLeon seemed to be enough for the naysayers, and the meeting broke up.

Was it over? Was OMB just looking for a face-saver? Was it trying to get a concession on the dead satellites to show that something was done and Motorola didn't get everything it wanted? One thing that everyone knew was that in the absence of firm decisive action, Iridium was a black hole that would suck in new complications.

And no, it was not over. The next day OMB remained fixated on the dead satellites. After fielding several calls from the White House, all of them related to OMB problems, Mark Adams called Colussy to say that a dangerous game was going on here. Because OMB came in so late, their policy wonks seemed oblivious to the fact that Motorola had its finger on the trigger. Motorola was not going to be swayed by any fine-point "moral hazard" arguments, no matter how commonsensical they were. OMB was in the way. They needed to be stifled.

"We have to get OMB off of this," said Adams. "Time to play the Africa card with Clinton."

And Colussy agreed. Calls went out to Herb Wilkins, then to Ty Brown, then to Jesse Jackson and Bob Johnson. Targets: Sperling and Chief of Staff John Podesta. Message: time to ram this thing through. Within twenty-four hours Ty Brown called back to say that Jesse Jackson didn't want to talk to Sperling or Podesta; he was going straight to the President. The FOBs were attacking en masse.

The next day, Podesta was asking the Pentagon for a briefing.

Keith McNerney, Chief Financial Officer of Gene Curcio's company, called Iridium's lawyers to say he thought the fix was in at the bankruptcy hearing and "I'm filing fraud charges and you're all going to jail."

Dave Oliver called Colussy to say, "The OMB is still holding everything up." It wasn't a face-saver issue at all. The "dead satellite issue" was now its main focus, and Oliver had gone back to his statistical arguments, bringing in a personal injury lawyer and computing the total risk to the government at $70,000 for the working constellation and $16,000 for the dead satellites, even using the NASA study. That wasn't the point, though. OMB was irked by Motorola expecting the government to cover the dead satellites at all. Dorothy Robyn agreed that it didn't make sense, but she knew how prickly Motorola had become and warned everyone: "This might be a deal killer."

James Swartz called Colussy to say the Saudis would invest, but only with a written guarantee that no executives from the old Iridium were allowed to work for the new company.

And the argument kept swirling higher and higher, getting very close to the President. "Rudy deLeon is on his way to talk to the national security advisor, Sandy Berger," Mark Adams told Colussy. "Apparently it couldn't be worked out without Rudy going in person. The OMB wants the National Security Council to tell them that the national interest is at stake."

Dorothy Robyn called to say, "Representative Lewis is refusing to meet with Rudy."

Finally, on November 22, the day Iridium Space Vehicle 79 was scheduled to crash into the ocean, the powers that be at the Pentagon made a collective decision to spend the entire day ignoring the problems of the rest of the planet so that Iridium could finally be put to rest in a "nobody leaves till we're done" session. Assembling in the largest SCIF in the Building were the four most powerful people in the Pentagon after Secretary Cohen: Deputy Secretary Rudy deLeon, Pentagon acquisition czar Jacques Gansler, Pentagon general counsel Douglas

Dworkin, and Bill Lynn, Under Secretary for Fiscal Matters, popularly regarded as "the CFO of the Pentagon." They had drawn up what they called a "term sheet" for getting things done, and most of the items were requirements that had to be approved by Motorola, represented at the meeting by Ted Schaffner and in-house attorney Bruce Ramo. Also attending were Colussy, Isaac Neuberger, Boeing lobbyist Shep Hill, Dave Oliver, Pentagon lawyer Harvey Nathan (the in-house expert on foreign ownership), and what Neuberger later called "untold legions of others, all of whom were tripping over themselves."

The big issues were settled in the first couple of hours. The DISA contract was to state clearly that Iridium would receive $72 million for twenty thousand handsets over a two-year period, with options to extend the contract in the future. DeLeon was even more effusive about the Iridium system than he had been ten weeks earlier, stating that the total $112 million "investment" was something he regarded as "equity" that the government was putting into a private company because it would be impossible for the United States to duplicate what Iridium provided, and even if the government could build its own stand-alone system, it would cost $9 billion. Therefore, in the interests of the United States, the government would invoke Public Law 85-804 to indemnify Motorola for the sixty-six satellites in the constellation, the eight spares in orbit, and the five dead satellites expected to fall to Earth in the next ten years.

But that wasn't all the satellites. Motorola had seven satellites that were sitting in the Chandler Lab and hadn't been launched yet, and they wanted those added to the indemnification deal. Dorothy Robyn responded with some variation of "Are you insane?" and that part of the equation was pushed aside "for later consideration." Attention then shifted to what Dave Oliver called "those troublesome dead satellites." Six were in orbits that would not reenter the atmosphere for at least a hundred years, and after some brief discussion, everyone decided just to ignore those. That left four satellites—the ones that would be crashing to Earth later than ten years hence but earlier than twenty years hence—and for some reason Motorola was paranoid about those. Schaffner and Ramo kept insisting that the government should indemnify those as well, but the Pentagon officials kept pushing back, saying they shouldn't be expected to insure satellites that far into the future. After ten hours of wrangling, Schaffner said he wasn't able to commit to the deal and suggested that deLeon call Keith Bane directly. Everyone was stunned. After all this time, this thing was going to fall apart over *four dead satellites?* It was the last day before the Thanksgiving break, so there was no way to ask anyone to stay late. The meeting finally broke up at 6:00 P.M. with everyone frustrated and angry, and with all the anger directed at Motorola.

After the long holiday weekend, everyone returned to work, only to find out the deal was being held up yet again, this time by Rudy deLeon.

"Rudy wants to get this past Representative Lewis," said Oliver in a call to Colussy. "He wants Lewis to sign off on it. It's bigger than Iridium. Lewis controls spending for a lot of other stuff the Pentagon needs."

But Jerry Lewis was refusing to meet with him.

"That's what's making it hard. Tomorrow for sure. Come to the Building. We'll do the deal."

So on November 28, Colussy and Neuberger went to the Building and were ushered into a SCIF, where they waited for Oliver, deLeon, and the others involved to come by with the various contracts that needed to be signed. At the end of the day, they were informed by a polite enlisted man that no one would be coming to see them this day, so they could go home. Apparently John Podesta was being briefed again, and everyone was waiting to see what he said. DeLeon was also drafting separate letters to Senator Stevens and Representative Lewis, explaining why the government needed Iridium. (*Wasn't that something we already did a month ago?*) At the Pentagon, the policy for sending a letter to Congress was to call the congressman first and tell him everything that was in the letter in advance—which begs the question as to why the letter was being sent in the first place—but, at any rate, that was why it took all day.

The dead-satellite issue persisted. The next day, a conference call with deLeon, Sperling, Rudman, Lew, Bane, and Schaffner resulted in more confusion. NASA could locate only two of the four dummy satellites. The agency needed to know how many were up there so they could be identified in the indemnification contract. Were there ninety satellites up there or ninety-two? And in the middle of the debate about it, the number changed again, because word came from Cheyenne Mountain that Iridium Space Vehicle 79 had descended through the atmosphere, narrowly missed the Northwest Territories, and disintegrated into the Arctic Ocean at 4:44 A.M. Eastern time. Not a single major media outlet noticed.

On the last day of November, Dave Oliver and Keith Bane got into yet another heated discussion, with Bane saying, "Why is this taking so long?" and Oliver explaining about the need to debrief Representative Lewis. Oliver explained the Pentagon letter-writing as "laying down a soft gauntlet," but Bane didn't want to hear it and eventually said, "The decommissioning of the constellation will begin at noon on Monday, December 4"—in four days. Oliver called Colussy to say that Bane was a crazy man. "But it doesn't matter anyway," said Oliver. "I still have to clean up a few things over on the Hill tomorrow morning, but be prepared to sign tomorrow afternoon."

Once again, Oliver had been optimistic. Jerry Lewis's staff members called later that night to say they were "dropping our objection," but four days later DISA still hadn't signed the contract, and Oliver still didn't have a formal okay from Senator Stevens.

Finally, on Wednesday, December 6, everyone gathered in Washington to go over the final drafts of all the contracts and, as Keith Bane liked to put it, "stay put till finished." Motorola's team had already been camped out for a week at the Ritz-Carlton in Pentagon City. Colussy's people checked into the Lansdowne Resort, just a stone's throw from the SNOC, for the first meetings of the management team. Colussy had designated Thursday as the day of the capital call, meaning everyone had to write checks—and surprisingly, Michael Boyd, calling from Canberra, asked for an option for an additional $20 million.

The other surprise that day was a new clause in the indemnification agreement. The government was insisting on a paragraph stating that Motorola would promise not to launch another low-earth-orbit satellite system. Bane and Schaffner were highly amused by that particular provision and said, "How fast can we agree?"

In an attempt to close the deal before the weekend, everyone stayed up most of Thursday night wrangling over details of the indemnification document, which had to be signed by Motorola, the Pentagon, Boeing, and Iridium Satellite LLC, which was the name chosen to replace "Newco." Apparently nerves were so frayed that several e-mails were exchanged the next day apologizing "for the way things ended last night," especially "inappropriate tones and statements that added to the unpleasantness." Despite their best efforts, the deal didn't close on Friday, because Ann Kurinskas, the Chase Manhattan collection agent, said she couldn't allow the sale before 10:30 A.M. on Monday. She had sent out letters to the thirty banks in her consortium, and she had to give them time to object to the deal. If she didn't hear from them, she would approve the sale.

On Sunday, December 10, Secretary of Defense Cohen finally signed a letter saying a contract for Iridium would "facilitate the national defense," and that Public Law 85-804 was being invoked to indemnify Motorola for any lawsuits that still remained after all of Colussy's insurance had been used up, after all of Motorola's current ten-year policies had been used up, and after all of Boeing's insurance had been used up—and the indemnification applied only to lawsuits against Motorola, not Iridium or Boeing or anybody else. This was essentially an endorsement of all the "ridiculous scenarios" that had been raised by the Motorola legal team over the past several months, including indemnification for *the seven satellites that hadn't yet been launched*. Motorola, like the bratty child who constantly throws temper tantrums to get his way, was finally muzzled, isolated, and appeased.

It was not until Monday, December 11, after six hours of waiting in a Pentagon SCIF, that Colussy and Neuberger were introduced to the infamous Sharlene Capobianco, contract officer for the DISA contract, along with Dave Oliver, an assistant secretary, two Pentagon lawyers, and a controller. The precious signature was duly obtained, and an impenetrable seventy-page document called a "source selection sensitive" contract—Pentagonese for a contract you're not supposed to discuss publicly—was handed over. Neuberger took photos to memorialize the occasion, reporting later to the Iridium investors that most of the six hours was spent "watching several Motorolans cringe and cry."

Shortly before 8:00 P.M., Colussy had his $72 million deal, which took the form of a comprehensive agreement among the Department of Defense, Motorola, Boeing, and Iridium. It was one of the few joyous moments of the year, and yet there was no champagne celebration because everyone was exhausted by the equivalent of a fifteen-round heavyweight fight. Mark Adams, Dave Oliver, Dorothy Robyn, and Dannie Stamp all seemed stunned, like they had finally slain a monster and were now watching the dead body, afraid it might come back to life. "When you serve a lifetime in government," said FCC chief of staff Kathy Brown, "you rarely have the chance to do something that you know is good, something that corrects a wrong, something you can be proud of. I was proud to be a part of this." Dave Beier, chief domestic policy advisor to Al Gore, told Dorothy Robyn, "Few will remember your persistence if this succeeds, but the national security (both in the defense sense and in terms of advancing the connection by telephone of the developing world) will be advanced by your work." Robyn's reply: "Thanks, but Dave Oliver is the real hero of this story."

The following day Boeing took over formal management of the SNOC. Unfortunately, many of the engineers had been so spooked by the previous year of uncertainty that they had either quit or made plans to leave. Boeing didn't want to run the risk of losing people who had intimate knowledge of all the peculiarities of the system, so all salaries were doubled and the workforce was largely retained.

By December 12, it was over. Oliver conducted a Pentagon press conference. Colussy conducted a conference call with the media. And almost all the news accounts of the sale were wrong. The London media said the Pentagon had stepped in to avoid mass panic from falling satellites. The Dow Jones News Service implied that the Colussy group was being uncommonly secretive when a reporter tried to find out the names of all the investors and couldn't—owing to the Prince's obsession with privacy. Roger Rusch, now president of a satellite consulting firm called TelAstra in Palos Verdes, California, told *Satellite News* that the Pentagon contract was woefully inadequate, that the satellites were going to fall out of the sky in five years anyway, and asked the rhetorical question: "Who would invest in another Iridium constellation?"[48] Most of the analysts who bothered to look at the deal—and there weren't many, since Iridium was old news by now—said it was a strange, pathetic ending to a really bad idea.

Dannie Stamp didn't think so. On December 11 he turned in his badge at the Chandler Lab. On the morning of the twelfth he reported to the gateway in Tempe along with two other employees, including the janitor. On the way to his new office, he stopped at a convenience store to buy toilet paper, Scott towels, and coffee. Once again, he had the greatest job in the world.

At the Battle of Lepanto, the outnumbered Holy League destroyed 210 Ottoman ships and killed thirty thousand sailors, ending Turkish dominance of the Mediterranean and leading to a new age in which the nimble Venetian fleet would become the model for navies everywhere.

The Iridium deal inspired historical comparisons. Three years after it was concluded, a writer for *USA Today* was casting about for some parallel in business history, but he couldn't find a precise equivalent for a $6.5 million cash outlay that resulted in possession of a resource that would probably never again be available in the whole future history of the United States. The one he came up with was from the nineteenth century, when a Virginian short of cash had to borrow $15 million to satisfy a disagreeable French dictator who controlled a lot of real estate. That deal was called the Louisiana Purchase.

Dan Colussy, the little guy, had won.

Chapter 15

MOOSE HUNTERS, MARINES, AND THUGS

MAY 9, 2004
SATELLITE NETWORK OPERATIONS CENTER,
LEESBURG, VIRGINIA

The story of Iridium wouldn't be complete without describing what happened to the satellites after the Colussy group took over. It took a little over three years, but on a day in May 2004 his instincts were all confirmed—the world couldn't live without the system. The day started with a technical problem at the SNOC. About halfway through the 7:00 A.M. to 7:00 P.M. shift, a technician on the third console received an inquiry from the military gateway in Wahiawa, Hawaii. It seemed that the system was experiencing a high level of dropped calls—or, more precisely, failures to connect—and an extraordinary level of traffic in general. Shortly thereafter, a similar call came in from the civilian gateway in Tempe. The system was jamming up; it was overloaded. The technician scanned his CRT screen. Everything looked normal. Most of the twelve hundred "care and feeding" tasks had been completed for the day. It wasn't an "anomaly" or a "bump"—the most feared words among the nine men on duty, meaning something in space has ceased working—and it wasn't an "exception," which would mean some part of the constellation was working in an unexpected way.

Any of those situations would have caused the whole room to scramble into emergency mode. It wasn't a weather problem—sunspots and solar eclipses had been known to affect the system in the past. No, this was something else. It seemed that satellites in one of the six orbital planes had been stressed beyond capacity.

But that was impossible. The Iridium system could handle 98,586 calls at a time—the math had been worked out by Ken Peterson long ago—and the company barely had 140,000 phones in service. Could there be a glitch in the pattern of cell reuse? Could the phased-array antennas be duplicating the same call or failing to pass it off to the next satellite? The technician dug deeper into the numbers scrolling rapidly across his screen, zeroed in on the heavy traffic, and then realized all the calls were originating in the same fifty-thousand-square-mile zone. It *was* possible to overload one small part of the system if, by some odd set of circumstances, massive numbers of people were all sending signals to the same satellite.

And that's what was happening. It was a holiday in the United States, and there were 146,000 American troops in Iraq, and there was only one phone they could all depend on. Technically you weren't supposed to use an Iridium phone for a "morale call," but when you absolutely, positively had to get your call through at a certain time to a certain person, every soldier, sailor, Marine, and airman knew there was only one solution. The only situation that could max out the Iridium system was Mother's Day in a combat zone.

Mother's Day 2004 was not only a turning point for Iridium—around that time the company reached cash-flow break-even and started making money—but it was the first time the rest of the world realized there was a phone that worked everywhere on the surface of the planet. For cutting-edge professionals, there were actually three events that made Iridium famous. The first was 9/11. The second was the war in Afghanistan, even more than Iraq, since the terrain was so rugged and the infrastructure so primitive. And the third game-changing event was Hurricane Katrina, which was not the first natural disaster where relief workers used the Iridium phone as a lifeline, but was definitely the highest profile.

The new company had started off with exuberance and great promise. On January 17, 2001, Dannie Stamp dressed up in a spaceman costume for the victory celebration at the Metropolitan Club in Washington, and people let their hair down after almost a full year of constant stress. The champagne flowed, the room was filled with warmth, and there was one brief moment when it appeared that Colussy's ragged coalition could form itself into a family of dedicated satellite operators. But it wasn't to be. With just a tiny marketing budget, the company struggled from day to day, surviving hand to mouth, trying to overcome the infamy of the name "Iridium" while the boardroom descended into squabbling and dysfunction. Personalities and cultural backgrounds grated on one another. Prince Khalid assigned five different employees to oversee his investment, and their constant requests for microscopic details on every decision, combined with their refusal to make timely decisions themselves, annoyed the laid-back owners of Syncom even as the Brazilians and the Australians ran into money problems and started missing their cash calls. The very fact that the Prince never introduced himself to the other partners, much less showed up for a board meeting, was regarded as insufferably insulting.

But there was a more fundamental problem. Most of the sixty-three thousand original subscribers never came back. In fact, Iridium would come very close to disappearing altogether during those first three precarious years after the Pentagon deal. Colussy began the year 2001 with just $14 million in the bank—enough to run the constellation for two months—but that wasn't the worst part. He also had no choice but to break his vow to Helene and take over as CEO. The Saudis wouldn't accept Ed Staiano and there weren't a lot of telecom executives out there who wanted to associate themselves with Iridium.

Still, it was gratifying to see that the small number of people who did use the phone were fanatical about it. Every time Colussy encountered an active-duty serviceman at an airport, he would ask if he was familiar with the Iridium phone, and the response was always immediate and ebullient. *We love the Iridium phone,* the war-fighters told him. *We can't live without the Iridium phone,* they said. But awareness by the general public came much more slowly. Iridium was still mostly associated with a failed venture by Motorola, not a marvel of modern engineering. Even when the phone started turning up in popular culture—Robert Downey Jr. uses one in the Iron Man movies, Dirk Pitt runs his secret operations with an Iridium phone in Clive Cussler's adventure novels, Brad Pitt saves the world with an Iridium handset in *World War Z*—it blended into all the other devices, most of them fictional, that superheroes tend to make use of. The first time most people noticed an Iridium phone in a movie was in 2014 when Bradley Cooper, playing Navy SEAL marksman Chris Kyle in *American Sniper*, executed the longest kill shot in history, then extended the antenna on his Iridium handset so he could call his wife eight thousand miles away to say, "Baby, I'm ready to come home."

It was John Castle, of all people, who made the first Iridium phone call from New York after the terrorist attacks of September 11, 2001. With downtown Manhattan in chaos and the Pentagon on fire, Castle couldn't make any of the phones work in the East 58th Street skyscraper where Castle Harlan was headquartered, so he took out his Iridium handset to call Colussy. Castle's first thought had been, *I wonder if Iridium still works.* And, of course, it did, on the day when there was no other service in Manhattan.

Unfortunately, Iridium executives soon became alarmed that the guys responsible for 9/11 may have been using Iridium as well. In late September Colussy got a troubling report from the SNOC and the Tempe gateway. Technicians reported that two or three days prior to 9/11, the dormant Jeddah gateway in Saudi Arabia had suddenly been activated after eighteen months of being shut down. Two or three days *after* the attacks, the gateway ceased operations again. Iridium was using only two official gateways—the one in Hawaii for military calls and the one in Arizona for civilian calls—but for some reason the old Saudi gateway was still tied into the network. Colussy called a meeting of top executives to decide what to do. Should it be reported to the government? How many calls went in and out of Saudi Arabia during the activation? Was the company in any kind of peril because of the usage? Fortunately Mark Adams was at the meeting, and he said, cryptically, "I'll take care of it." So the report was made through channels, the FBI was informed, and no one heard anything else about it—for five years. Then, when former Director of Homeland Security Tom Ridge was asked to join the Iridium board in 2006, Ridge insisted that the matter be reopened.

This time Fulbright & Jaworski was asked to investigate a number of issues involving the Jeddah gateway, including whether it had been turned on around 9/11, whether gateway manager Tom Alabakis had any contact with United Arab Emirates intelligence agencies from his office in Dubai, and why Mark Adams made two visits to Saudi Arabia around the same time. Alabakis' laptop computer was seized, all call records were examined, documents were accumulated, employees were interviewed, and Fulbright & Jaworski gave its findings to Michael Deutschman, Iridium's general counsel. That report concluded that the Jeddah gateway had never been turned off because Iridium still had customers in Turkey and Egypt whose phones were "homed" on the gateway. These customers were complaining of spotty service because one of the two Jeddah antennas was defunct and the other one only worked sporadically. As a result, Alabakis had hired a company called Scientific Atlanta to do antenna repair work between September 7 and September 14, 2001, and test calls were being made during that time. The report also determined that UAE intelligence agents contacted Alabakis in October and asked for data on a list of Iridium phone numbers, but apparently the information was never turned over because the UAE had no court order. The visits of Mark Adams were meetings with the

Saudi Ministry of Defense and Ministry of the Interior, and they were both deemed to be routine sales visits and nothing more, as Adams solicited business for his various ventures. Deutschman recommended to the Iridium board that no further action be taken, and Tom Ridge then agreed to serve as a board member. Of course, the backdrop to the whole 9/11 query was that Prince Khalid, majority owner of the Jeddah gateway, was a 24 percent owner of Iridium (soon to rise to 31 percent as the Australians and Brazilians had their shares diluted), and various low-level government functionaries had been raising foreign ownership issues even before the Prince invested. It also didn't help that the minority owner of the Jeddah gateway was the Saudi Binladin Group, especially once it became known that 19 of the 20 hijackers came from Saudi Arabia.[49]

Ironically, it was President Bush's actions after 9/11 that ended the foreign ownership issue forever. The President had just announced a "coalition of the willing" to fight the war in Afghanistan, and he was desperate for Middle Eastern allies. Envoys had been sent to Saudi Arabia to get the blessing of the royal family for the military action—and Prince Khalid was married to the sister of King Fahd. "We won't be issuing any edicts against Middle Eastern companies while this is going on," reported Dave Oliver.

Meanwhile, Mark Adams continued his double life through a unique employment contract. He received a salary as Iridium's Chief Technology Officer, but he also had a "minority interest" in a company called NexGen Communications that would be developing top secret Iridium applications that could be used by the military and other government agencies. (The other owners of NexGen were all engineers who worked together at MITRE Corporation.) And perhaps Adams had a third job as well: one of the conditions of his employment was that six government operatives would be stationed at the SNOC. (Colussy didn't know exactly what they did, and didn't want to know.) Still, it was not exactly clear where Adams' primary loyalties lay—with Iridium, with the government, or with the economic interests of his own highly privileged company.

In December 2001 Adams suddenly turned up in London, having a face-to-face meeting with Prince Khalid. Reports of the meeting startled Colussy, since no one else at Iridium had ever met the Prince or even spoken to him. It was an especially vulnerable time for the company, with cash on hand running low and Craig McCaw hovering around again, hoping to get a bargain if Iridium ran into trouble. When Adams returned to the States, he explained that the meeting was no big deal—the Prince simply wanted to plead for the Jeddah gateway being kept open. James Swartz backed up Adams' account, saying the meeting was "just a coincidence." But how could any meeting with the reclusive Prince be a coincidence? And if it was a coincidence, what was Mark Adams doing at the Prince's town house in the first place? What were the Saudis attempting to do? What was Mark Adams attempting to do? The deeper you got into Iridium politics, the more paranoid you became.

And then there were those ten stolen phones.

In January 2001 it was discovered that ten Iridium phones had been in constant use, running up millions of minutes in airtime, without ever being registered to any particular user. Once the handsets were identified, they were blocked from using the system—but *who were those people and what were they doing?* No one ever found out.

So there was this dark underside to Iridium usage—bad guys seemed to know about it, and bad guys seemed to like it—but at the same time it was becoming the favorite handset of the American military. The Colussy investment group had taken over Iridium on December 12, 2000, and less than twenty-four hours later the first six orders for handsets came in—from the U.S. Army's Battle Command Battle Lab in Fort Gordon, Georgia. By April 2003, when the invasion of Iraq began, it was clear that Iridium was not just a great add-on for small units but the *only* handheld phone that worked in Middle East combat zones. And because it worked all the time, under all conditions, civilian contracts started to roll in as well, especially from Halliburton, Blackwater USA, and other companies hired to provide services in Iraq and Kuwait. By August 2003, the popularity of the phone had spread to other armies—those of the United Kingdom, Australia, New Zealand, and Canada—but the ultimate badge of approval came when the Israel Defense Forces ordered five hundred handsets, since Israel was notoriously hard to please when it came to military equipment. "We have some forces that can't get along today without the Iridium system," said Colonel Barry Patterson, chief of the U.S. Air Force Satellite Communications Division.

It was ultimately the Marines who would become the most ardent supporters of Iridium. They had always relied on Navy satellites for on-the-move communications, but that system remained 250 percent over capacity through the first decade of the twenty-first century. Just as Jack Woodward had predicted, Iridium was perfect for platoons in motion. The Marine Corps Warfighting Laboratory started working with Mark Adams' company beginning in 2004 to develop a "netted" system that allowed every Marine to be issued an Iridium phone with a "push-to-talk" button on it—no dialing numbers, no delayed connection, everyone can hear everyone else—and after being tested at the Combat Development Command in Quantico, Virginia, the new system was launched. It was called the Expeditionary Tactical Communications System (ETCS), and it became a $70 million addition to the DISA contract in 2009. Netted Iridium was, in fact, the most dramatic advance in combat communications since Motorola's invention of the Walkie-Talkie during World War II. For the first time, every Marine in a unit could communicate in real time with every other Marine in that unit. Since its introduction, Netted Iridium has been continuously improved and expanded at the Naval Surface Warfare Center in Dahlgren, Virginia, with a view toward achieving the military nirvana of "global push-to-talk." (Currently it only works within a 250-square-mile area.)

And yet, even as the reputation of Iridium grew, Colussy was frustrated. The company was growing, but it could have been growing faster if it had more capital to introduce new services, especially asset-tracking applications. "This is not a game for individual investors doing a typical venture capital deal," he complained to the board. "This is for deep-pocket strategic investors looking out twenty years. We've only invested $131 million at this point and we need much more to build the customer list and get ready for the future-generation broadband solution, not to mention replacing the whole satellite constellation in ten years." He pleaded with the board to let him set up a merger or partnership with a company that could finance expansion. But after months of wrangling and the dilution of both the Australians and Brazilians, other companies with inferior products started winning contracts because they were better capitalized. The Federal Air Marshal Service was on the verge of ordering Iridium phones for every officer, but at the last minute the lobbyists for Verizon stepped in and got the government to choose a ground-based system instead. The government of Iraq was about to place a massive order for Iridium phones, only to be talked out of it by Boeing, which had an ownership stake in the new Thuraya phones, a regional service that operated two GEOs from a gateway in Sharjah, United Arab Emirates, and had an annoying half-second signal delay. (The Pentagon eventually banned the use of Thuraya phones in Iraq because they were not secure. It would have been possible to use them to track American troop movements.) Even the DISA contract came under scrutiny by the Pentagon, with several bureaucrats saying, "Maybe we should give part of this to Globalstar"—after Globalstar lobbying, of course, and at a time when the Globalstar constellation was literally falling apart. Just as the Iridium inventors had predicted, the altitude chosen by Globalstar forced it to fly through the South Atlantic Anomaly of the Van Allen belt, where the antennas took so much radiation punishment that by the year 2007 they started failing.

"We're getting outgunned at every point," Colussy told his board. "We need a strategic partner with significant resources, even if it means consolidating with one of these inferior satellite companies."

Colussy stayed on as chairman of the board as the company went through a series of CEOs, including General Electric veteran Gino Picasso, who butted heads so often with the Saudis over their micromanagement that he eventually resigned under pressure. His successor was an executive named Stephen Carroll, whose previous experience had been at a London company called Storm Telecommunications. Colussy made no secret of the fact that he considered Carroll a "lightweight" who was alienating Boeing and Motorola at a time when Iridium needed debt forbearance from both—so eventually Carroll was replaced with Carmen Lloyd, who resigned as CEO of Stratos Global to run Iridium. Lloyd had an extensive background in electronics, especially aircraft electronics, but he, too, started to wear out his welcome when he repeatedly failed to raise money needed for expansion. Meanwhile, the partners resisted Colussy's attempts to set up a merger with a larger player. Wasn't the company growing? Wasn't Iridium the only global satellite service? And besides, didn't they already have the most powerful lobbyist possible?

Yes, in a manner of speaking, they did. A few months after the company launched, Colussy had received a call from Bob Johnson. Johnson wanted to have lunch—and, by the way, Bill Cohen would be joining them. Cohen had left the Pentagon at the end of Clinton's presidency, of course, but he had now set up a lobbying firm called the Cohen Group. Wouldn't the Cohen Group be a great asset for Iridium? Shouldn't Iridium have a lobbyist? suggested Johnson. Colussy got the message and agreed it would probably be a smart move. By May 2001 he was signing up Bill Cohen as Iridium's man in Washington.

Hurricane Katrina was the event that cemented Iridium's reputation. Mark Adams had first realized what a powerful search-and-rescue tool the phone was when he received a call at home late one night in 2003. An emergency had been declared by the Alaska Rescue Coordination Center. A small private plane had gone down somewhere in the heavily forested northern regions of the state, but no one knew where. There was an Iridium phone on board the plane. Was there any way to figure out where the last transmission from that phone had come from? Adams got busy and did figure out how to do that, and the plane was located within several hours. (Unfortunately the pilot didn't survive.) It was an emotionally wrenching experience that made Adams realize just how important these situations would be in the future, and for the next two years he remained actively involved with emergencies and rescue efforts, especially in rugged areas like Alaska where he dealt with every-thing from lost backpackers to a hunter who stabbed himself in the leg while dressing a moose.

Meanwhile, stories were trickling in from the field indicating that Iridium phones were already saving lives every day, with or without the help of company headquarters. On April 11, 2001, just two weeks after the system was switched back on, the physician on duty at the Amundsen-Scott South Pole Station, Dr. Ron Shemenski, passed a gall-stone and diagnosed himself with acute pancreatitis. After consultations via Iridium phone with specialists in the States, it was decided that he needed to be evacuated. But with winter coming on, the rescue required coordination on three continents involving the military, the National Science Foundation, and a Canadian airline that leased the de Havilland Canada DHC-6, known as the "Twin Otter," because that plane could land on skis. The South Pole runway was normally closed during the six months of night, with the last plane departing in February, leaving only forty-seven personnel to fend for themselves until October. It took two weeks, but eventually two Twin Otters did manage to land at the Pole with a replacement doctor aboard and to get Shemenski to his home in Denver, where he was diagnosed with heart problems as well as pancreatitis and underwent open-heart surgery—but his life was saved.

Iridium phones were used at every stage of that rescue—on the ground, in the air, at every way station, remotely by physicians—and it was one of a thousand similar stories that began to pour into Iridium headquarters as soon as service was enabled. By 2005, the year of Katrina, Iridium was set up to mobilize for disaster relief and rescue efforts wherever they were called for, having already scrambled for the much smaller Hurricane Frances in 2004. The first impact from Katrina killed 1,833 people, drove 1.5 million people from their homes—and destroyed the telecommunications infrastructure of southern Louisiana.[50] The demand for Iridium phones was overwhelming—from FEMA, from law enforcement agencies, from relief workers, from everyone who had to function in the New Orleans area. Thousands of handsets were moved into place—so many, in fact, that the last of the old Kyocera units were finally sold and the company's new handset factory in Malaysia had to start running around the clock.

In the aftermath of Katrina, relief agencies all over the world started stocking Iridium phones. In 2006 thousands of phones were sent to Taiwan after the typhoons that knocked out telecommunications. In 2008 Hurricanes Gustav and Ike hit back-to-back in late August and early September, and the company sent sixty-two hundred phones to the Gulf Coast in a three-week period. Fortunately, by the time a 7.0-magnitude earthquake hit near the capital of Haiti in 2010, many of the first responders from the United States and Mexico were already outfitted with Iridium handsets kept in readiness for just such a situation. In fact, it could be said that the Haiti earthquake was the first time Iridium became standard operating equipment for virtually all disaster relief workers. More than fifty organizations used Iridium, including the media and Project HOPE, the international health care organization that employed the phones so that local doctors could consult with specialists in acute-care cases such as pediatric burn victims.

But once again, the growing admiration for the phone itself didn't translate into respect from potential partners. Colussy spent years trying to motivate various CEOs to orchestrate a merger with a larger corporation, but the talks would never get very far. At various times he opened discussions with Inmarsat, AT&T Wireless, Boeing, General Electric, General Dynamics, several defense companies, and even Globalstar once Bernie Schwartz ceased to be part of its management in 2003. He went to Bob Johnson one more time. He had meetings with deep-pocket investors like Paladin Capital Group, the Carlyle Group, Miami tobacco magnate Bennett LeBow, the Pivotal Group of Phoenix, and even his old nemesis Gene Curcio, who said he had $450 million from South Korea and wanted to spend some of it on Iridium. Colussy went back to Eagle River more than once, but the tenor of those conversations changed dramatically after Nextlink Communications, McCaw's broadband provider, lost 97 percent of its value in a sixteen-month period in the year 2001. Nextel, McCaw's wireless network, was down 75 percent in the same time period. No wonder he was talking about launching ICO-Global Teledesic "in about four years." By then McCaw had picked up yet another company—Ellipso, the joint venture of Boeing and Fairchild—and folded yet another batch of intellectual property into Teledesic. "Our problem now became how are we gonna raise $9 billion?" recalled Dennis Weibling, Eagle River's CEO. "With Ellipso and Odyssey and ICO we were taking over spectrum allocations, but I had to be up and running in order to keep that spectrum." In the end, the skepticism of the Iridium/Celestri team proved correct—McCaw couldn't raise the capital.

One of the ironies of Iridium's history was that it was evaluated by some of the wealthiest investment funds and telecommunications companies in the world, and none of them could ever come to terms with Colussy, who considered himself a motivated seller. By the fall of 2002 Bear Stearns already valued the company at $300 million, indicating that a money-losing company had actually become $275 million more valuable during the eighteen months it was losing money. But one by one, the potential investors bowed out. It was too small for some, too speculative for others. Colussy kept plugging, but when he checked with his friends at Bear Stearns, they gave him some sobering news: they'd taken this investment to twenty-six companies and been turned down twenty-six times.

When you go through dozens of potential investors, all excited about an Iridium investment, who then withdraw after going through due diligence, there's a tendency to start questioning yourself, and this is what happened with the Prince's advisors, who would periodically suggest that the company simply give up, revert to the Finite Life Project, and see how much money the owners could squeeze out of the constellation before it crumbled and decayed. This infuriated Colussy. In fact, the life span of the satellites had been greatly expanded, mostly through the efforts of Mark Adams and Dannie Stamp, who were now predicting that the constellation could survive until the year 2020 by using such simple measures as letting some of the batteries hibernate during certain points in their orbits. *All the better,* said the consultants. *We can run it until the year 2020 without ever having the expense of a second generation.* It was a position that caused Colussy to almost lose his temper on several occasions, and that was something that had rarely occurred during his six decades in the business world.

During the first few months of 2003, Colussy got a temporary respite from constant squabbling with Prince Khalid's employees as one of the Prince's Thoroughbred racehorses, a three-year-old named Empire Maker, swept to victory in several stakes races and became the odds-on favorite to win the Triple Crown.[51] With most of the Prince's army preoccupied with horse racing during May and June, Colussy had time to reflect. He was now six years into his retirement, and four of those years had been spent dealing with the daily drama of Iridium. It was time to back off and enjoy his life with Helene. They had been married in the Coast Guard Academy chapel the day after his graduation in 1953, and with their golden wedding anniversary approaching, Colussy decided to transition to less stressful pursuits. That summer he invited his two daughters, his grandchildren, and forty-one other people to a villa near the little Tuscan village of Pienza, and there he and Helene were remarried under the boughs of an eight-hundred-year-old olive tree by a jolly Catholic priest who was willing to ignore the fact that they were Episcopalians. The party continued for a month, with hot-air balloon rides every day, creating the kind of lifetime memories that the

Colussys had hoped to make for themselves now that they were selling their Maryland home and moving full-time to Florida. The experience made Colussy take stock of which aspects of his life were most important. Constant bickering with Iridium board members was not high on his list.

Fortunately, while the board continued to argue over CEOs, FCC filings, advisors leaking inside information, exit strategies, and the like, the Iridium system was gradually becoming indispensable to several industries—and one of the main reasons was the emergence of the asset-tracking business. What started in the eighties as a tool for the three million long-haul trucks in the United States had since become an industry considered vital for any kind of large mobile equipment. What Colussy's team had recognized—and Motorola scoffed at—was that satellite data speeds could be extremely slow and still be extremely useful. That's not to say it wouldn't be possible someday to achieve the Bill Gates/Craig McCaw dream of broadband from space. A District of Columbia firm called Sky Station International was launched in 1996 with a $4.2 billion plan to float seventeen-ton stationary platforms eighteen miles above the ground, each one held in place by two balloons and a solar-powered thruster. This would reduce the length of the transmission path so drastically that about 250 platforms could serve the world—one for each major metropolitan area—without interfering with any of the other spectrum users. Sky Station operated for four years but eventually foundered on the regulatory front. Since it functioned so close to the ground, it required airspace approvals from every nation in the world.[52] So in a world where high-speed data didn't yet exist, Iridium offered the next best thing.

As early as June 2001, Mark Adams had introduced the first Iridium data device, working at the painfully slow rate of 2.4 kilobytes per second but still representing the first time the Internet was available globally. Adams admitted it was primitive, but pointed out that it was intended for parts of the planet where the only other option was carrier pigeon.[53] A few months later Adams had jury-rigged the system to bump that rate up by a factor of ten, enabling the first Iridium modem for asset tracking. It was huge and retailed at $850, so the only people who used it were those who needed to track very expensive equipment (helicopters, aircraft, ships) in the Arctic and Antarctic regions. In order to get the price down to $100, thereby becoming competitive with the Orbcomm Little LEOs, it took Iridium management five years and seven trips to the board to get the development money. The device that emerged in 2006 allowed short-burst machine-to-machine messaging between any two points on the globe within five seconds. That was good enough for ExxonMobil, which instantly bought three hundred units, creating publicity that made it possible for Iridium to start capturing huge contracts from trucking companies, then power companies, then pipeline companies. (The most important contract of all didn't come until 2013, when Caterpillar—the industry leader in heavy equipment—started using Iridium as standard equipment on all its machinery.)

The next frontier, Colussy thought, would be airlines, and a few small ones were indeed signed up—Aloha Airlines of Hawaii, the regional Frontier Airlines, the private-jet leasing firm NetJets, and two cargo carriers—but it was the shipping magnates who came aboard in vast numbers. By 2004 the maritime uses of Iridium had become 50 percent of the system's total usage, mainly because Iridium was the first cheap handheld device to be available in places operating far from a landmass. Iridium calling cards were introduced so that crew members, frequently at sea for weeks at a time, could remain in contact with their families at rates that were affordable to a sailor. On smaller ships, the Iridium phone started gradually replacing the SSB radio, which was used in the same way truckers use CBs—socially as well as for safety and business—but the SSB system was intricate and hard to master, whereas the Iridium handset worked just like any other phone. The company's biggest leap forward came when Iridium introduced a device for ships called OpenPort. At that time Inmarsat's cheapest device for ships was $18,000. OpenPort retailed for $3,500, and despite everything

Inmarsat attempted in an effort to match Iridium, it could only get its price point down to $7,500. By this time a British technological design firm called Cambridge Consultants had worked out a compression solution that raised the data rate on Iridium to 128 kilobytes per second, or the equivalent of traditional copper-wire dial-up services. Slowly Iridium started to make inroads into the $400 million maritime market that had been dominated by Inmarsat for decades.

By the end of 2005 Iridium had shown its first small profit ($30 million). But that was also the point at which Colussy, with board approval, asked for the resignation of Carmen Lloyd, the company's fourth CEO, because Lloyd had been at the helm during a disastrous failed merger with Inmarsat that ate up nine months of due diligence time and, in the opinion of some, gave away trade secrets to the enemy. Lloyd had also tried to raise $100 million in a Bank of America bond offering—and failed.

In early 2006 Colussy apologized to his wife once again and told her that he was going to spend his seventy-fifth birthday functioning as a full-time CEO. Four months later, Iridium had $210 million in interim financing. Colussy had done a one-man road show and come home with the bridge money the company would need.

But Colussy had now interrupted his retirement for six full years, and he wanted out. He wanted to be building his final home in Florida and resuming his golf lessons. But he vowed, before he quit forever, that he would find a competent CEO who believed in the future of the constellation—no more "empty suits" backed by the Saudis. At the end of 2006, the company had obviously entered an explosive growth stage, with around 250,000 subscribers, $200 million in revenues, and $50 million in earnings. That curve would continue all through the decade, with Iridium eventually passing Inmarsat in number of subscribers and being declared "the market leader and industry standard" by the consulting firm Frost & Sullivan. The board members thought this was reason to celebrate, and to start talking about distributing the wealth.

But the success of the company was actually costing them money. Since Iridium was set up as a limited-liability corporation, each partner had to pay huge tax bills whenever there was an operating profit. Colussy, Syncom, and the Saudis all believed *in theory* that they should be building the next generation of satellites, but not if it meant never getting back their original investment. The dilemma had less meaning for the Saudis, simply because they had deeper pockets than Syncom or Colussy, but Colussy kept looking for a way to do both—get the partners out of debt and fund the next generation at the same time.

Another person who believed that Iridium had to be saved and preserved at all costs was Ray Leopold, the most public of the three Iridium inventors, who was still often asked to speak about the system. "In the history of mankind," said Leopold, "Iridium is the first and only worldwide communications system. We've had radios since Marconi, but it's the first common carrier for anyone to use. When Iridium first went into bankruptcy, of course I was shocked like everyone else. But then I read about the building of the railroads in the nineteenth century. All of those companies went out of business, just like all the satellite companies went out of business. The people who built the infrastructure were never the same people who were able to figure out how to use the infrastructure. It had to be consolidated and preserved."

And indeed, Iridium was proving day by day, month by month, year by year, that the world couldn't live without it. It wasn't going too far to say, in fact, that Iridium's effect on business and government was revolutionary. Beyond the asset-tracking business, it had become essential by the end of the decade for fisheries management, man-overboard alarm systems, location detectors used to map seventy thousand ships on the high seas, the SSAS silent-alarm system to combat piracy, flight following systems, airline cockpit communications, over-the-horizon military communications, the Blue Force Tracking systems to identify the positions of military forces in real time, oceanographic data sensing, the tracking of dog sleds in the 1,150-mile Iditarod race through the Alaskan wilderness (making it a spectator sport for the first time), early-warning systems for Pacific tsunamis, devices to monitor migrating whales, and cell service for people in Newfoundland, Greenland, Siberia, Iceland, the Yukon, northern Scandinavia, the southern Atlantic, and the Indian Ocean who otherwise wouldn't have mobile phones at all.

When the latest Iridium data device was introduced in late 2009, its small size and cheap price made the market explode. Suddenly Iridium transceivers were being used by the oil, gas, chemical, rail, cargo, and shipping industries for machine-to-machine transactions in which no human being was ever involved. Navibulgar, a Bulgarian shipping company, used Iridium to monitor all sixty of its heavy vessels, while the Albanian government employed it to monitor that nation's commercial fishing fleet, thereby ensuring compliance with complicated European Union regulations. When the last ice blockage in the Laptev Sea melted in 2008, the Northeast Passage between Europe and Asia was finally opened, allowing ships to travel north of Russia instead of using the Suez Canal. The first regularly scheduled vessels to use that route were the heavy ships of Bremen-based Beluga Shipping, which were outfitted with Iridium because it was the only communications system that worked that far north. More mundanely, the iCone was introduced at the 2010 Rose Bowl, using Iridium to measure traffic patterns and bottlenecks in Pasadena, sensing what happened on the road, counting the cars, transmitting the speeds and road temperatures, and then sending short-burst messages to technicians who could reroute vehicles.

Then there were the humanitarian uses of the phone, which so far surpassed everyone's expectations that Iridium was starting to gain the status of standard safety equipment, like having life preservers aboard a ship or a second parachute. In fact, some government agencies started treating it just that way, such as the Alaska Department of the Interior, which required that every plane carrying government employees be equipped with Iridium in the cockpit. GlobalMedic, a Canadian agency known for its fast-acting paramedics, considered the Iridium phone standard equipment for its personnel in Angola, Cambodia, Iraq, Somalia, Sri Lanka, Chad, and Gaza. MedStar Health maintained 165 Iridium phones in its Washington, D.C., medical facilities and its helicopter fleet in case of terrorism, war, or natural disasters. Iridium was used to monitor the health of small Beechcraft airplanes and, in case of trouble, allow the pilot to speak to an expert on the ground. When the Tohoku earthquake and tsunami hit Japan in 2011, it was Iridium buoys that first detected the tidal waves and Iridium phones that were used for disaster relief.

And then there was the Iridium market that no one ever anticipated: daredevils and thrill seekers. Trekkers, mountain climbers, competitive yacht racers, wilderness campers, and adventurers of all sorts started packing Iridium phones right alongside their bungee cords and first-aid kits. John Castle thought he was making history when he stood on the South Pole in 2003 and placed an Iridium call to Colussy, but he was actually doing the same thing everyone did when he or she got to either of the Poles for the first time—call someone on an Iridium phone and say "Guess where I am."[54] Most adventurers use Iridium for safety, not novelty, although a team of whitewater kayakers used it to chronicle their journey on Twitter from wilderness areas on the Kamchatka Peninsula. Others think they're going to use it for filing reports on the Internet and end up using it to save their lives—like Cathy O'Dowd, the South African climber famous for being the first woman to reach the summit of Everest from both the north and the south faces, but whose ascent of the east face in 2003 had to be abandoned. Then there was Henri Chorosz, better known as "the mad Frenchman," a sixty-eight-year-old retired IBM executive who tried to fly around the world via the Poles in a home-built single-engine plane called a Glosair. He narrowly escaped death after leaving Cape Town bound for the South Pole, when he suddenly lost speed, then lost altitude, then ended up flying between icebergs and barely making it to Marion Island, where he flipped over and was pinned inside the plane. Fortunately someone was monitoring his

Iridium tracking device and noted a 180-degree turn just prior to his disappearance. He was rescued and, true to his nickname, returned later to salvage the plane and reconstruct it for a later attempt.

Ironically, Iridium inventor Ray Leopold was a Boy Scout leader who had participated in wilderness expeditions in remote regions of northwestern Montana in which the organizers refused to carry Iridium phones. "The thinking is that it's supposed to be a survival hike, and Iridium is an unfair advantage," said Leopold. Many outdoorsmen, pilots, sailors, and hermits would disagree with that gamble—like the flight instructor and student who crash-landed in a mountain range south of Auckland, New Zealand, but whose lives were saved by Iridium; or the Seattle couple stranded in the Pacific with no propeller, no sails, no engine, and no communications equipment (just an Iridium phone that eventually got a lifesaving message to the Colombian armada); or Cliff Glover, the man who lived so far out in the boondocks of British Columbia that the only way to make sure he didn't miss his transplant call was to carry an Iridium phone—he has two new lungs today as a result.

Rear Admiral Hugh D. Wisely assessed the Iridium system halfway through the Iraq War and concluded that it had been "transformational." With reviews like that, law enforcement and security agencies took notice. In 2004 the Republican Party was paranoid about terrorism at its national convention in New York City, so they ordered 230 Iridium phones in advance. Orders came in from the UK Defense Ministry, the national police of Colombia (which needed Iridium phones to keep up with the Iridium-linked drug traffickers in the mountains), the government of Alberta, the Spanish special forces, and the government of Poland, which became so enamored of Iridium that at one point it inquired about what it would take to install its own military gateway. In 2006 Boeing and Iridium teamed up to create a "virtual fence" along the U.S.-Mexico border, and Homeland Security chief Tom Ridge ordered hundreds of phones for various first-responder personnel. There was an Iridium device in every bus, van, or sedan used to transport illegal immigrants, under the theory that it was the only way to guarantee officer safety when one man was in charge of dozens of detainees. DARPA developed an Iridium app to be used in concert with the GPS constellation to create a highly classified 3-D mapping service in outer space. (Whereas GPS is accurate to within several meters, the DARPA application is accurate to within centimeters.) In 2011 Navy SEAL Team Six carried an Iridium phone into the house in Abbottabad, Pakistan, where Osama bin Laden drew his last breath.

But when the final accounting comes of all the ways Iridium has changed the world, no doubt the primary focus will be on its contributions to science. In some cases Iridium simply provided contact with the outside world for scientists who had no other option, such as the team from the University of New South Wales that settled in at the most remote place in the world: Dome C in Antarctica, where France and Italy have run Concordia Station since 2005. The darkest place on the planet, and therefore the best place in the world for ground-based telescopes, this was where the Iridium-connected operation built the Italian Robotic Antarctic Infrared Telescope (IRAIT). Likewise, the International Space Station had an Iridium phone aboard (in the TMA-4 descent module), and there were several at the Institut Polaire Français, where fourteen scientists worked in temperatures as cold as 112 below, relying on Iridium for an audiovisual application that made it possible for station physician Alexander Kumar to have consultations with doctors in the United Kingdom.

In other cases, Iridium was part of the scientific device itself. Seaglider, for example, is a bright pink torpedo-shaped drone created by the Applied Physics Laboratory at the University of Washington. It gathers data from the ocean, then pokes its nose up out of the water to transmit Iridium messages detailing what it has found. After the Deepwater Horizon oil-rig explosion in April 2010, seven helicopters were equipped with an Iridium application that allowed biologists in the air to tell biologists on the water the exact location of birds in the most imminent danger. The Pelagos Sanctuary for Mediterranean Marine Mammals used an

Iridium device to record the locations of whales in real time and transmit this information to high-speed ferries so that courses could be altered when collisions were likely.

At the beginning of the second millennial decade, Iridium applications and machine-to-machine uses started growing at such a rapid rate that the company ceased to be primarily a phone service and became more of a communications platform that could be accessed and used by the app developers of the world, especially those in the asset-tracking fields. As early as 1997, MIT's Nick Negroponte, guru of the digital revolution and a Motorola board member, had said, "The phone should answer the phone, like a secretary does. I wish Motorola did not make handsets, but phones which behaved liked well-trained English butlers." And that's essentially what the Iridium system had become.

Oddly, many of the uses Colussy had predicted in that first meeting in John Castle's Palm Beach house—the most obvious uses—never worked out. Year after year, Colussy thought the major airlines of the world would adopt Iridium as their first choice in cockpit communications and air traffic control, but time and again they passed on the mere $30,000 per plane that would have been required to employ Honeywell's AirSat system. To demonstrate just how reliable it was, Honeywell installed it in an F-16 and had the pilot do three- to five-second aileron rolls at Mach 1.6 in an effort to lose the signal. The best the pilot could ever do was a slight fade when the plane was upside down.[55] Nevertheless, the airline industry as a whole was continuing to use Inmarsat, despite it being triple the cost of Iridium and despite it being worthless beyond the 65th parallels, as late as 2011. That was the year that Iridium was finally certified by the FAA for aviation safety services, and slowly the airlines started coming around to AirSat.

Likewise, the plan to provide village phone service to the most disadvantaged parts of the Third World never panned out, despite concerted efforts by Syncom. Shortly after the new Iridium launched, a special WorldTel division was created for western Africa. Ty Brown made several trips to Ghana, Senegal, and Nigeria in 2001 and 2002 to secure licenses but could never line up any African partners. A pilot project in Senegal started with great promise, including the blessing of Iridium supporter Nelson Mandela, but there were practical problems. With only one phone in each village, determinations had to be made about when it should be turned on, who was appointed to receive calls, how a call intended for a particular person would be handed off to that person, who would be available to translate from the national language to the tribal language— in other words, a single phone per village provided an administrative problem, in that you ended up needing two to four employees to operate the one Iridium handset. But that was not even the main roadblock. "We were threatening British Telecom and France Télécom," said Brown. "They wanted to develop something by themselves. If we developed an alternative to their land-based trunk lines, we could possibly get bigger contracts later and supplant them. And when you have a battle like that, it's usually the legacy company that wins out." In the early days of the project, Brown expected to get support from the World Bank and other funding organizations devoted to the welfare of the developing world, but that never worked out either, primarily because the potential African partners couldn't get their act together to apply for the funding.

Eventually Brown and Wilkins gave up on that particular concept, but the idealistic goal of "global connectivity" for the Third World was still alive eight years later when Google announced it was backing an Internet-in-the-sky project called O3b, supposedly to provide Internet access to "the other three billion" without any connection to the Web. Ten other partners joined with Google to raise $1.2 billion for 8 MEOs and 180 LEOs. Strangely, O3b would cover only 70 percent of the world's population. The announcement of O3b sounded, in fact, a lot like a revamped version of Teledesic, with "the other three billion" replacing "universal dial tone" as the goal of all right-minded technology enterprises. Then, in 2014, several O3b employees left Google to form yet another broadband-in-space company called WorldVu, with backing from Qualcomm and Richard Branson's

Virgin Group. WorldVu announced a system called OneWeb that would consist of 700 LEOs traveling through the Van Allen belt at 750 miles up. Then, as though the satellite craze of the early nineties had never abated, yet another new company—LeoSat of Arlington, Virginia—announced a 120-satellite constellation that would provide global broadband. ("We are a commercial company," said CEO Vern Fotheringham in a *Wired* interview. "We'll stick with the top 3,000 rather than the other three billion.") By the end of 2015 there was once again a full-fledged battle to be the first broadband operator in space, with Elon Musk, the entrepreneur best known for founding SpaceX, Tesla Motors, and PayPal, announcing a constellation that would make Craig McCaw dizzy—a constellation of 4,425 LEOs designed to provide broadband not just to our planet but eventually to Martian colonists as well. With big satellite systems suddenly becoming "cool" again after fifteen years of neglect, it's odd that most of the emerging companies are choosing bent-pipe transmission, eschewing the intersatellite switching in space pioneered by Iridium, and all of them are choosing to fly through the Van Allen belt that destroyed the first Globalstar constellation. All of the systems would also require extensive infrastructure on the ground.

Since no Mega LEO system has ever been built, some of the old low-tech solutions started to revive in popularity as well. Google resurrected the abandoned Sky Station International business plan in 2013 and launched thirty experimental balloons in the Canterbury area of New Zealand, hoping to eventually supply the world with broadband through what it was calling Project Loon. (Interestingly, each balloon carries an Iridium device—a "command and control" transceiver that tells the balloons to climb, descend, move into the wind, or change direction.) And in 2014 Facebook founder Mark Zuckerberg dusted off the old HALOSTAR project and announced he was backing development of a solar-powered drone that would provide broadband access for "suburban areas" around the world that weren't yet wired. A few weeks later Google acquired its own high-altitude drone company, Titan Aerospace. When you try to chart the objectives of all these various Ka-band and Ku-band projects, you see the investors running up against the same problem Craig McCaw had—namely, that they're not sure who the customers are, not sure what those customers want, and not entirely certain that people won't be able to get it some other way before the system is launched. None of the balloon or drone ventures have explained how they expect to conquer the fundamental problem of licensing and government approval, since all would be flying below the Kármán line and would therefore be within the national airspace of every country in the world.[56] The answer seems to be that none of the systems would be truly global.

The executive business traveler is another species that never really took to the Iridium phone, even after it became smaller and capable of doing what Ed Staiano had promised it could do. Just a few months after taking over the company, Iridium management had already figured out a way to get rid of the old cassette-based roaming system. By accessing terrestrial repeaters of the sort used by Sirius and XM to deliver music to car radios, Iridium could have been used as a universal phone that would seek out the strongest signal wherever it was—no cassettes or attachments needed—but the FCC wouldn't allow it. After two years of rule-making hearings, during which Craig McCaw sent a small army of lawyers to Washington to try to get approval for what was called the "Ancillary Terrestrial Component" (systems to use satellite signals through land-based relays), the FCC denied every application, calling it a "murky" area where two technologies converge. The Iridium and Globalstar spectrums were allocated for satellite use only, and the FCC was uncomfortable with the concept of a hybrid satellite/terrestrial phone.

Still, with engineers all over the world choosing Iridium every day because it's an easy, reliable connection in the sky, and therefore free of all the infrastructure you need for any other communications system, the question is worth asking: Why didn't the original Motorola plan work? Why isn't everyone in the world walking around today with satellite phones in their hands instead of phones dependent on land-based cell towers?

Most of the analysis of the first Iridium said it failed because the phone was too big, the phone was too expensive, the phone had limited capacity for data, and the link margin was weak. But all these were the kinds of first-generation problems that eventually would have been solved—and, in fact, were mostly solved by the end of 2001. Judge James M. Peck, who studied eighteen years of Iridium data and listened to numerous expert witnesses during the bankruptcy trial, eventually concluded that "handset size was not a material factor in Iridium's lack of success in attracting subscribers." At any rate, the phone got smaller. The price came down dramatically for industrial users, and if Iridium had been signing up millions of consumer users, it would have been even cheaper still. Leopold, Peterson, and Bertiger had worked out the logistics of expanding system capacity by layering new planes of satellites between the existing planes, so the number of calls that could be handled by Iridium was essentially infinite. The second-generation Celestri system, which would have been launched in 2003 to interface with Iridium, had a data rate 11,000 percent faster. As for the link margin, the terrestrial repeater had already solved the strength-of-signal problem. Even if the FCC continued to block the use of terrestrial repeaters, economies of scale would have allowed for larger satellites, more durable batteries, and stronger signals.

The bottom line is everything about a world using satellite phones would have been safer and probably cheaper as well. The average consumer believes that cell service is much more pervasive than it really is. Only about 12 percent of Earth will ever be covered by terrestrial cell towers. The five million towers that exist today were erected at a cost of about $1 trillion, and they periodically have to be replaced, moved, torn down, or altered at a huge continuing cost not just in money but in lives. (Working on cell towers is among the most dangerous construction jobs in the world.) Systems like OnStar, the General Motors auto safety and navigation system, work only within range of a tower, whereas Iridium keeps you safe in places that don't even show up on the map. Every time disaster strikes, in the form of earthquakes or fires or hurricanes or bombings, the first thing that goes down is the cellular phone system. During an electrical blackout, most of the panic among the citizenry is not caused by "Will I be able to stay warm?" but "Will my smartphone work?" And one reason companies like AT&T, Verizon, and T-Mobile spend hundreds of millions of advertising dollars promoting their coverage maps is that, even in the United States, none of them is able to provide universal coverage. Dropped calls remain common even in the most heavily populated urban areas, and anyone who has traveled on a long-distance train knows that there are extensive dead zones all across the continent that has the best coverage in the world. Fishermen who never leave the continental United States sometimes buy Iridium phones simply because they spend long periods on lakes where cell coverage disappears as soon as you move away from the shoreline.

The answer seems to be that Iridium's backers, frightened by the debt burden, simply gave up too soon, and then Motorola failed to show the kind of leadership that would have carried the company through a Chapter 11 reorganization. "After they fired Ed," said Leo Mondale, "we still had enough money to run the company for two years, provided we could defer those unreasonable Motorola payments. It didn't seem like a fatal situation. All we were asking Motorola for was time. But we could never get extensions from them." Even Ted Schaffner, who examined Iridium repeatedly over the years with the most skeptical of eyes, said the project should have succeeded. "I don't think the size of the phone had much to do with the decline of Iridium," recalled Schaffner, "and I don't think the link margin had much to do with it. I think there was not enough runway for the company to succeed." The crucial moment came during the first six months of 1999, when Motorola could have essentially taken over Iridium with a cash infusion at a time when everyone else wanted to sell. Why not redeem twelve years of work and preserve the technology for the future? Instead, Motorola issued a press release: "Motorola will not provide any further support beyond existing contractual commitments unless there is substantial participation in the Iridium LLC restructuring from all parties with a significant financial interest." In other words, the gateways and the bondholders and the banks and the service providers and the stockholders—everybody involved in the giant, multinational mess that Motorola had created—needed to dig deep down into their pockets one more time and prop up a system designed to funnel fees to Motorola. To use just one example, Motorola slashed its monthly operations-and-maintenance charge from $45 million to $8.8 million once the company entered bankruptcy and McCaw was doing due diligence—the same deal Staiano had asked for a year earlier and been denied.

Then, once Iridium so publicly crashed and burned, there was the perception by every other company that "if Motorola couldn't make that thing work, then nobody can." And all the mythology surrounding the company—it was a fancy, wrongheaded science project; it was analog; it was worthless for data; it was too elaborate and expensive to keep flying—not only enveloped the inexperienced partners in gloom but stuck to the company for years to come. If Motorola had simply adjusted its expectations, written off the debt, and taken control of the situation—in other words, chosen a middle course, like Dan Colussy, who was constantly saying "the data is not great but it's good enough," "the name is not perfect but it's good enough," "the link margin is faulty but it's good enough"—it would have ended up with a perpetual multibillion-dollar asset that would set Motorola up for all kinds of other applications in every country of the world. And it wouldn't have taken Motorola four or five years to turn the corner either, because its international marketing and sales force would have already been in place.

But the solution didn't have to be Motorola. If any other large corporation had decided to buy Iridium, it could have had an even faster turnaround and an even more lucrative asset, because it wouldn't have had to deal with all the lawsuits against Motorola. That's the genius of the American bankruptcy system: the asset is preserved even if the founders lose all their money. In later years, as people started to realize the value of Iridium, they tended to say, "Well, that's the American bankruptcy system in action. That's the way things are supposed to work." Not really. The bankruptcy system in action would have had fifty deep-pocket bidders at an Iridium auction. In this case, there were six shallow-pocket bidders and fifteen empty-pocket bidders, and only one of them was prepared to write a check. The most complicated satellite constellation ever devised was saved because of the persistence of a single man.

And now, several years later, the same man was still fighting his own board, the telecommunications industry as a whole, and sometimes even the government to try to preserve the constellation and get it to the next level. He drew up deals. He drew up partnership agreements. And every time Colussy got to the key meeting with potential investors, their response was always the same: "Aren't those analog devices? Aren't those satellites going to fail two years from now? Hasn't the world moved on from satellite phones?"[57] The one company Colussy kept coming back to was Inmarsat. Inmarsat was the world leader in fixed-terminal satellite service. Iridium was the world leader in handheld satellite service. Didn't it make sense to merge them? Surely the powers that be at Inmarsat knew their business model was an endangered species. After twenty-five years in business Inmarsat was still charging $8 to $10 per minute for airplane calls, whereas the Iridium price averaged $1.50 and the installation was much cheaper. As early as August 2004 there were ten teams working on an Inmarsat-Iridium deal, but by October everything had blown up. Inmarsat executives professed to be "shocked" by the valuation of Iridium, which was now $400 million, saying they were thinking more along the lines of $125 million and 14 percent of Iridium's stock. Later attempts to revive that deal were equally frustrating.

And so, after five years of trying to partner with a larger entity, everything fizzled. The board members were all fatigued. They were starting to get tired of Colussy's "dream vision" of a new constellation.

And then Matt Desch showed up.

When Desch was interviewed in 2006 to be the fifth Iridium CEO in a little over five years—the sixth if you counted Colussy twice, like Grover Cleveland—Colussy worked quietly with Ty Brown to line up candidates and stage-manage interviews, mainly to limit the influence of the Saudis. Colussy's first impression was favorable—Desch seemed scrappy, opinionated, confident. He was a restless, fast-talking tough guy from Dayton, Ohio, a computer scientist who had been on the front lines during the past two decades of telecom wars, first at Bell Labs, then as head of wireless at Nortel Networks, and most recently as the CEO of Telcordia Technologies, which was the descendant of Bell Labs after the breakup of AT&T in 1982.

Colussy told Desch he liked his energy and attitude but would vote against him unless he agreed to one nonnegotiable condition: "You must believe that there can be a next-generation Iridium system. You will need to find $3 billion for the next generation of satellites. And that may be conservative. Some people are telling me $4 billion." The current constellation might last five years or it might last fifteen years or it might, if you believed Iridium's detractors, last only three years. But before the satellites started to die, there had to be new satellites, with high-speed data capacity, in their place. *You will have no allies,* he told Desch. *The board of directors is tired of talking about it, and the partners would prefer to take their profits and go home. They're probably perfectly willing at this point to let the constellation coast into history, taking out a few hundred million in profit as it dies. Everyone you talk to will tell you you can't raise the money and that Iridium is hopelessly outdated technology. The system that runs the constellation uses the same 32-bit processors that were used in the Commodore Amiga personal computer. Remember the Commodore Amiga? Neither does anyone else. I realize I'm asking you to do the impossible, but that's what I'm asking you to do.*

Desch listened carefully, studied the system, talked to other board members (many of whom disagreed with Colussy), took a weekend to think about it, and called Colussy back on Monday morning.

"Okay," he said, "I drank the Kool-Aid."

In fact, Desch wanted to shake up his life. He had grown bored with the wireless business. That market was mature. The cell business was all about who could sign up the most service plans and hawk the most gadget-heavy handsets—a pure marketing game. "It went from a techno-logical innovation to a mere commodity," he said. "Wireless companies are all the same. So what do you do? Become faster? Become cheaper? The world I was part of, the technology end of the industry, was coming out of a six-year downturn. It was a declining business. Everything was moving to China, becoming commoditized, breaking up, pieces were being sold off. But there's still that 91 percent of the world that doesn't have cell coverage. And there are high barriers to entry with satellite. And if there's one thing I learned from twenty years in wireless, it's that the first three things you need are coverage, coverage, and coverage. So I felt I had a lot to work with." He was also intrigued by Iridium because it was an underdog company with the potential to break through one final frontier and become the dominant satellite communications company in the world. Or at least that's how he saw it. And that's how Colussy saw it as well. But a lot of what Colussy was telling him still seemed like wish-ful thinking. Since the company was only on target to earn $75 million in the year 2007, you could get discouraged by the math—it would take forty years at that rate to pay for the new constellation—so obviously Desch was going to have to get creative.

From his first day on the job, in September 2006, Desch started making plans for what was being called "Iridium NEXT." When the first constellation was launched in 1997, Motorola intentionally avoided any secondary payloads like cameras or sensing devices, because the company wanted to avoid appearing to be an instrument of the U.S. government.[58] The new Iridium had no such limitations, and Iridium engineers started suggesting government uses that would help fund the next generation, everything from cameras to telescopes to weather sensors. In each case, the appropriate government agency would express initial interest; then the proposal would fall into an endless bureaucratic dead zone, delaying decisions and forcing the company to look elsewhere. Colussy thought the most promising secondary payload would be air traffic control, but he also became fascinated by the idea of a camera that was pointed not down at Earth, but up at the rest of the space traffic. Since Iridium had the lowest orbit of all the big space systems, it was in perfect position to photograph the space vehicles and space junk above. For a while Colussy thought the idea might pass muster with the spy-versus-spy guys in the Pentagon, but the company never gained any traction. Iridium executive Don Thoma also had a high-level meeting in Mountain View, California, with the three gurus at Google Earth, including one of the "Fathers of the Internet," Vinton Cerf, who admitted that Google's database consisted of imagery that was four years old, bought from DigitalGlobe, the Colorado-based satellite company. But there was no follow-up on Iridium's offer to provide close-up real-time global photography.

Meanwhile, all the board members were tired of writing big tax checks and started pushing for an exit strategy—meaning either selling the company outright or merging with a larger company in a deal involving cash and stock. Colussy still believed—and Desch agreed with him—that the solution was a strong partner. "We needed a Blackstone or a Carlyle," said Desch, "someone with deeper pockets." The Saudis, on the other hand, believed the company shouldn't be wasting more time after seven years spent looking for a partner and not finding one. The Prince's team wanted to simply take the company public with an initial public offering and realize its true value. Colussy was extremely skeptical of IPOs, however, having run small-cap companies for much of his life and suffered the slings and arrows of day trading and short selling.

After a full year of arguments, Desch took action in 2008, hiring Evercore Partners, a boutique investment bank in New York, to find a deal. Evercore came up with fifty-four possible buyers for Iridium, including several hedge funds and special-purpose acquisition companies, or

SPACs. Colussy was steeped in the old-school ethics of Harvard Business, where trading for a living was considered sinful, and he hated hedge funds on principle. "Any group of smart people that devote 100 percent of their energy and time to hedge securities and manipulate markets—at the end of the day, what have you done for the nation or the economy?" he told Herb Wilkins. "They say, 'We're providing liquidity.' Bullshit! Commodities markets were created for users and producers. The brightest minds in our country should be building companies, building our nation." Colussy was not quite so hostile toward SPACs, although he regarded them with suspicion. SPACs were corporations set up specifically to acquire a company and take it public, thereby realizing a profit from the IPO. The concept was pioneered in the nineties, waxing and waning in popularity as a last resort for companies that couldn't raise money from banks or institutional investors, or companies that were unable to launch IPOs on their own. From the investor's point of view, it's essentially a bet on the SPAC management team—that it will be able to identify an acquisition target and shrewdly increase its value fairly quickly. The SPAC will also end up owning warrants to buy shares of the new stock at a low price—a factor in the deal that often comes back to haunt the company. For the target company, a SPAC provides a source of quick cash and a motivated partner to set up the IPO.

The board members made it clear to Desch that they wanted "a clean buyout deal," not a SPAC, because they didn't want to end up owning stock in a company that may or may not have their best interests at heart. Besides, SPACs had a reputation for being less than they were cracked up to be from the point of view of the company being acquired. But Desch had a good reason for pushing a SPAC. Of the fifty-four companies that had entertained the Evercore road show, only one was left standing—GHL Acquisition, a SPAC formed by Greenhill & Company, the New York investment bank cofounded by former Smith Barney chairman Robert F. Greenhill in 1996. GHL was willing to value Iridium at $650 million, to be paid partly in stock and partly in cash, which would mean the Iridium partners would earn back—depending on how the stock performed—up to fifty times their initial investment.

Colussy was still skeptical but didn't want to waste any more time on a marketplace that was obviously less enamored of Iridium's future than he was. Just to be sure the company had more than one bidder, he called John Castle to tell him he had another chance to own Iridium. "You missed out in 2000 because you wanted immediate cash flow," Colussy told his friend. "You've followed the company from the beginning. You've called me from the South Pole on your beloved Iridium phone. Now you can still get in for $650 million, but we need your money now."

Surprisingly, Castle said yes this time, and Colussy flew to New York to try to work out a deal. Castle said he could put in up to $100 million in cash and work out the rest of the deal for a valuation of $600 million. Colussy thought a Castle Harlan buyout would be cleaner than a SPAC deal, because he thought it would provide more liquidity for the cash-strapped partners, but Desch wasn't that crazy about working with Castle. Desch believed that a private equity fund like Castle's would be self-interested, looking after its own profit before the long-term needs of the company, and he wanted to go public sooner rather than later. Castle also had a reputation for nitpicking his investments. He was notorious for knowing arcane details about the various restaurant chains he owned, including the fact that Morton's The Steakhouse always had between forty-five and fifty thousand steaks in the refrigerators of sixty-four restaurants, and that Marie Callender's sold twenty-one million pieces of pie per year in the state of California. Desch frankly said he feared that Castle would be "domineering," and Desch didn't like people looking over his shoulder. One of the advantages of going public was that Desch would finally be free of what he called "this ragtag group of investors who, after all, were not going to come up with the money themselves"—but he wasn't sure John Castle as the owner would be a better trade. For a while Colussy tried to work out a combination deal—Castle Harlan and Greenhill working together—but after a month of meetings, he gave it up. Greenhill was simply not interested in partnering with Castle, whose ardor was cooling as well, once he heard the company would have to immediately raise $3 billion for the next generation of satellites. In September 2008 Colussy told Castle that if he wanted in, he would have to put up the whole amount and he only had a couple of days to make up his mind. Four days later Castle called to say he was still considering the investment. Colussy told him, "Too late." The deal with Greenhill had been concluded.[59]

The merger between GHL Acquisition and Iridium was publicly announced on September 23, 2008—ten years to the day from Ed Staiano's "can't miss" start date—with the company being sold for $591 million, allowing the partners to receive $102 million plus 45 percent of the new company. Greenhill ended up with 17.5 percent of the stock. But this was exactly eight days after Lehman Brothers became the biggest bankruptcy in U.S. history, presaging a collapse of the world economy, leaving half-finished deals in peril and throwing any proposed IPO into jeopardy. By December Desch was telling Colussy that Greenhill wanted to "reprice" the deal. When Bob Niehaus, the banker who cofounded Greenhill Capital Partners in 2000, ran into the CFO of Inmarsat at a technology conference, the CFO openly mocked his deal, telling him it would never close given the dire straits of the economy, and, even if it did, Desch would never be able to raise the money he needed for the next generation. Then, as if Desch needed any more headaches, a defunct Russian Kosmos spy satellite slammed into Iridium Space Vehicle 33 in February 2009, smashing it to smithereens somewhere over northern Siberia. (If anything, the one-in-a-million accident demonstrated the resiliency of the system. A spare satellite was moved from its storage orbit to replace the dead bird in less than a month.)

The Greenhill deal turned out to be both a blessing and a curse. By the time Iridium Communications Inc. was launched on the NASDAQ in September 2009, early trading valued the company at around a billion dollars, but that valuation didn't hold up. Greenhill & Company had overexpanded and ended up getting battered by the bear markets of 2009 and 2010, ultimately losing half its value. To assuage angry investors, the Greenhill board agreed to dump all its Iridium stock. That meant regularly scheduled sales of large blocks of stock every month for three years, depressing the price and the valuation. There were also millions of warrants in circulation, most of them held by Greenhill, allowing the bearer to buy the stock at a steep discount. Then, even after Greenhill had sold off its holdings, Niehaus remained chairman and Greenhill kept its seats on the board of directors, angering big shareholders who now regarded them as meddlers and speculators. At the same time, short sellers were saying that Desch would never be able to raise what was now an estimated $2.7 billion to launch the second generation of satellites, leading to analysts' reports that portrayed the stock as highly risky. Then, as though the past decade had never happened, all the old whispers returned—the satellites would all fail at the same time because the momentum wheels were faulty, the link margin was bad, the system was analog, the company was too small to compete—and institutional investors stayed away while day traders piled on, enjoying the up-and-down bumps that were only possible to control when it was a small-cap company. Adding to the misery was a sudden lawsuit filed in Chicago by Motorola, rising zombielike from the Iridium past to claim that it was owed $24.68 million because of a "change of control" clause inserted into that $30 million loan agreement that Keith Bane had grudgingly approved for Colussy back in 2000. The Greenhill deal had stimulated yet another Motorola performance reminiscent of George C. Scott in *The Hustler*: "You owe me money!"

Matt Desch ignored all the ambient noise and, during the years 2009 and 2010, positioned himself to launch the next generation. First he invited bidders for the $2.7 billion contract. Then he narrowed those down to two: Lockheed Martin and Thales Alenia Space, owned by the French electronics company Thales with a minority interest held by Italy's Finmeccanica. He told both companies that the contract would go to whoever helped him raise the most money—and a year and a half later, the answer was clear. The French had wanted to be major players in space ever since launching the first European satellite in the seventies, and Thales Alenia Space had spent four decades becoming the preeminent satellite source for most of the civilian and military systems of Europe. The company also had the backing of the French government, the French banking system, and, most important, Compagnie Française d'Assurance pour le Commerce Extérieur, better known as COFACE, which was similar to the Export-Import Bank in the United States and existed to promote French industry. A consortium of six French banks, headed by Société Générale, were willing to set up a $1.8 billion credit facility for Iridium's next generation, and that credit facility was guaranteed by COFACE up to 95 percent. The rest of the eventual $2.9 billion cost of the new eighty-one-satellite system would be covered by Iridium cash flow, which was still growing with each passing year, and by secondary payloads. Desch was so happy with the French deal that he took advantage of the low price of Iridium stock and loaded up on it himself.

It's hard to imagine a more complete French corporate victory over America. Lockheed Martin would have been equally capable of building the satellites and managing the launches, but Lockheed's entreaties to America's Export-Import Bank, begging it to match the COFACE offer, fell on deaf ears. The Ex-Im Bank was set up to help American companies exporting products overseas, so it was not that interested in helping an American company invest in another American company. The prospect of Iridium NEXT being designed in Toulouse and Cannes was full of irony. Thales Alenia Space was not only the company that had built the Globalstar satellites, but some of its executives were the same fierce opponents of Iridium who were accused of rifling through the hotel rooms of Motorola executives in the early nineties, and then challenging Iridium's European patent through its subsidiary Alcatel-Lucent. In the final analysis, it was a no-brainer for Desch. All systems were go for a constellation to be launched in 2015 that would double the subscriber capacity of the old system, have a much higher data speed, and contain secondary payloads that would help defray the cost. On January 14, 2017, Iridium successfully launched the first ten satellites of Iridium NEXT, with plans in place to complete the constellation by early 2018. The suicide software would then be used to de-orbit the Chandler satellites—and, yes, Public Law 85-804 would still be in effect. The annual cost of running Iridium NEXT was expected to be only about $35 million a year, meaning that finally, after twenty years, it would be a "mature" system. In fact, Iridium after 2018 was expected to be such a cash cow that a third generation around the year 2033 was projected to be funded entirely out of the company's cash flow. Iridium satellites would fly into perpetuity.

Epilogue

TWILIGHT OF THE WARRIORS

By the time the last of the old Motorola satellites are switched off and the new French satellites switched on, it will be thirty years since Ray Leopold, Bary Bertiger, and Ken Peterson commandeered the whiteboard in the parking lot of the Chandler Lab and came up with the blueprint for Iridium. During that time the three inventors have gone through mixed emotions as their brainchild was questioned, then celebrated, then cursed, then abandoned, then laughed at, then reborn in the care of people they never knew or met. They were warned from the beginning: "You're betting your careers on this." They were warned so many times that they started saying it to one another as an ongoing joke. It became a source of pride. "Let's create some miracles today," Leopold would say at the beginning of each workday. They were, in fact, proud to be betting their careers. And all three of them say today that if they had to do it over again, they would once again bet their careers, even in the face of those who believed the project was folly. It should also be noted that, prior to 1997, before the first satellite had even been launched, they had already corrected all the first-generation problems that resulted from locking in the technology of 1995. Their next-generation system, which would have been called Celestri, or M-Star, or Macrocell, or Teledesic, depending on which business plan prevailed, would have had broadband data capacity. In May 1997 Leopold gave the commencement address to the graduates of the South Dakota School of Mines and Technology, and he said, "I have five children—a little girl nine years old, and she has four younger brothers, ages one, three, five, and seven. In ten years, when they are teenagers, they and their friends will be able to create applications for Iridium which the several thousand people now bringing it into being cannot even imagine. That's the vision of Iridium."

In other words, they knew. They knew they were creating a framework for something larger than a mere telephone system. "Other than the atomic bomb project," said Bertiger, "there has never been an engineered system more complex than Iridium. In fact, it may be even more complex than Los Alamos, because they were only doing one thing and we were doing many."

Peterson was the first of the three men to leave Motorola—but not because he wanted to. A month after the SNOC was turned over to Boeing, he was told that "a big boss from the Chicago office" needed him to come to a meeting. There he was told by the head of the satellite division, "Our department has been dissolved and your services will no longer be needed." There would be no more hiding him in marketing. Apparently one of the nation's top mathematicians was no longer Galvinized. So on February 28, 2001, at age sixty-two, after thirty years at Motorola, Peterson spent his last day in the Chandler Lab. He found work soon enough, at the Advanced Technology Center of Rockwell Collins in Cedar Rapids, Iowa, but it didn't have the same pizzazz as the Motorola years. Eventually he moved home to Lake Mills, Iowa, the town settled by his Norwegian great-grandfather, and became a farmer again, on the same land he'd been working when he first got the idea of going off to Iowa State in 1956. In his spare time he worked for the FAA on frequency-interference issues, solving the algorithms that would be used for the next generation of air-traffic-control technology.

Once it was obvious that the Celestri system would never see the light of day, Bertiger and Leopold moved on to new jobs in Motorola's Global Telecommunications Sector, designing cellular infrastructure systems for companies all over the world. Bertiger was General Manager, Leopold was Chief Technology Officer. But Bertiger, who held thirteen patents for his work at Motorola, left the company in 2003 after getting into a philosophical battle with Chris Galvin over whether the company should pursue the consumer market or the infrastructure market. "I told Galvin, 'We're terrible in the consumer business,'" he said. "And we got totally sideways. They sold off the military business and the semiconductor business, but they still had the best systems house in the world. So why not invest in it? One day a very junior person in the human resources department came to my office and said, 'Maybe you oughta think about a retirement package.' I kicked him out of my office." Bertiger nevertheless did negotiate that package and left after thirty years at the company. He served for a time as Managing Director at Grayhawk Capital, a Phoenix venture capital firm, and eventually relocated to Vero Beach, Florida.

Ray Leopold left the same year, taking an early retirement and moving his large family—now six children—to a patch of land between Kalispell and Whitefish, Montana, an area he chose because he considered it the most beautiful place in the world. "Have you ever wondered where water begins?" he would ask visitors, and then take them to the sheer cliffs in Glacier National Park, where waterfalls appear to be springing directly out of the rock. "This is where water begins." In retirement Leopold made frequent appearances at universities, becoming a senior lecturer at the Massachusetts Institute of Technology, and was asked to deliver the prestigious Minta Martin lecture there in 2004. Leopold chose to speak about Iridium.

Everyone involved with the early days of Iridium looked back on the experience with a mixture of pride and regret, like a man who marries the most beautiful woman in the world and becomes a widower a few months later. Bob Kinzie would have preferred to stay with the new company, but there was no longer any need for the old business model of seeking liaisons with foreign governments and state-owned telecommunications companies. He retired to an upscale subdivision in Bethesda, Maryland, but got bored with being inactive. While recovering from carotid surgery, the power in his house went out—so he started an emergency-generator business for suburban Maryland, which he runs to this day.

Dave Oliver left the Pentagon and was appointed to the ultimate "fixer" job. He was the Minister of Finance in Iraq during the interim government run by American diplomat Paul Bremer in 2003 and 2004. He later became Chief Operating Officer for the North American division of EADS, the European defense contractor, and got into a brutal public battle with Boeing over a $35 billion contract for refueling tanker jets that he first won, then lost, after the political establishment of the Pacific Northwest ganged up on him.

Dorothy Robyn left the White House to work for the Brookings Institution and then the Brattle Group, an economics consulting firm, before going to the Pentagon in the first year of the Obama administration to work under Defense Secretary Robert Gates as Deputy Under Secretary of Defense for Installations and Environment. After a brief stint as Commissioner of Public Buildings at the General Services Administration beginning in 2012, she left government service to devote herself to consulting and writing for think tanks, primarily on energy issues, spectrum policy, and her pet project, the privatization of air traffic control. Her old boss, Gene Sperling, became a consultant for NBC's hit series *The West Wing*, then a spokesman for Goldman Sachs before rejoining government in 2009 as the counselor to Treasury Secretary Timothy Geithner. In 2011 he was promoted to Director of the National Economic Council, the same job he had under Clinton, making him President Obama's top economic advisor. He resigned in early 2014.

Ted Schaffner left The Tower to run Motorola operations in China before retiring in June 2008. Six months later he was brought to

Washington by Secretary of the Treasury Henry Paulson to be program manager of the Troubled Assets Recovery Program, better known as TARP, literally writing the checks to troubled banks. As head of the Capital Purchase Program, Schaffner built a team to regulate who got money and how much they got. When he closed out the program two years later, he was proud of the fact that he had loaned $206 billion and gotten back $140 billion, with the other $66 billion earning 10 percent interest for the taxpayer.

Keith Bane retired from Motorola in 2003 and started dividing his time among three homes—an estate in the Chicago suburb of Algonquin; a winter vacation home in Longboat Key, Florida; and the Bane family's traditional summer vacation home in Lake Minocqua, in the Wisconsin Northwoods, where in 2006 he raised the money to convert an old Catholic church into the Campanile Center for the Arts.

Rudy deLeon left the Pentagon in 2001 to take a job with Boeing. In 2006 he joined the Center for American Progress, a Washington think tank founded by Clinton chief of staff John Podesta. DeLeon became a frequent guest on news-channel talk shows, specializing in national security and international policy issues.

Castle Harlan continued to expand, reaching $6 billion in assets even after the rest of the financial world crashed in 2008. The *Marianne* completed a second circumnavigation of Earth, and then John Castle turned to the rivers of America, traversing the nation from New York Harbor to the Pacific by using the Erie Canal, the Ohio River system, and the historic route of Lewis and Clark. He also landed a Twin Otter at the North Pole—and called Colussy on his Iridium phone. In May 2014 he put the Kennedy Winter White House up for sale. Asking price: $38.5 million, or a premium of 680 percent over what he paid for it in 1995.

Bill Cohen continued to build the Cohen Group into one of the most powerful international lobbying firms in Washington and a first stop for many senior-level undersecretaries, generals, and admirals when leaving government, including retired Lieutenant General Harry D. Raduege Jr., former head of DISA, who turned out to be a three-star, not a two-star. Cohen developed a special expertise in companies doing business in the People's Republic of China, where he established two branch offices. Janet Langhart Cohen published her memoir, *From Rage to Reason: My Life in Two Americas*, in 2004, changing her designation from "First Lady of the USO" to "First Lady of the Pentagon." Three years later the Cohens collaborated on *Love in Black and White*, their joint memoir, which includes reflections on race and religion. The Cohen Group remains on the Iridium payroll.

Kathy Brown left the FCC to work at the Washington offices of the prestigious law firm Wilmer Cutler Pickering Hale & Dorr, then joined Verizon as Senior Vice President/Public Policy Development and Corporate Responsibility. She left that job at the end of 2013 to become CEO of the Internet Society, the organization that governs the Internet worldwide through its Internet Engineering Task Force.

Bill Kennard left the FCC at the end of the Clinton administration to become Managing Director of the Carlyle Group, where he handled telecommunications and media, at one time engaging Colussy in investment talks when Iridium was looking for a strategic partner. Kennard then returned to government in 2009 when he was sent to Brussels by President Obama as the first U.S. ambassador to the European Union, returning to Carlyle in 2013.

Roy Grant, the Chief Financial Officer whose decisions were still being debated in the court system for several years after the collapse of the original Iridium, opened a bicycle shop in Half Moon Bay, California.

John Richardson left Iridium in February 2000 and founded two cement companies in Kazakhstan, then semiretired to a life of safaris, mountain climbing, and consulting for banks.

As soon as Mauro Sentinelli returned to Telecom Italia in 1999, he was knighted by the President of the Italian Republic. He ran the wireless division until 2005 and then retired with the largest government pension in the history of the nation: $1.6 million per year. Iridium was never able to make a deal to use Fucino as a backup gateway. Sentinelli's official biography nowhere mentions Iridium, listing the years 1997 to 1999 as "an experience in the United States."

Aside from that one secretive meeting with Mark Adams, no one at Iridium ever saw Prince Khalid, although Ty Brown exchanged words with him during a chance encounter at the London townhouse, thereby proving his existence to the partners. The Prince continually increased his investment in Iridium, first in 2002 when he joined with Syncom to fill the gap caused by the Australians and Brazilians dropping out, then again in 2009 when the company needed an additional $10 million to complete the Greenhill SPAC transaction. As of 2016, he remained the largest single Iridium shareholder, holding 18 percent of the stock. By the end of the decade, the Prince's sons had taken over most of the day-to-day management of the Prince's interests, although the Prince himself still managed Juddmonte Farms, his Thoroughbred bloodstock operation. The Prince retired to his six homes, including a town house on the Parc Monceau in Paris and the thousand-acre Fairlawne Estate in Kent where, oddly, he drilled for oil on the property. When he closed off a footpath that had been used since Roman times, he ended up facing the wrath of the seven hundred villagers of Shipbourne, who raised a ruckus with the Planning Inspectorate and had it reopened.

Craig McCaw's ICO Global, after all those years of angst and promises, ended up launching just three space vehicles. Teledesic 1, first of the 288 satellites that would supposedly provide Internet connections for the planet, was launched in early 1998 and went dead shortly after achieving orbit. A second satellite in 2001 lasted less than three years, then ceased working and went into a decaying orbit. The company was on life support in 2008 when McCaw tried to revive it as a mobile television service, with grand plans to deliver video signals to automobile-based entertainment centers. A GEO launched that year was heralded as "the largest satellite ever launched" at 14,625 pounds—a curious claim in the age of increasing miniaturization of parts—but the business plan never proceeded beyond alpha trials. In 2011 ICO Global changed its name to Pendrell Corporation and began functioning primarily as an investment group for patents and intellectual property rights. In 2012 the company's one operating satellite was sold to the Dish Network, and McCaw's eighteen-year flirtation with communications satellites came to an end.[60] That same year, McCaw made news by buying the most expensive car ever sold, a 1962 Ferrari 250 GTO built for Sir Stirling Moss, for which he paid $35 million.

In 2001, around the time Ken Peterson was fired and the satellite division was shut down, Motorola was still one of the largest electronics companies in the world. But Chris Galvin was already in a death struggle with the Motorola board. During the first quarter of 2001, Motorola recorded its first operating loss in sixteen years, and this came at a time when its direct competitors—Nokia, Qualcomm, and Ericsson—were all showing record profits. In a single fourteen-month period, from May 2000 to July 2001, the company lost 72 percent of its market value as the stock plummeted from $60 to $16.75. The company's share of the wireless phone market declined to just 13 percent, compared with Nokia's 35 percent, and every article in the business press cited Iridium as one of the main reasons that Galvin had no credibility.

And yet it's hard to see how Motorola suffered at all from the demise of Iridium. The original contracts, drawn up around 1992, were so one-sided in Motorola's favor that Iridium had already paid $3.65 billion to Motorola by the time of the bankruptcy. During the frenzied years of 1997 and 1998, when Iridium was selling junk bonds and raising money in the stock market, Ed Staiano borrowed a total of $1.4 billion—and $1.038 billion of it went directly to Motorola. The company also ended up with a thousand patents related to satellite technology that were enormously valuable a decade later. Bary Bertiger, who was General Manager of the satellite division after Durrell Hillis left, said the division always made money and that Iridium was profitable for Motorola. It's true that Motorola claimed $2.6 billion in write-downs that were blamed on Iridium, but it's hard to say where that money was lost. An analyst for Hambrecht & Quist, the San Francisco investment bank, calculated that Motorola's entire Iridium investment had been written off prior to May 1999—long before the highly publicized write-downs—and Bertiger confirms that. "Motorola definitely made a profit on Iridium. The expenses were less than $100 million and that had already been written off long before. The company even received a research-and-development tax credit for it. So we never lost money."

In other words, Motorola's exposure to the Iridium bankruptcy was zero. Ida Picker, a business analyst for Bloomberg News, compared Motorola to Tom Sawyer, who convinced all his friends to pay for the privilege of painting his fence. Indeed, the reason it was so easy to attack Motorola in the courts was that it seemed that Motorola was the *only* company that ever got paid. You would have to be a forensic accountant to track that $2.6 billion it claimed to have lost, but could it be possible that it was easier to blame a write-down on Iridium than on the more visible reasons for Motorola's declining fortunes? Motorola had once controlled 63 percent of the world's cell-phone business. As that number started sinking toward the single digits, some speculated that Motorola simply dumped all its losses into a trash bin called "Iridium." By way of comparison, Motorola loaned $2 billion in cash and equipment in the late nineties to a company controlled by Istanbul business magnate Cem Uzan so that Uzan could build out the Turkish cell system. Uzan turned out to be a con man, and even after receiving a judgment of $4.2 billion from a New York court, Motorola was reduced to seizing Uzan's foreign residences and a private jet in order to recover fractions of a penny on the dollar. This misadventure resulted in twice as much wasted cash as Iridium—even accepting Motorola's own numbers—and yet it was never mentioned as a reason for Motorola's decline under Galvin.

Still, three years after Motorola turned over the SNOC to the Colussy group, Iridium continued to haunt Galvin's résumé. The financial press increasingly portrayed Galvin as remote, slow-acting, out of touch. His close friends advised him to resign the presidency and become a ceremonial chairman, in the manner of William Ford Jr. at Ford Motors. "You would be revered if you did that," one of his managers told him. But Galvin wanted to govern, not preside. The end came on January 4, 2004, when the Motorola board forced Galvin to resign after failing to be impressed by three months of arguments as to why he should keep his job. The following day was the first in Motorola's seventy-six-year history in which no member of the Galvin family was employed by the company. Chris Galvin would spend much of the next decade trying to repair his reputation, maintaining, with some justice, that his three-year restructuring of the company had just started to bear fruit at the moment of his firing. (Indeed, Motorola showed a 454 percent increase in operating earnings for the year 2004—a result the new CEO could have had nothing to do with.)

If the Motorola board was looking for a new type of CEO, it got one. Ed Zander, a goofy guy who liked to do stunts like riding a bicycle onto the stage at the Consumer Electronics Show, was brought in from Sun Microsystems, and even though he was a Brooklyn native, he couldn't have been more emblematic of a rock-and-roll California guy taking on the crusty old Chicago stuffed shirts. He immediately got into a culture war with the Motorola engineer-managers as he moved them away from the company's traditional strengths, slashing all research-and-development budgets, and brought in a marketing guy from Nike. The result: Motorola ended up in the cutthroat business of trying to create the newest, coolest handset each year. The strategy worked for a while, especially with the popular Razr phone, an ultra-slim lightweight flip phone that had an innovative nickel-plated keypad and enjoyed a couple of years of immense popularity stoked by its "Hello Moto" ad campaign. Zander's taking credit for the success of the Razr was especially galling to Galvin, since the phone had been created on Galvin's watch. Motorola sold more than one hundred million Razrs, but the folly of depending on consumer tastes emerged on a single day—June 29, 2007—the day Apple introduced the iPhone.

The iPhone was made possible by the latest generation of Motorola microprocessors, but it was also made possible because Zander had shared Motorola hardware technology with Steve Jobs during a 2005 collaboration on the ill-fated Rokr, which was the first phone to feature iTunes. Jobs was regarded among Motorola veterans as "a software genius and a hardware thief," so the decision to share with him was very controversial among the old-timers. Zander proved why when he essentially gave Jobs the keys to the kingdom: Apple ended up contributing minor expertise to the Rokr while using the same hardware to introduce the iPod Nano, and Jobs even had the effrontery to release the Nano at the same time as the Rokr. Jobs had insisted by contract that Motorola's Rokr would be limited to one hundred iTunes songs, whereas the iPod was virtually unlimited. Once Zander figured out what had happened, he was furious—"Screw the Nano! What the hell does the Nano do? Who listens to a thousand songs?"—but Galvin was even more furious. As soon as his separation contract allowed him to speak publicly, in 2008, Galvin pointed to that moment as the beginning of the end. It had always been Motorola policy to keep Silicon Valley software guys away from the hardware secrets. "Why let his really thoughtful, imaginative nose under the camel's tent?" said Galvin. At any rate, Zander paid for the misstep. Motorola's stock price plunged, and then corporate raider Carl Icahn bought up the cheap stock and demanded a management change. Zander was not only fired but would end up on *Business Insider*'s 2009 list of "The Worst CEOs Ever" for his repeated public gaffes, such as telling the *Wall Street Journal*, "I love my job. I hate my customers," and for describing Chicago as a boring place where "people think about what kind of picnic to go to this weekend." Zander's cluelessness was especially evident in his refusal to admit that the smartphone would catch on. He was still insisting that the iPhone was no big deal when he was being escorted out of Schaumburg.

One of the few positive developments from Zander's four years at the helm was a $1 billion sale to Comcast of the state-of-the-art Motorola cable modem set-top box—which, by the way, was developed by Galvin. The handset division recorded a loss of $1.2 billion in the fourth quarter of 2007—with the company falling behind not only Apple but Samsung, Nokia, Sanyo, LG, and Research in Motion's BlackBerry—and Motorola became a ghost of its former self. After several rounds of layoffs and a huge exodus of executives to work for Apple in 2008, there were attempts to sell the handset division, but Motorola couldn't find any serious buyers. Chris Galvin continued to brood for years about being displaced by the board at the moment of the turnaround, until one day when he was talking to his father, on the front porch of Bob Galvin's home in Barrington, Illinois.

"I'm not going to let the Motorola board ruin the rest of my life," said the elder Galvin to his son. "Nor should you."

Soon thereafter, father and son created Harrison Street Real Estate Capital, named after the original 1928 location of Galvin Manufacturing Company, and started building what would become a $5.8 billion investment fund.

By that time John Mitchell, last of the old-time Motorola bosses and widely regarded as the man who made cell phones possible, had retired for the third and last time to the village of Inverness, just eight minutes from his old office in The Tower. He insisted on using his ancient analog Motorola car phone until Verizon told him it was switching off service because no one used analog anymore, and he was in the process of writing Verizon a letter in protest shortly before his death in 2009, at the age of eighty-one.

In 2011 Motorola split into two separate companies: Motorola Mobility (for the handset operation in Libertyville) and Motorola Solutions (for the police radio and commercial operations in Schaumburg). A few months later Motorola Mobility was sold to Google for $12.5 billion, and the company that had invented the cell phone officially left the cell-phone business. It was never clear that Google really wanted to be in the handset business, and the speculation was that it was really buying

Motorola's seventeen thousand patents. Sure enough, Google would keep Motorola Mobility for less than three years, then sell it to Lenovo, a personal-computer company based in China, where the very word "Motorola" had once been a synonym for "mobile phone." Bob Galvin, who passed away two days after his eighty-ninth birthday in 2011, didn't live to see his China strategy come full circle, as the company that once dominated telecommunications in Asia was sold at a discount price in a market now dominated by Asian firms. In 2016, Motorola Solutions announced plans to subdivide the Schaumburg campus and sell it off piece by piece to other corporations. As much as it pained the Galvin heirs to admit it, Motorola had joined Zenith, Westinghouse, Magnavox, and Philco as once great icons of the American consumer that were no longer with us.

Lawsuits against Iridium and Motorola continued for years. Probably the most important was over by 2002, when New York federal district judge Alvin Hellerstein ruled that Motorola had to pay the Chase banking consortium $300 million for being complicit in the "fantasies" of Ed Staiano and Roy Grant that had been put forward as real income projections. (Hellerstein rarely spoke from the bench when a case was being presented, but his first question to attorneys was "What are gateways again?") In the Iridium bankruptcy case, Judge Blackshear ruled that the creditors could go after Motorola separately from Iridium because, in his opinion, Motorola controlled much of what Iridium did. Blackshear retired in 2005, before that case was complete, so it was passed on to Judge James M. Peck, who presided over the testimony of fifty-two witnesses during fifty trial days stretched over eight months, admitting 866 exhibits into evidence—and that was only the first of what was expected to be two phases of litigation. Since all parties were exhausted by then, they came to the bargaining table, and Peck approved a final settlement in May 2008. The bondholders received $16 million from Motorola—one penny per dollar of what they had loaned to Iridium.

The most remarkable Motorola case—one for the law books—occurred in Washington, D.C., where eighteen class-action suits brought by plaintiffs who had bought Iridium stocks and bonds were consolidated in the court of Judge Thomas Penfield Jackson. Jackson was best known for calling Microsoft "stubborn mules who should be walloped with a two-by-four" after ruling against Bill Gates in the government's landmark antitrust case, and he was equally withering toward Motorola. After Jackson's retirement in 2004, the cases passed to visiting judge Nanette K. Laughrey from Missouri, and she surprisingly denied all Motorola's motions to dismiss. Cases of this sort almost never reach trial because the standard of proof is so difficult. The plaintiff has to prove that the seller of stock either spoke untruthfully or said things that were technically true that would lead to an untrue conclusion, and that those false or reckless statements directly caused people to buy the stocks or bonds. That's why only thirteen class-action shareholder cases *in history* have ever gone to trial. Iridium was one of them. After nine years of litigation, the production of 1.5 million documents, and what one lawyer called a "five-ring circus" of depositions, the parties finally settled in October 2008. Motorola was ordered to pay $20 million. Ed Staiano and Roy Grant were ordered to pay $14.89 million. And five underwriters of Iridium stocks were ordered to pay a total of $8.25 million—Merrill Lynch, Goldman Sachs, NationsBanc, Salomon Smith Barney, and SoundView Technology Group.

Ed Staiano didn't like retirement any more than he liked vacations, so he started his own venture capital fund, the Sorrento Group, named after his favorite part of the planet. Sorrento funded fifteen companies, including a start-up telecommunications firm in Italy. Staiano continued to be active on the board of trustees of his alma mater, Bucknell University, where he had started his career as a professor of engineering, founding the computer program there. He also bounced around the semiconductor world, ending up as chairman and CEO of a Texas-based company called Quickfilter Technologies.

The legacy of Motorola in Arizona was devastated by post-Iridium events. For decades Motorola had been the largest employer in the state, with thirteen plants in the Phoenix area and the original head-quarters building on 48th Street in Scottsdale, at the foot of McDowell Mountain. But beginning in the 1990s, newspaper reports blamed Motorola for TCE contamination of the Indian Bend Wash aquifer in Scottsdale. The source of the pollution was never proven, but the city of Scottsdale and several individuals ended up suing the company for more than a decade. TCE, an industrial solvent, had never been used in any substantial quantities at any Motorola plant, and since Motorola had some of the top environmental engineers in the world, they did an in-house study that concluded the source of the contamination was probably not Motorola but some local laundries. Motorola won all the lawsuits but, in the interests of being a good corporate citizen, agreed to clean up the TCE anyway. Ever since Dan Noble had started building transistors and developing semiconductors in the 1940s, Motorola had been a contributor to major Arizona philanthropies and a champion for the state, but something about this controversy left a sour taste in everyone's mouth. Motorola was being portrayed as a dirty company and a bad neighbor. For the first time, Motorola was dealing with people who didn't know who Dan Noble was and didn't care. It was as though, after all the years of prosperity, the local leaders wanted to kick them when they were down. So they left. They closed every facility, sold off the buildings, and pulled out of Arizona entirely. Today General Dynamics occupies the Scottsdale building where Noble's little team built the first transistors. But the most visible symbol of Motorola's vanished grandeur is the Chandler Lab, the center of the world's space industry just one decade before, now abandoned and derelict, looking even bleaker because of the few ducks that still manage to survive in the landscaped ponds near the entrance, the same entrance through which the greatest scientists in the world once arrived to marvel at the "next best thing" called Iridium.

In 2008 Dan and Helene Colussy were finally able to start their retirement in earnest, moving into an Italianate mansion they designed themselves in the exclusive Bear's Club subdivision founded by Jack Nicklaus on ecologically protected land in Jupiter, Florida. The Colussys returned to their life of dinner parties and the hosting of visitors from faraway places, many of them titans of industry, government leaders, and, of course, Colussy's colleagues from the U.S. Coast Guard Academy, class of 1953. As late as 2012—almost six decades later—Colussy's fellow crew members on the Coast Guard training ship the *Eagle* were still arguing about whether their commanding officer had surreptitiously ordered the engines to be turned on in Copenhagen Harbor while the crew was backing the foresails against a light dead-ahead wind. A crew of Danish cadets had just brought its own ship effortlessly into port, and the Danes stood on deck watching the Americans' floundering attempts to cast off from the wharf. Five of the cadets on the *Eagle* that day had become admirals, so their honor was at stake. Colussy played the bad guy in the argument, claiming he was standing on the bridge and heard the captain give the command for a brief engine thrust, whereas his fellow shipmates denied that any such thing would ever be possible, as it would forever sully the reputation of the U.S. Coast Guard.

Colussy continued to go to work every day in his library, a spacious wood-paneled study featuring a ceiling fresco depicting his three favorite assignments: the *Eagle*, Pan American World Airways, and Iridium. After his final resignation from the board of directors in 2009, he mostly kept up with his former Iridium colleagues by phone—and many of them were upset with Matt Desch.

Desch had decided not to use Isaac Neuberger as corporate counsel anymore—and Neuberger was hurt by that, after all those years of giving the struggling company such a discount legal rate. Nor did Desch have much use for Mark Adams, now that the company was developing its own applications and had its own research-and-development goals. Adams had given up his job as Chief Technology Officer in 2006, after Colussy discovered he was making more money through NexGen than he was revealing to the company, and in some cases taking Department of Defense business without telling anyone. Adams eventually sold his company to EDO Corporation—yet another "gray ops" firm that had worked for the Pentagon and CIA for years, mostly doing electronic warfare projects—but remained as President of his division. EDO was in turn bought by ITT Corporation in 2007, and that company was acquired by military supplier Harris Corporation in 2015. Adams continued to run his own subsidiary of Harris from an unmarked Pentagon-vetted office overlooking the runways at Dulles Airport.

One by one, the various handpicked employees who helped Colussy win the DISA contract and build the company were being let go by Desch. Also feeling slighted were the Iridium board members, who were used to being consulted on every decision of the company. Some of them, like Herb Wilkins, felt frozen out by the new management, and they called Colussy to reminisce or share stories about family. Wilkins and Terry Jones and their families became frequent visitors to the Colussy home, and Colussy was often asked to speak at meetings of African American businessmen's associations where he was the only Caucasian present. When Wilkins died in December 2013, Bob Johnson gave the eulogy at St. John Baptist Church in Columbia, Maryland, and in the course of his remarks he singled out Colussy as "the only white man Herb ever trusted." Tyrone Brown served on the Iridium board for several years and then left private law practice to head up the nonprofit Media Access Project, arguing before the FCC for fair-use doctrines such as net neutrality. In 2013 he returned to the Washington, D.C., law firm Wiley Rein, where he had worked during two previous stints.

By that time Bob Johnson had sold Black Entertainment Television to Viacom for $3 billion, thereby becoming the first African American billionaire, but he continued to dabble in restaurants, hotels, music, film, financial services, and a string of car dealerships he partnered on with Mack McLarty, former chief of staff for President Clinton. Most visibly, Johnson became the first black owner of a major sports franchise in 2002 when he bought the Charlotte Bobcats. Every time Johnson saw

Jones and Wilkins, he ribbed them about "not letting me in on the best deal you guys ever put together"—meaning Iridium.

But perhaps the loneliest guy of all in the post-Greenhill years was Dannie Stamp. Stamp had organized the launches of seven spare satellites during the first three years of the new Iridium. The first was a flawless five-bird launch on a Delta II from Vandenberg. The next was supposed to be a Long March, but the Chinese by that time had raised their prices to exorbitant levels. Stamp went back to his Russian friends at Khrunichev, who had become firmly devoted to commercial space launches thanks to their new partnership with the largest European aerospace company, Munich-based Astrium. The two companies, working together under the corporate name Eurockot, pulled an SS-19 "Stiletto" missile out of an active ICBM silo so that Iridium could use it for a two-satellite launch from Plesetsk Cosmodrome. Stamp was also part of the early planning for Iridium NEXT, but he butted heads with Desch over several issues and Desch decided "his vision for the program was inconsistent with how we wanted to build it." Stamp left the company in early 2007 and returned to agriculture, farming land in Iowa and the Texas Panhandle, while occasionally doing consulting work for a small Tempe company that was involved in scientific missions to Venus and Pluto. Stamp was especially proud of his cotton, corn, and milo farms in the Llano Estacado of Texas, where he installed state-of-the-art drip irrigation technology that he thinks might be the long-term solution for the rapidly depleting Ogalalla Aquifer.

The reason Stamp was supplanted had to do, once again, with Desch going in new directions. In June 2010 Desch announced that the rocket used to launch the new satellites would be the Falcon 9, which would have scared Stamp to death since the Falcon 9 had had only a single successful launch at that point and Stamp always used rockets that had been proven through years of use. The Falcon 9 had been developed by SpaceX, but analysts were skeptical of a rocket that had misfired the first three times it was tested and then been delayed for years. Desch thought the risk was minimal, though. By the time it was used for Iridium, the Falcon 9 would have been tested via twenty-five other launches, including six for NASA to resupply the International Space Station. And the best thing about the Falcon 9 was that the nose cone could handle ten satellites, so the entire constellation could be completed with a mere eight launches.

But the most impressive move Desch made wasn't the rocket but the creation of his own secondary payload in the form of a subsidiary called Aireon. Aireon would be the first truly global air-traffic-control system. Most people don't realize that airplanes fly out of range of traffic controllers all the time, especially over the heavily traveled North Atlantic, where planes have to be kept five miles apart in all directions until they return to the grid. There are even huge landmasses, notably in Africa, where radar systems are spaced so far apart that airplanes often can't be seen by controllers. Much of the world was shocked in March 2014 when a Boeing 777 operated by Malaysia Airlines apparently disappeared off the face of the earth, but it wasn't the first time that had happened since Amelia Earhart's ill-fated flight in 1937. (The doomed Malaysian airliner was using Inmarsat.) The Aireon system would make it possible for aircraft to be tracked 100 percent of the time, which means planes could fly much closer together, travel times could be decreased, airports could handle more traffic at peak times, search-and-rescue efforts could begin immediately, and billions in fuel costs could be saved.

Although a system of this type seems obvious, it had actually failed on three previous tries. INTELSAT proposed a system in 1962 but failed to get the support of the airlines. The Russians and the U.S. Air Force came up with a plan to combine their global positioning systems in 1987, but that cooperative effort fell apart as well. Boeing announced a system of air-traffic-control MEOs in 2001 but never built it.[61] This time, to make sure the system would be adopted, Desch sold off an equity position in Aireon to NAV CANADA, the world's busiest air-traffic-control system, covering eighteen countries, and, more to the point, a fully privatized system, meaning CEO John Crichton could work with Iridium without all the interminable delays Desch faced at the FAA. Desch then sold off positions to ENAV (the Italian agency that controls much of Europe's airspace), the Irish Aviation Authority (which controls much of the North Atlantic), and Naviair (the Danish agency that controls much of the Arctic). And since the U.S. government agencies could never get their act together to approve any other secondary payloads, Desch also created a program called Iridium Prime. If any company or agency needs to lock into the Iridium constellation at any time in the future, Desch will launch a satellite specific to that purpose. (There's plenty of space on those eleven-satellite planes.) Meanwhile, the company continued to create new apps and devices that made inroads into the consumer markets envisioned by the original inventors—notably the first portable hotspot that makes it possible for you to use your smartphone or tablet anywhere in the world.

After Colussy's retirement party in 2009, he told Desch not to share anything with him unless it was public. He was just a shareholder now, and when he would get calls from any of his old colleagues from the cramped offices at the SNOC, he would sympathize with their plight. *Yes, Matt marches to his own drummer,* he would tell them. *Yes, it's a shame he didn't see the value of your experience and commitment,* he would say. But inwardly he couldn't have been more thrilled. Desch was running the company exactly as he would have run it himself, with his own team and his own goals. Colussy could finally relax. Desch had done everything Colussy hoped he would do in 2006 and more. He had even figured out a way to get the hardheaded airline executives to enter the twenty-first century.

For the first time since 1999, Colussy started working seriously on his golf game again, and golf at his new residence—the Bear's Club—was a very serious thing. Some of the top golfers in the world either lived or played there, including Rory McIlroy, Michelle Wie, Luke Donald, Ernie Els, Gary Player, Ian Baker-Finch, Mark Calcavecchia, and Jack Nicklaus himself, as well as celebrities from other sports like Michael Jordan, who built a palatial estate three doors down. Colussy delighted in competing against old business friends like Ross Johnson of RJR Nabisco, Ed Artzt of Procter & Gamble, and Jim Perrella of Ingersoll-Rand, as well as legendary New York Giants running back Tucker Frederickson, since they all tended to exaggerate their handicaps. Colussy had never played the game when he was young, so he was able to win round after round simply by using his real handicap. Like everything else he did, he was competitive and thorough. When Helene developed a respiratory condition and had to give up the game, he played less often, but he still practiced most days, at least until Helene took a turn for the worse in the summer of 2012 and he became her caregiver. She finally passed away at home in the summer of 2013, and a week later 120 people showed up at the residence for a memorial service for the woman he had married sixty years and twelve days earlier. On that particular day, Colussy recalled their first date, when, as a Coast Guard cadet, he'd tried to impress the beautiful New England girl by taking her sailing on the Thames River—only to embarrass himself when the wind died and the boat got hung up on the Gold Star Memorial Bridge.

It was the kind of memory that could make you feel wistful for a vanished past. Then, a few days later, his old friend Russell Chew called. The two men had met when Chew was at the FAA and remained close when he became the COO of JetBlue. Chew told Colussy he had found this new company and he thought maybe they could work together to get control of it. Chew would be needing a board chairman and a fundraiser. Colussy listened intently as his friend ticked off the advantages of the deal—it was a new technology, it revolutionized ground maintenance time for big aircraft, there were possibilities in other industries as well—and as he listened, Colussy felt his whole being starting to stir. After all, he was only eighty-two.

I just want to make one thing clear, he told Chew. *When we do it, you'll need to find someone else to be the CEO.*

AUTHOR'S NOTE

Eccentric Orbits was written over a period of several years using materials from many sources, including personal interviews, government archives, private archives, court records, correspondence, personal journals, videotapes of rocket launches, corporate publications, visits to various facilities and homes mentioned in the narrative, and contemporary media accounts. Due to the passage of time, some of the participants couldn't be found or had passed away, and a few chose not to speak to me. Three people agreed to be interviewed only under the condition of anonymity. Two principals chose to communicate with me through their law firm.

For the most part, I researched the book with face-to-face interviews, letting people tell their own stories, and then filled in the rest of the narrative with archival material. The most extensive interviews were with Dan Colussy at his home in Jupiter, Florida, at the Coast Guard Academy in New London, Connecticut, at the Four Seasons restaurant and the Harvard Club in New York, and by phone; Ray Leopold at various locations around Glacier National Park, Kalispell, Montana, and his home in Whitefish, Montana; Dorothy Robyn at the Sofitel Hotel in

Washington, D.C.; and Dannie Stamp at various locations in Phoenix, Tempe, and Chandler, Arizona. Additional interviews were conducted with Bary Bertiger at his home in Vero Beach, Florida; Ken Peterson at his home and at the Grand Cafe in Lake Mills, Iowa; Ed Staiano in his Scottsdale, Arizona, office; Randy Brouckman at restaurants in Alexandria, Virginia, and Philadelphia; Isaac Neuberger in his downtown Baltimore law office; General Jack Woodward at a restaurant in Reston, Virginia; John Richardson by phone from London and at the Hay-Adams Hotel in Washington, D.C.; Mark Adams at his ITT Corporation office at Washington Dulles International Airport; John Castle at the New York headquarters of Castle Harlan and by phone; Herb and Sheran Wilkins at their home in Ellicott City, Maryland; Dave Oliver at the headquarters of EADS North America in Rosslyn, Virginia; Ted Schaffner at a restaurant in Hoffman Estates, Illinois; Bob Kinzie at his home in Bethesda, Maryland; Rudy deLeon at the Center for American Progress in Washington, D.C.; Matt Desch at Iridium headquarters in Tyson's Corner, Virginia; Steve Pfeiffer and Larry Franceski at the offices of Fulbright & Jaworski and Norton Rose Fulbright in Washington, D.C., and at the Yale Club in New York City; Terry Jones at Syncom headquarters in Silver Spring, Maryland; Ty Brown at a hotel in Washington, D.C., and at Herb Wilkins' residence in Ellicott City, Maryland; Kathy Brown at Verizon headquarters in Washington, D.C.; Dennis Weibling at the Loews Regency Hotel in New York City; Steve Miles at the Satellite Network Operations Center in Leesburg, Virginia; General Carl O'Berry at his home in Scottsdale, Arizona; Jennifer Colussy Grace at her home in Jupiter, Florida, and at the Waldorf Astoria Hotel in New York City; Helene Colussy at her home in Jupiter, Florida; Durrell Hillis at the Paradise Valley Country Club and at his home in Paradise Valley, Arizona; Mike Fisher at his office in Tempe, Arizona; Ted O'Brien, Don Thoma, and Greg Ewert, all at Iridium headquarters in Tyson's Corner, Virginia.

Phone interviews were conducted with Leo Mondale from his Inmarsat office in Eysins, Switzerland; Rick Stephens from his home in

Alpine, Texas; Carlton Jennings from his home in Greenwood Village, Colorado; Jim Walz from his Alliance Data office in Columbus, Ohio; and Bill Kennard from his Carlyle Group office in Washington, D.C.

I spent several weeks at the William J. Clinton Presidential Library and Museum in Little Rock, Arkansas, where my Freedom of Information Act request had resulted in the discovery of 16,343 pages of archival documents related to Iridium. Of those, 1,912 pages were redacted either in whole or in part. The largest number of withheld pages (973) were suppressed for reasons of national security. Documents were also withheld because they would reveal corporate trade secrets (691 pages), because of invasion-of-privacy issues (232 pages), because the release would violate a federal statute (47 pages), because the documents would reveal law-enforcement techniques (21 pages), and because the life or physical safety of someone could be endangered (12 pages). Forty-six pages were withheld because they contained confidential advice between the President and his advisors, but I filed an appeal based on President Obama's Executive Order 13526, issued in 2009 to create more transparency and quicker release of government records. After this appeal, one three-page diplomatic cable and 23 of the 46 "confidential advice" pages were released. After the first edition of this book was published, the government released 53 additional pages of redacted government cables, without affecting any of my conclusions. All of the Iridium records may be examined at the library by accessing Freedom of Information Act request 2010-0706-F. I want to give special thanks to Adam Bergfeld, the archivist in charge of electronic records, who unearthed hundreds of pages of handwritten notes, allowing me to pinpoint what was said during meetings and phone calls. I also want to thank Dana Simmons, the supervisory archivist; archivist Rhonda Young; archive technician Jennifer Caddell; Kelly Hendren, mandatory review archivist; and Rob Seibert, mandatory classification review archivist.

The records of the United States Bankruptcy Court for the Southern District of New York were reviewed on-site at One Bowling Green in Manhattan. Depositions in the Iridium bankruptcy (case number 99-45005) had been sealed, but some of the attorneys involved assisted me with the general narrative of the case. The records of *Chase v. Motorola,* including a bench trial in November 2001, were reviewed on-site at the Daniel Patrick Moynihan United States Courthouse in lower Manhattan. The records of the securities case, *Freeland et al. vs. Iridium World Communications et al.,* were reviewed on-site at the U.S. District Court for the District of Columbia, on Constitution Avenue in Washington, D.C. Most of the records in this case had been sealed by Judge Nanette Kay Laughrey at the request of Motorola.

Dan Colussy maintains a private archive of documents relating to his professional career at his home in Jupiter, Florida. Several thousand documents relating to Iridium, including Colussy's handwritten notes from every phone call and meeting, were pulled out of that archive and digitized by his archivist, Cari Ann Gowin, who organized them for me. The process of going through this material was expedited by Janet Coffey, his assistant.

Public-domain documents relating to Iridium and the history of communications satellites amounted to more than 100,000 pages, including dozens of academic studies and monographs, and these were all assembled by my longtime research assistant, Rebecca Brock, Director of the Chapmanville Public Library in Chapmanville, West Virginia. The detailed research involved in locating and assembling photographs was carried out by Rebecca Brock and my personal assistant, Tracy Vonder Brink of Loveland, Ohio.

Tyler Love, an archivist at the Udvar-Hazy Center in Chantilly, Virginia, part of the National Air and Space Museum, assisted with photograph collection and access to the personal correspondence of Arthur C. Clarke, and read an early version of the manuscript. Also helping with early manuscript corrections were Andrew Stuttaford, head of the New York office of the Norwegian investment bank ABG Sundal Collier and a frequent commentator for the *Wall Street Journal* and *National Review* on complex international financial issues; Jennifer Morrow, an editor at the Routledge academic publishing firm in New York City; Renner Jo

St. John, an attorney at Brock & Scott in Charlotte who had been the pilot for General Norman Schwarzkopf in Desert Storm; and Jerry Flieger, Professor of French and Comparative Literature at Rutgers University.

Diane Hockenberry, Director of Communications and Public Relations at Iridium, helped ensure access to photographs and Iridium personnel. Her predecessor, Marie Knowles, provided interview assistance and access to the Satellite Network Operations Center.

Special thanks to Tom and Chris Casey, who purchased Dan Colussy's house in the Glen Oban section of Arnold, Maryland, and opened it to me so that I could research physical details.

Several people have asked about the genesis of this project. I was introduced socially to Dan Colussy by Ted Malloch, at that time a professor at Yale University. Colussy gave me a two-minute version of the Iridium story and was surprised and skeptical when I told him it sounded like a book. I met with Colussy a few days later at the Harvard Club in New York City, where he gave me a longer version of the story. The following week I had lunch with Morgan Entrekin, president and publisher of Grove Atlantic, at the Union Square Café in New York City, and he took a chance on the story by agreeing to commission the book. It turned out to be much more complex than it originally appeared, since several of the principal figures in the narrative refused to be interviewed and the government took forty-nine months to provide the necessary clearances for the release of documents. Nevertheless, the book was shepherded to completion by my editor, Jamison Stoltz, who suggested many changes that were incorporated into the final work, and his two assistants, Allison Malecha and Nicole Nyhan.

The other question I'm asked by manuscript readers is what happened to Dan Colussy after his wife's death. The answer is that he quit trying to retire and, as this book went to press, was eighty-five years old and operating three "green" companies that he hopes will be revolutionary in their beneficial effect on the environment.

—John Bloom, New York City, February 10, 2017

NOTES

Chapter 1: THE CONSPIRATORS

1. The "Winter White House" had lost the aura of Camelot over the years thanks to highly publicized drinking binges by Senator Ted Kennedy and, most notorious of all, as the place where William Kennedy Smith had sex on the lawn with a woman he picked up in a local bar. (He was acquitted on rape charges in a 1991 trial that attracted national attention.) The structure was a wreck when Castle bought it in 1995, having been left in disrepair by successive waves of Kennedys who were so unsentimental about their famous relative that they hired lawyers to prevent the Palm Beach Town Council from declaring the house a national landmark and making it hard to sell. "It had become almost like a time-share for the Kennedy children," Castle said. "Seven of them divided it up during the season—the Lawfords, the Shrivers, Caroline, Rosemary." That meant none of them wanted the restoration expenses attendant on a building declared historic, so they fought the city for fifteen years. When family matriarch Rose Kennedy died in early 1995, the house was placed on the market, which was actually a relief to the city, since that was a better fate than demolition. Castle had seen all the drama within the family as a buying opportunity, telling the Kennedys he would be happy to restore the place to the standards of the city, but only for $4.9 million, not the $7.6 million they were asking for, and the Kennedys eventually took the deal. He then spent five years restoring the crumbling Mediterranean Revival to the way it looked in 1923, when celebrated architect Addison Mizner designed it for Rodman Wanamaker, the Philadelphia department store magnate. It was not one of Mizner's better efforts, and if it had not been used so often by the First Family, it would no doubt have been torn down long ago. But Castle plunged into the project, surviving some three hundred inspections by the city,

even going so far as to leave the presidential bedroom exactly as Jackie had decorated it and to preserve such tiny details as dresser drawer labels handwritten by Rose Kennedy.

2. It would be eclipsed a year and a half later by Enron, which listed $63.4 billion in assets at the time of filing but was still insolvent. The valuations of bankrupt companies are so dicey—do you count what was invested, what was spent, what was owed, or assets at the time of filing?—that several different firms have been cited as the largest flameouts in history. In the 1990s those companies' valuations were counted in the tens of billions, but in the following decade those numbers would be counted in the hundreds of billions, culminating in the 2008 bankruptcy of Lehman Brothers, widely reported as a $691 billion loss.

Chapter 2: NERDS, NAZIS, AND NUKES

3. Despite the constant rhetoric about satellites as instruments of peace, suspicions that they were actually intended to be used as military weapons pervaded the government. It didn't help the image of the American program that, at a conference held at the National Academy of Sciences, Vanguard was referred to by the director of the program, John P. Hagen, as the "Naval Research Lab satellite," only to be corrected by a very junior member of the IGY staff: "The National Academy's satellite, Dr. Hagen." Nor did it help that whenever reporters showed up at Cape Canaveral, the result was headlines like "Devoted Navy Men Work Around the Clock at Cape Canaveral to Put Up Vanguard Vehicles"—when, in fact, the vast number of workers in the Vanguard Operations Group were the civilian engineers of the Glenn L. Martin Company, who bristled at the idea that they were hired hands for the military.

Chapter 3: THE SPOOKS

4. This was where the confusingly named Naval Computer and Telecommunications Area Master Station Pacific had been the communications base for the Navy ever since a few days after the Pearl Harbor attacks of 1941, when commanders in the field realized they needed to move their vulnerable radar and radio links into more isolated areas, away from potential combat zones. Wahiawa means "place of noise" in Hawaiian but is actually one of the quietest places in the islands, surrounded by Lake Wilson on three sides, buried between two volcanic mountains, and five miles from the largest Dole pineapple plantation. The only noise occurs in the headphones of the technicians, who listen to coded messages from all over the world, as well as encrypted Iridium phone calls that are routed through a tracking device next door to a cattle ranch into what is officially known as the Enhanced Mobile Satellite Services Gateway.

5. Woodward was really comparing apples to oranges. The UHF Follow-On system, tagged with the sardonic acronym UFO, consisted of eight satellites in geosynchronous

orbit and required terminals and ground stations. UFO was indeed designed for "transit and off-station communications," meaning when ships, planes, and infantry units were on the move, but it was not nearly as mobile as Iridium.

6. The nuclear reactor plant is today the Mohegan Sun resort and casino. When Colussy shut down the plant in the eighties, he was unable to sell the building even after it was declared uncontaminated and cleared for unrestricted uses. Eventually he approached the local Indian tribe, the Mohegans, and asked if it would be interested in using the building for a casino. The chief, a boatbuilder, told him the tribe didn't qualify for casino ownership. At the time the Mohegans owned an old church but had no money and no land, having recently been turned down by the Bureau of Indian Affairs for classification as a federally designated tribe. So Colussy hired a Washington, D.C., law firm to resubmit their application, noting that one of their burial grounds was on the 240 acres then owned by UNC, and eventually they were approved. He then introduced them to Sol Kerzner, the gambling entrepreneur best known for Sun City in South Africa and the Atlantis development in the Bahamas, and Kerzner worked out a deal with the state of Connecticut to build the Mohegans a highway interchange. Colussy sold Kerzner the building, and it stands today as the main structure of Mohegan Sun. The 240 acres once designated as "Dean's Chicken Farm" constitutes the Mohegan reservation.

7. The Marine mobile devices in use were the SMART-T (Secure Mobile Anti-Jam Reliable Tactical-Terminal), the UHF TACSAT (a satellite phone that worked off the top-secret Pentagon GEOs), and a high-frequency radio device. All of them had to be activated by the platoon's radio officer.

Chapter 4: THE DREAMERS

8. HALOSTAR used a manned aircraft flying at fifty-two thousand feet in patterns designed to cover an area of fifty to seventy-five miles in diameter. The test plane, called the High Altitude Long Operation Proteus, resembled a flying catamaran, with the communications payload slung low under the body, and its flights in 2000 set three world altitude records for piloted aircraft, eventually reaching 63,245 feet, or almost twelve miles above sea level. Unfortunately, Angel Technologies never raised the $700 million it would have taken to launch.

9. Aerospace worked closely with the astronauts on Mercury, Gemini, and Apollo, developing data on how the human body reacts to space, and had access to the expertise of Dr. Hubertus Strughold, the "father of space medicine," one of Werner von Braun's colleagues brought to the United States after the war. Strughold had access to data derived from Dachau inmates who had been frozen and put into pressure chambers.

Chapter 5: TREASONS, STRATAGEMS, AND SPOILS

10. Odyssey would become notorious for what was considered one of the most ridiculous patent filings in history. TRW claimed ownership of the medium earth orbit, defined as any trajectory between 5,600 and 10,000 nautical miles from Earth's surface, and insisted that no one could use that orbit without infringing on its intellectual property. The only other company planning to use MEOs at the time of the 1992 filing was Inmarsat, which was planning to launch ICO Global, and the two companies had been feuding over other issues. Amazingly, the patent was actually granted in 1995, then revoked the same day when Patent and Trademark Office Commissioner Bruce Lehman received a call from Air Force One. To this day no one knows how TRW managed to get even remotely close to patenting a portion of outer space, or who called the President to complain about it, but it was one of only two cases in a hundred years in which a patent was issued and revoked on the same day.

11. To this day the Iridium patent is invalid in Europe, although it's in full force and effect everywhere else in the world.

12. Teller was an advocate for environmental reform and later asked if he could put a secondary payload on the Iridium constellation—for global ocean-temperature sensing— but the design team turned him down. "We didn't need an additional bundle of complications," said Bertiger. "We had to stay focused. What if the secondary payload started sucking power? It could end up screwing up the primary mission."

13. A quarter of a century later, the $6.1 billion Galileo navigation system was still not operational. The first two satellites were launched in 2011, but the entire thirty-satellite constellation was not expected to be complete until 2019.

14. Asset tracking had been introduced in 1983 by a company called Geostar. The Geostar inventor was a Princeton University physics professor and futurist named Gerard K. O'Neill who had no interest in assets—he envisioned a device that would identify your precise location and send short messages to the police if you were attacked by muggers. Fortunately Qualcomm knew what to do with the technology, introducing OmniTRACS in 1988, a system that could store and forward messages through chips installed in the cabs of eighteen-wheelers. This allowed truck fleets to be monitored so that owners could make sure the trucks never traveled empty. Geostar eventually went bankrupt, and its licenses were purchased by Iridium in 1992.

Chapter 6: ROCKET MAN

15. "Phased array" simply means a radar unit composed of many elements which, when combined electronically, allow it to send and receive signals from all directions without the physical rotation of the antenna. The most famous device equipped with phased-array

antennas is the Patriot missile (Patriot is an acronym that stands for "Phased Array Tracking Radar to Intercept of Target"), which was used in the Gulf War to shoot down Scud missiles in flight. The Patriot missile was built by Raytheon, the same company that built the Iridium antennas.

16. Gary Powers' capture resulted in a sharp curtailment of the CIA's use of manned aircraft, although the U-2 would remain in service for special assignments into the twenty-first century. Within two months after Powers's capture, photography surveillance missions had been reassigned to cruise missiles that effectively worked as drones. (Before the missile crashed into the ocean, its photographic film would be ejected and retrieved in midair by a JC-119 aircraft.) Primitive drones had been used in both world wars, but the first modern one was an aircraft that began its life in 1945 when the Army added wings to a Nazi V-2 missile. After many design changes it became the XSM-64, known as the Navaho, but it had so many disastrous tests during 1956 and 1957 that the "Never-go Navaho" was canceled after the nation had spent $700 million on less than one hour of flight time. Even more infamous was the Snark, built by Northrop and introduced in 1946. The Snark could supposedly fly at 150,000 feet at Mach 0.9 speeds, but it had one test failure after another, including an infamous launch in 1956 in which the drone was lost somewhere in the Amazon rainforest and never recovered. With Air Force personnel literally enacting "The Hunting of the Snark," it was destined to be the last missile of any kind named after a Lewis Carroll creature, especially after techies started referring to the Atlantic coast off Cape Canaveral as "Snark-infested waters." (The errant Snark was finally found by a Brazilian farmer in 1982.) The Snark program was canceled by President Kennedy in 1961, but it was doomed anyway, because the general staff at the Strategic Air Command hated it. Even after drones proved themselves in combat—especially the so-called Lightning Bugs that were launched from C-130s over Vietnam—the commanders in charge of them continued to despise them. The only drone that ever got any Pentagon support was the AQM-34 Ryan Firebee, and that's because its only purpose was to be shot down by other planes. The Firebee had thirty-four thousand successful flights in the 1960s—as a gunnery target.

17. For cost reasons, Motorola determined that the Chandler satellite factory should be a 5.3-Sigma operation instead of a full Six Sigma facility. Given what we now know about the durability of the satellites, even this standard was beyond what was needed.

18. An ill-fated INTELSAT launch on February 15, 1996, did end up with a Long March 3B careening into a mountain village, but the extent of the damage was never reported to the outside world. The launch was managed by Loral Space, which was later hauled before the Senate Select Committee on Intelligence and raked over the coals for allegedly sharing sensitive missile information with the Chinese during the post-crash investigation. Loral's sin: teaching the Chinese how to avoid crashes in the future by "improving launch reliability." The committee, chaired by Senator Richard Shelby of Alabama, was

equally upset about a post-crash investigation carried out by Hughes Aircraft after a failed Long March 2E mission carrying an Apstar communications satellite in 1995, but that rocket apparently didn't hit any population centers.

19. Hedy Lamarr, the inventor of CDMA, was an Austrian Jew who was married to the munitions manufacturer Friedrich Mandl when she left him for a career in Hollywood in the 1930s. A decade later she took a 1903 patent by Nikola Tesla describing "spread spectrum" and updated it as a means of wirelessly guiding torpedoes in a manner intended to confuse enemy jamming. She worked out the system on twelve synchronized player pianos with her friend George Antheil, an avant-garde composer, then applied for and received a patent in 1942 for what she called a "Secret Communications System." Her letters to the U.S. Navy offering it as part of the war effort were ignored, and it wasn't really employed until 1962, for covert ship-to-ship communications during the Cuban Missile Crisis. It was later used in military satellites but wasn't used commercially until Qualcomm introduced it in 1988 for its OmniTRACS truck-tracking system. Then, when the first 3G wireless systems were introduced in the late 1990s, they all used CDMA, the "Secret Communications System."

20. This didn't mean the other potential competitors went away. A Torrance, California, project called Celstar was a planned $600 million system consisting of a single GEO that would provide bandwidth to existing land-based operators so that their GSM systems would work anywhere. When the Celstar investors were frozen out of the FCC licensing process, they complained loud and long, backed up by small wireless companies that had no other way to achieve 100 percent coverage in the United States. Ultimately they lost their appeals because their project was not global and the WARC frequency allocation had been specifically for global firms.

21. Even though 1610 to 1626.5 MHz was allocated, Motorola built the satellites to use only 1616 to 1626.5, mainly because of WARC footnotes that resulted in tougher restrictions for the lower frequencies. Then, when Globalstar became operational, Iridium had to split that 10.5 MHz of spectrum, so in most of the world the system only uses 1621.25 to 1626.5 MHz. In regions where Globalstar doesn't operate, Iridium uses the full 1616 to 1626.5.

22. Stamp's confidence in the Proton proved correct as it became the heavy-lifting workhorse of the entire space industry over the next two decades, resulting in Khrunichev setting up a company in Reston, Virginia, called International Launch Services that eventually captured 30 percent of the world's business. Oddly, space cooperation between the United States and Russia started to disintegrate in 2014 after NASA cited Russia's annexation of Crimea as a reason for ending several agreements between the two countries, including agreements that had been in place since 1975. Even more devastating, the Senate passed a bill banning the purchase of Russian-made rocket engines,

thereby causing havoc with the launch schedules of Boeing, Lockheed Martin, and the U.S. Air Force. The congressional action essentially ended the life of the American-made Atlas V rocket, because it depended on the RD-180 engine, manufactured by NPO Energomash of Belgorod, Russia. It's ironic that the "handshake in space" brokered by President Ford could survive an era in which twenty thousand ICBMs were on high alert but couldn't survive much less dangerous regional disputes over Ukraine. In its half century of service, the Proton has had more than four hundred successful launches.

Chapter 7: FAST EDDIE

23. The FBI was being disingenuous about wiretap arrangements, since it was well known within law enforcement circles that nations within "the family"—England, Australia, and Canada—were often allowed to break the rules. In fact, there were already satellite calls originating in the United States and being processed through a foreign gateway—calls from the American Mobile Satellite Corporation's "briefcase phone" terminals, which came to Earth in Ottawa. The emphasis on Canada in the FBI legal filings is baffling anyway, since the Canadian gateway would be less dangerous to national security than almost any of the other ten Iridium gateways, and the FBI's position is even more puzzling now that we know the agency had formed an organization called ILETS (International Law Enforcement Telecommunications Seminars) in 1993 to foster wiretapping across borders among the twenty-one member nations, including Canada. ILETS resulted in "cross-border interception arrangements" that were primarily directed at Internet e-mail but also included satellite phones. This was the first government initiative that routinely downloaded subscriber information, regardless of whether the user was suspected of a crime or not, and that policy would become common in the first decade of the twenty-first century.

24. Four months later Odyssey "merged" with ICO by selling its patents for 7 percent of the company. Roger Rusch, the inventor of Odyssey, would later say the patents were valued at $150 million.

25. Before Celestri, there were two other proposed Motorola broadband systems, one called Millennium, which would have used four Ka-band GEOs, and another called M-Star, which would have used seventy-two LEOs in the EHF (Extremely High Frequency) spectrum. Celestri included both systems and added sixty-three Ka-band LEOs in a massive constellation that would provide every kind of broadband service between the 60th parallels.

26. A Monte Carlo simulation is an algorithm that uses random sampling to take into account every possible outcome of open-ended situations. Through years of experience, Motorola knew that Monte Carlo simulations were always better predictors of cost overruns or schedule overruns than any other method.

27. The combined Teledesic-Celestri system coined a new term. Unlike the Little LEOs, such as Orbcomm, and the Big LEOs, like Iridium, this was a "Mega LEO." Shortly after the new Teledesic-Celestri constellation was mapped, France's Alcatel responded with a Mega LEO plan of its own, called SkyBridge, consisting of eighty LEOs at 905 miles up that would offer broadband to fixed terminals. Now that the ACTS satellite had proven how versatile the Ka-band could be, Ka-band projects were sprouting up all over the place. Hughes Aircraft announced it was building a $4.5 billion Ka-band project called SpaceWay as a rival to Teledesic, and AOL would soon join Hughes as a $1 billion partner. Not to be outdone, Lockheed Martin chimed in with a nine-GEO system called Astrolink, budgeted at $3.7 billion, in partnership with Telespazio and TRW. General Electric started looking for partners for a $2.5 billion constellation of seven satellites to be called GE Star, and Loral announced a $2 billion three-GEO system called CyberStar. Thanks to the events of 1998 and 1999, none of these broadband systems would ever be built.

Chapter 8: MAN OVERBOARD

28. The Iridium Annual Report for 1997 actually lists $4.7 billion of "financing," but it's unclear whether the word "financing" implies debt. If it does, then the total start-up cost for Iridium would be $7.061 billion.

29. The minimalistic logo, created by Landor Associates of San Francisco, is still in use today.

30. Claircom was a partnership of Craig McCaw and Hughes Aircraft that was acquired by AT&T in 1994 when McCaw sold his cellular empire, and it had captured 39 percent of the market by building out a system of 151 ground stations that could be accessed from the air in North America. In Europe the system used both ground stations and Inmarsat.

31. Iridium was increasingly regarded as a first-aid kit in the event of apocalyptic situations. The Office of Management and Budget studied the Iridium phone throughout 1999 and recommended its use as the "last resort" backup communications device for all government agencies on New Year's Eve 1999, when the nation was poised for the possibility of massive infrastructure breakdowns. Similar conclusions were reached by the President's Commission on Critical Infrastructure Protection, which had been meeting for years to deal with the question "What if a cyber attack brings down an essential service like fire, police, medical, banking, or energy distribution?" The answer: use Iridium phones until the services are restored.

32. The Motorola Timeport was quickly followed by the Bosch World Phone and the Ericsson I888, both worldwide roaming phones operating on the GSM standard and retailing for under $300.

Chapter 9: THE WONKS SCRAMBLE

33. Fifteen years later, Lockheed Martin was still building Buchanan's vaunted Mobile User Objective System. It turned out to be more expensive than Iridium at $7.3 billion and rising.

34. This is not quite as cynical as it sounds. Seven-eighths of the world's land area is in the Northern Hemisphere.

35. The idea of adapting Iridium for use in parts of the world that were otherwise without telephony stirred the imaginations of futurists and intellectuals who weighed in on an e-mail thread initiated by distinguished science fiction author David Brin. Among them were Bangladeshi telephone visionary Iqbal Quadir, IBM executive Michael R. Nelson, Microsoft chief strategist Nathan Myhrvold, British science writer Oliver Morton, "Father of the Internet" Vinton Cerf, *Whole Earth Catalog* creator Stewart Brand, and World Bank economists Carlos Braga and Andrew H. W. Stone. The high level of interest is not that surprising since sociologists, anthropologists, political scientists, and economists had been writing about the projected impact of Iridium throughout the 1990s.

Chapter 11: *ET TU?*

36. Scott Adams, creator of the *Dilbert* comic strip, was following the drama that summer and penned two panels about Iridium in late July 2000. In one, Dilbert raises his hand to ask a question of the executive announcing the Iridium de-orbit. "Question: Wouldn't that create dozens of deadly flame balls speeding toward Earth?" asks Dilbert. The answer from the executive: "That's why we're aiming for cities that have lots of swimming pools."

37. The one time Neuberger did work on the Sabbath was the exception that proved the rule. When Colussy and Neuberger were in the process of selling UNC to General Electric, Colussy was suddenly asked to an "urgent" negotiating session with GE chairman Jack Welch on a Friday night. Not certain what to do, and uncomfortable with the idea of negotiating a nine-figure deal with Welch alone, Colussy took the risk of calling Neuberger's Miami hotel room. After several rings, Neuberger picked up the phone but didn't speak. Colussy explained the situation. There was a long pause. Finally Neuberger said, "Come to my room, knock three times on the door, and say you have an emergency. Repeat this three times." Then he abruptly hung up. Colussy did as he was told, going to the room, knocking three times, declaring an emergency, and then repeating the ritualized request for assistance. After the third request, Neuberger cracked open the door and gave Colussy some advice, warning him not to agree to anything Welch wanted. An hour later, Colussy was seated in the hotel lobby, speaking directly to Welch, when he suddenly noticed something out of the corner of his eye. A man was hiding behind a potted plant, making hand gestures to Colussy. It was Neuberger, still clad in

his pajamas, on a mission of business assistance that somehow conformed to the rules of the Sabbath but didn't allow getting dressed.

38. EZ Bank had essentially the same business plan as PayPal, which had been launched in 1999 but didn't truly take off until 2001. PayPal worked over common Internet connections available to everyone, but Maruani apparently believed that a more secure network was required for banking operations.

Chapter 12: THE FIXER

39. "Syncom" was also the name of a satellite launched by Hughes Aircraft in 1963. The outer space Syncom stood for "Synchronous Communication," whereas the Wilkins-Jones partnership stood for "Syndicated Communications," but choosing the namesake of the most powerful of all the first-generation communications satellites would later seem prescient.

40. Wilkins was not alone. Al Gore, while a Senator, once called Malone "the Darth Vader of the cable industry" for his hardball tactics and anti-consumer policies.

41. The first space junk to plop down in a populated area was a 20-pound piece of Sputnik 4, which landed on the center line of North 8th Street in Manitowoc, Wisconsin, on September 5, 1960. Nine years later, five Japanese seamen were hit by falling debris near Sakhalin Island, and that debris was assumed to be from space, but the origin of the objects was never determined. The only verified instance of a human being struck by space debris occurred on January 22, 1997, when Lottie Williams of Tulsa, Oklahoma, was struck on the shoulder by a rumpled metal shard about twice the size of a human hand while walking her dog in O'Brien Park. The debris turned out to be part of a Delta II upper-stage booster. Williams was unharmed, but, more ominously, a 507-pound piece of stainless steel from the propellant tank of that same booster hit the ground near Georgetown, Texas, and a 66-pound piece of the titanium pressurant sphere landed near Seguin, Texas.

Chapter 14: FOUR DEAD BIRDS IN THE SITUATION ROOM

42. Orbital debris is monitored by NORAD because the primary reason for tracking objects in space is to make sure they're not mistaken for nuclear warheads. The United States and Russia both notify the other country whenever any object reenters Earth's atmosphere.

43. The main concern for the future was the expected life of the batteries, which was anywhere from eight to twenty-five years, depending on whose data you were using and how many phones were in service.

44. In fact, the island nations of the South Pacific *had* gotten into the satellite business, but only to apply to the ITU for orbital "slots" that they could then resell. Princess

Salote Mafile'o Pilolevu Tuita, sister of the king of Tonga, went so far as to buy a twenty-two-year-old decommissioned COMSAT satellite that had been used for AT&T phone service but had long since become obsolete. The salvaged satellite was never used, as far as anyone could tell, and her company—TONGASAT—ended up at the center of a national scandal, owing $32.3 million she somehow managed to borrow from the national treasury.

45 These recollections by Dan Colussy and Isaac Neuberger were disputed by the Saudis, who say they were never influenced by the prospect of Craig McCaw investing.

46. Money eventually produced a one-page document called "Compelling Reasons Supporting DoD Use of Iridium," and it circulated throughout Washington as the answer whenever anyone asked, "Why do we need these phones?" It was basically a rearranged version of Dave Oliver's reasons: 1) the military's satcom system was maxed out, 2) this was the only system with global pole-to-pole coverage, 3) there was a need to preserve the government's $200 million investment, and 4) Iridium was the only handheld device that offered Type 1 security, the highest level of encryption.

47. Robyn's answer put her in the curious position of defending Motorola's motives to OMB. "The precedent issue concerns me," she wrote, "but I don't consider this a 'bail out' really. Rather it's a way to counter Motorola's incentive to destroy a very valuable asset rather than face the prospect of frivolous (but potentially successful) lawsuits for things for which it is not responsible. . . . Surely it cannot be good public policy to have the creating company destroy an asset for fear that, in our overly litigious society, some clever lawyer is going to figure out a way to hold the company liable for something that it cannot control if the asset is sold."

48. Rusch was constantly quoted in the press, always identified as a "satellite industry consultant" who believed Iridium was a virtually worthless system that would soon wear out, but the reporters never noted that he was the inventor of TRW's Odyssey system, a competitor of Iridium that was never built.

Chapter 15: MOOSE HUNTERS, MARINES, AND THUGS

49. The Prince had formed a Curaçao corporation called Baralonco NV in the 1980s to handle investments in the United States, and he chose to switch his ownership to that company instead of Mawarid Overseas Investments Ltd., which had made the original Iridium investment. Curaçao is part of the Netherlands, and the Netherlands is a World Trade Organization partner that is immune from some of the foreign-ownership rules. But the FCC refused to accept Baralonco as a "transparent" entity and kept pressing the company for information, suggesting that the transfer of licenses from Motorola to the new Iridium might be held up, even after Baralonco made all the appropriate disclosures.

50. Among the displaced residents was Globalstar CEO Jay Monroe, whose New Orleans home was directly in the path of the hurricane. Monroe was unable to get any information on the condition of his property, a fact that Iridium chose not to emphasize in its public relations despite Globalstar's less-than-charitable comments on Iridium's service reliability four years earlier.

51. The 2003 Kentucky Derby turned out to be legendary in the world of Thoroughbred racing. As Derby day approached, Empire Maker seemed to be the culmination of two and a half decades and hundreds of millions of dollars that Prince Khalid had poured into his Juddmonte Farms, which consisted of eight breeding operations in Kentucky, Ireland, and England. Although Juddmonte had earned the industry's highest awards for its broodmare stock, the stable had never won a Triple Crown race, much less the Kentucky Derby. So when race day arrived, a small legion of the Prince's employees showed up in Louisville to watch Empire Maker go off as the solid 6-to-5 favorite in a field of sixteen, with U.S. Racing Hall of Fame jockey Jerry Bailey aboard—only to be defeated by the most unlikely winner of the Derby in its 129-year history. Funny Cide, a gelding purchased for $75,000 by an unknown trainer named Barclay Tagg, was owned by some high school buddies from the little Upstate New York town of Sackets Harbor—their stable was playfully called "Sackatoga"—and they had pooled their resources to rent a school bus to take them the 750 miles to Churchill Downs. Going off at 15-to-1 and winning by almost two lengths, Funny Cide's victory was considered one of the greatest upsets in Derby history, and the gentle gelding soon became a national favorite two weeks later when he defied the odds again, winning the second leg of the Triple Crown, the Preakness, just a few miles up I-95 from Iridium headquarters. It must have rankled the highest-paid breeders, trainers, and jockeys in the world to be beaten by a horse considered so unlikely to be worth breeding that he was gelded before he ran his first race. The Prince's racing team pulled Empire Maker out of the Preakness in order to rest him for the final Triple Crown race, the Belmont Stakes, and as race day approached for that grandest of old-money New York pageants, the matchup between Funny Cide and Empire Maker was one of the most anticipated Belmont had ever hosted: the "people's horse," the gelding from nowhere, the first New York–bred horse to win the Kentucky Derby, the cheapest horse to win any major stakes race in decades, trying to become the first Triple Crown winner since 1978 by running against the slick multimillion-dollar Arab-owned dark bay colt descended from the best bloodstock in the world. The Sackatoga boys were superstitious, so they again rented a school bus to bring them to Belmont Park, but it rained all day before the race and left one of the muddiest tracks the Belmont railbirds had ever seen. When the horses paraded out of the paddock, the crowd broke into chants of "Funny Cide! Funny Cide!" and jockey José Santos raised his crop to acknowledge the cheers. But when the field of six broke from the starting gate on that cold, drizzly

day, Bailey aboard Empire Maker forced Funny Cide to the rail—where the mud was deepest—and stayed on his haunches all the way down the back stretch to make sure he had a dead ride. The mile-and-a-half race, the longest in the world on a dirt surface, took a toll on Funny Cide's legs, and as he entered the final stretch, he started losing the lead. Empire Maker, racing on the harder-packed dirt, easily swung around him and won the race, but when the jubilant Bailey entered the winner's circle, the partisan crowd responded with jeers and boos. When the mud-caked Funny Cide rode back into view a few seconds later, the boos turned to cheers. Empire Maker's trainer, Bobby Frankel, took credit for the race strategy and was honored eight years later when a bay foal sired at Barnstead Manor in England was named after him. That horse, Frankel, became the top rated Thoroughbred in history, winning fourteen races over three seasons in Europe and retiring to stud at the same farm where he was born, covering about 150 mares per season at the record fee of $190,000 per insemination. The next Triple Crown winner wouldn't come along until 2015, and it was a horse called American Pharaoh—Empire Maker's grandson.

52. Sky Station had the blessing of the ITU, two dozen strategic partners, and a blue-ribbon board that included former Secretary of State Alexander Haig and Martine Rothblatt, the same regulatory attorney who had been Martin Rothblatt at the time he founded Sirius Satellite Radio in 1990 but was now Martine after a sex-change operation. Another reason Sky Station may have had problems with investors was that it just sounded too darn freaky. While providing broadband, the platforms were also designed to suck up chlorine molecules, working as an environmental vacuum cleaner that "cleans the ozone layer." And cofounder Rothblatt was active in the cryogenics movement, advocating "xenotransplantation" as a way to achieve human immortality, as well as a scheme to preserve a person's brain data in digital files and upload it into clones.

53. The Bergen Linux User Group actually tested the Internet protocol for carrier pigeons. Called IPoAC (IP over Avian Carriers), it had an unfortunate 55 percent packet-loss rate. (This endnote will be humorous only if you're an engineer.)

54. The first polar phone call appears to have been made on May 7, 2001, by two trekkers claiming to be standing on the North Pole. But since the true North Pole is in the middle of the Arctic Ocean, amid constantly shifting sea ice, you can never be quite sure who's been there and who hasn't. The other place where a "Guess where I am" Iridium phone call has become traditional is the summit of Mount Everest, a tradition begun by mountaineer Byron Smith and captured in real time on *Good Morning Canada*.

55. Test pilots Steve Barter and Troy Pennington took off from the Lockheed Martin Tactical Aircraft Systems airport in Fort Worth, Texas, on the morning of August 27, 1999, and made continuous voice calls using a cockpit handset while executing vertical dives, barrel rolls, vertical climbs, and Mach 1.6 speeds at forty-two thousand feet.

56. Yael Maguire, engineering director at Facebook Connectivity Lab, appeared at the Social Good Summit in New York City in September 2014 and described the drones as "roughly the size of a commercial aircraft" flying between sixty thousand and ninety thousand feet. Maguire mentioned a "regulatory risk" in the course of his remarks but didn't go into detail.

57. The myth that the original Iridium phone was analog, not digital, stuck to the company throughout its life, mainly because every other phone released by Motorola was analog well into the late 1990s. Iridium was actually the first digital device released by Motorola, using a modulation scheme called "quadrature phase-shift keying." (Modulation is the process by which information is inserted into the radio wave. The original analog modulation schemes—amplitude modulation and frequency modulation—are the actual names of AM and FM. There are dozens of digital modulation schemes.) The confusion also stemmed from the fact that the Iridium phone was a narrowband device that used TDMA instead of CDMA, making it less versatile than some other terrestrial cellular devices.

58. In 1991 Ray Leopold met with Commerce Secretary Robert Mosbacher, the onetime Houston wildcatter and close friend of President Bush père, who was actively promoting initiatives by private industry. Mosbacher wanted to discuss NASA's Landsat program, consisting of two remote-sensing satellites that circled the planet at four hundred miles up, taking pictures of Earth that retailed to industry for $4,400 per photo. Mosbacher thought that was expensive, especially since the French had launched a satellite that could do the same thing for much less. Since Iridium would be at about the same altitude, couldn't they just take that over and do it more cheaply? Leopold told him it was a great idea, but Motorola had already made the decision not to put cameras on the satellites. Iridium had skeptical partners who thought the whole project was some kind of secret plot by the CIA, and cameras would only increase suspicion. Mosbacher nonetheless became a major ally of Iridium, helping Motorola later with permissions to cooperate with the Russians and Chinese.

59. Matt Desch disputed parts of this narrative, saying that none of the Iridium partners was that interested in the second generation when he was first interviewed for the job, and each was mainly concerned with tax bills, not the future of the constellation. This version is at variance with eight years of Colussy's documents, notes, speeches, and e-mails from the years 2000 to 2008, as well as recollections by the parties involved.

Epilogue: TWILIGHT OF THE WARRIORS

60. Several other companies ventured into the Ka-band business, but they all used bent-pipe GEOs and earth stations in the quest to supply the world's insatiable demand for data. Astrolink, the multibillion-dollar constellation planned by Lockheed Martin,

flamed out in 2001, never to be revived. But a Denver start-up called iSky launched a service called WildBlue on the first of two GEOs in 2004, offering Internet at about thirty times the speed of dial-up through small VSAT earth dishes that looked like the ones that delivered satellite television. WildBlue was partly financed by John Malone, the old nemesis of Herb Wilkins, and Malone still retained a 37 percent stake when the company was sold in 2009 for $568 million to ViaSat of Carlsbad, California. ViaSat built out an additional nineteen earth stations, bringing the total to thirty, and launched a third GEO offering service under the trade name Exede Internet in 2011, selling primarily to farmers and other rural customers in North America. Meanwhile, Hughes launched the long-delayed Spaceway project in 2005—two GEOs providing what it claimed to be broadband-on-demand—but before the service could ever be offered, the satellites were sold to News Corporation for its DirecTV service. A third Spaceway bird did finally launch in 2007, offering Internet access similar to the WildBlue system. On the other side of the Atlantic, Avanti Communications Group of London was formed in 2002 to provide broadband for Europe, and after spending $850 million, it launched a GEO called Hylas in 2010. Avanti's plan was similar to WildBlue, but after four years in business and two more launches, it was struggling to stay afloat, and increasingly its sales were to third-party cell systems trying to provide Internet for businesses forced to survive in areas like rural Libya and Zimbabwe. None of the Ka-band systems are global, and advertised maximum transmission speeds are four times slower than the rates achieved by the ACTS in 1998. Oddly, these systems are mostly directed at the most extensively wired parts of the planet. They're simply picking up customers stuck in random dead zones.

61. GLONASS, the Russian satellite navigation system, was designed in 1976 in an attempt to match the Pentagon's GPS, which was funded in 1973 and launched in 1978. GLONASS was finally completed in 1995, although it didn't always function properly and didn't attain full coverage of Russian territories until 2010. The twenty-four-satellite constellation was supposed to be part of a worldwide positioning system that would include GPS and Europe's Galileo system, but in the summer of 2014 Russian and American space officials started feuding over the arrangements for Russian ground stations in the United States and U.S. ground stations in Russia, and it appeared that the consortium was falling apart.

PHOTO CREDITS

Page 9: Dannie Stamp with Proton and Delta rockets, © 1997, Khrunichev State Research and Production Space Center; Taiyuan, © Yao Jianfeng/Xinhua Press/CORBIS.

Page 10: Iridium handset, Courtesy of the National Air and Space Museum.

Page 11: Mauro Sentinelli, Courtesy of Telecom Italia; Cornelius Blackshear, Courtesy of the National Conference of Bankruptcy Judges; Craig McCaw, Courtesy of Warren Mell; Doctor Fun, Courtesy of David Farley.

Page 12: John K. Castle, Courtesy of John K. Castle/Castle Harlan; Prince Khalid, Painting by Jemma Phipps, Courtesy of Prince Khalid; Terry Jones and Herb Wilkins, Courtesy of Terry Jones; Ty Brown and Jesse Jackson, Courtesy of John Mitchell.

Page 13: Motorola tower, Courtesy of Maggie Ragaisis; Old Executive Office Building, Ronald Reagan Presidential Library; Chris Galvin, Courtesy of Lisa Swann.

Page 14: Rudy de Leon, Courtesy of the Center for American Progress; Dorothy Robyn, Courtesy of Dorothy Robyn; Dave Oliver, Courtesy of Dave Oliver; Bill Cohen and Janet Langhart Cohen, Courtesy of Department of Defense.

Page 15: Ribbon cutting, courtesy of Dan Colussy; Jerry Lewis, © Chip Somodevilla/ Getty Images; Dan Colussy, Courtesy of Dan Colussy.

Page 16: Matt Desch, Courtesy of Iridium; Iridium Next, Courtesy of Iridium, U.S. Marine, Associated Press/Julie Jacobson

INDEX